Deep Learning Techniques for Medical Image Analysis

Deep Learning Techniques for Medical Image Analysis

Guest Editor

Zhuhuang Zhou

MDPI

Basel • Beijing • Wuhan • Barcelona • Belgrade • Novi Sad • Cluj • Manchester

Guest Editor
Zhuhuang Zhou
Department of Biomedical
Engineering
Beijing University of
Technology
Beijing
China

Editorial Office
MDPI AG
Grosspeteranlage 5
4052 Basel, Switzerland

This is a reprint of the Special Issue, published open access by the journal *Diagnostics* (ISSN 2075-4418), freely accessible at: https://www.mdpi.com/journal/diagnostics/special_issues/J476I9I3J9.

For citation purposes, cite each article independently as indicated on the article page online and as indicated below:

Lastname, A.A.; Lastname, B.B. Article Title. *Journal Name* **Year**, *Volume Number*, Page Range.

ISBN 978-3-7258-7763-8 (Hbk)
ISBN 978-3-7258-7764-5 (PDF)
https://doi.org/10.3390/books978-3-7258-7764-5

Contents

About the Editor

Zhuhuang Zhou

Zhuhuang Zhou is an associate professor at Beijing University of Technology. His research interests include biomedical ultrasonics, quantitative ultrasound for tissue characterization, ultrasound propagation in tissues, acoustic cavitation and bubble dynamics, biomedical signal/image processing, medical robotics, and artificial intelligence in medicine. He is an editorial board member of *Scientific Reports* (2024-present) and *Diagnostics* (2023-present). He has published 91 journal papers.

Preface

In recent years, deep learning techniques have been widely used in medical image analysis. These techniques employ deep neural networks to automatically extract multi-level, multi-scale, abundant information (features) from image data, which is hard for conventional machine learning techniques which use hand-crafted feature parameters, including supervised learning (with task-driven models), unsupervised or generative learning (with data-driven models), semi-supervised learning (with hybrid task-driven and data-driven models), reinforcement learning (with environment-driven models), and physics-informed learning (hybrid task-driven and physics-driven models). The vast applications of deep learning techniques in medical image analysis cover lesion detection and segmentation, disease diagnosis, treatment monitoring, efficacy evaluation, prognostic prediction, and even biomechanical analysis. In addition to medical image post-processing, deep learning techniques can also be applied to the front-end (e.g., image reconstruction) to enhance the quality of medical imaging.

Given the high level of research interest and clinical application prospects, deep learning techniques have continued to develop, especially in the field of medical image analysis. This Reprint aims to report on state-of-the-art deep learning techniques applied to medical image analysis, covering medical image based deep learning for automated object/lesion detection, segmentation, classification, measurement, and evaluation as well as treatment planning. The featured medical images include ultrasound, X-ray, CT, MRI, microscopy, dermoscopy, and auditory brainstem response images for different kinds of tissues.

Zhuhuang Zhou
Guest Editor

diagnostics

MDPI

Article

Development of an Artificial Intelligence-Based System for Evaluating Transthoracic Echocardiographic Imaging in Focus Cardiac Ultrasonography

Daigo Ikeda [1], Sanshiro Togo [1], Shogo Tsuge [1], Shu Ohya [1], Yuki Sugiura [1], Masaya Honda [1], Taiki Hosokawa [1], Kenshin Suzuki [1], Katsumasa Nakamura [2,3], Yuki Kurita [4,5], Kazuki Tamura [5,6,*] and Takeji Saitoh [2]

[1] Faculty of Medicine, Hamamatsu University School of Medicine, Hamamatsu 431-3192, Shizuoka, Japan; daigo-ikeda-doctengineer@outlook.jp (D.I.)
[2] Next Generation Creative Education Center for Medicine, Engineering and Informatics (Nx-CEC), Hamamatsu University School of Medicine, Hamamatsu 431-3192, Shizuoka, Japan
[3] Department of Radiation Oncology, Hamamatsu University Hospital, Hamamatsu 431-3192, Shizuoka, Japan
[4] Department of Regenerative and Infectious Pathology, Hamamatsu University School of Medicine, Hamamatsu 431-3192, Shizuoka, Japan
[5] Institute of Photonics Medicine, Hamamatsu University School of Medicine, Hamamatsu 431-3192, Shizuoka, Japan
[6] Graduate School of Engineering Science, Yokohama National University, Yokohama 240-8501, Kanagawa, Japan
* Correspondence: tamura-kazuki-nb@ynu.ac.jp; Tel.: +81-45-339-3014

Abstract

Background/Objectives: Transthoracic echocardiography (TTE) is a non-invasive tool for real-time assessment of cardiac motion and blood flow. It is widely used in emergency and bedside settings as a Focus Cardiac Ultrasound (FoCUS) device. However, standardized training methods and adequate educational environments are limited. **Methods:** A TTE image assessment artificial intelligence (AI) system was developed in this study, focusing on probe positioning and image quality for non-supervised TTE practice. **Results:** The view classification model achieved a high F1-score of 0.956. The position evaluation model achieved F1-scores of 0.678, 0.864, and 0.831 for the parasternal long-axis, parasternal short-axis, and apical four-chamber views, respectively. The quality evaluation model achieved F1-scores of 0.674, 0.845, and 0.746. Combining the position and quality models improved the F1-score for the parasternal long-axis view to 0.714, showing the benefit of integrating views with lower baseline performance. **Conclusions:** This study presents a novel AI-based educational system that assesses probe position and image quality in TTE. The model was developed using a custom dataset of healthy young adults that reflects beginner-level training scenarios, including many suboptimal images similar to those commonly acquired by novices. The proposed framework, which integrates position and quality models, lays the groundwork for future AI-assisted ultrasound training, particularly in unsupervised or resource-limited settings.

Keywords: transthoracic echocardiography; artificial intelligence; focus cardiac ultrasound; medical education

1. Introduction

Transthoracic echocardiography (TTE) is an essential diagnostic tool in cardiovascular medicine that enables the real-time assessment of cardiac motion and hemodynamics without causing pain or exposing patients to radiation. When TTE is performed by non-cardiology specialists for rapid diagnosis and initial treatment decision-making, it is termed Focus Cardiac Ultrasound (FoCUS). FoCUS is increasingly recognized as an essential skill for physicians, particularly in the emergency department, the intensive care unit (ICU), and primary care settings [1–5]. Globally, there is a growing emphasis on integrating ultrasound education, including TTE, into undergraduate medical training, with widespread adoption expected among medical students [6].

FoCUS is a brief bedside screening examination mainly conducted by non-cardiology specialists. Its primary goal is to prevent the oversight of common or potentially life-threatening conditions by focusing on a limited number of essential findings using simple and fundamental techniques [7]. Unlike comprehensive echocardiography conducted by ultrasound specialists, FoCUS is not designed to provide a detailed evaluation of the entire heart or quantify parameters, including left ventricular ejection fraction or transvalvular flow velocities.

A significant challenge for beginners in FoCUS training is determining whether the image they obtain represents an appropriate cross-sectional view. Typically, learners acquire this skill through hands-on sessions with ultrasound instructors, who provide direct feedback on image quality. However, tools that are currently available for evaluating image appropriateness are few, making it difficult for beginners to train effectively without the supervision of experienced practitioners. In Japan, many physicians receive FoCUS training during their residency; nevertheless, the limited availability of instructors or technicians in some institutions has led to junior residents completing their programs without acquiring adequate FoCUS skills [7].

Recently, advancements in artificial intelligence (AI) have improved image recognition technology. Notably, research has focused on tasks, including standard view classification, which involves identifying commonly depicted views on TTE, such as the parasternal long-axis view, parasternal short-axis view at the mitral valve level, and apical four-chamber view [8–15]. Additionally, ultrasound devices equipped with AI-guided scanning assistance have become available, helping to acquire appropriate cross-sectional views. However, the underlying algorithms differ among manufacturers and mostly depend on specific ultrasound systems.

There are two essential elements for acquiring optimal TTE images in FoCUS. The first is the position of the probe required to obtain the intended view. The second is whether the anatomical structures necessary for FoCUS were adequately visualized. Some factors, including respiratory motion, imaging artifacts, insufficient gel application, and inappropriate display of structures at their expected locations, can impair the second element. In this study, these two elements are referred to as "position" and "quality," respectively. Deficiencies in any of these elements can hinder rapid and accurate diagnosis in clinical settings. Skilled sonographers achieve optimal imaging by understanding the target view and continuously adjusting the position and orientation of the probe based on real-time image feedback. Trainees must learn to recognize optimal views and infer the necessary adjustments to the probe based on the images they observe to acquire this level of skill.

In this study, a TTE image assessment AI as an educational tool was developed to help beginners acquire FoCUS skills. A two-step framework was established for this purpose. Step 1 involves classifying the TTE images into three standard views. Step 2 involves evaluating whether each image corresponds to an optimal cross-sectional view and is of sufficient quality. Completing both steps enables the identification of images that are suitable for FoCUS.

https://doi.org/10.3390/diagnostics16071032

Various studies have addressed Step 1 (the prerequisite task), which involves classifying standard echocardiographic views [8–16]. From a methodological perspective, deep learning approaches—particularly convolutional neural networks (CNNs) such as ResNet, VGG, and Inception, as well as neural architecture search (NAS)—have become the standard for automated view classification. These models achieve high classification accuracies, often exceeding 95%, and have demonstrated robustness across diverse clinical scenarios, including point-of-care ultrasound (POCUS), contrast-enhanced echocardiography, and handling cardiac motion via spatial-temporal or graph-based constraints.

Despite these strengths, a key limitation of existing tools is their reliance on datasets composed primarily of clinically optimal, properly aligned standard views. Their primary objective is to identify which standard view is depicted, rather than to detect how the probe is misaligned. Consequently, they cannot evaluate the typical suboptimal images produced by novice trainees during hands-on practice. Some studies have also investigated the use of neural networks to evaluate the image quality in TTE as part of Step 2 (the primary task) [16–18]. However, estimating the position of the probe from an image requires a unique dataset that includes intentionally misaligned or suboptimal images. Because existing automated tools were not designed or trained to output fine-grained probe deviation classes (e.g., distinguishing between specific sliding, tilting, or rotating errors from a reference view), a direct empirical comparison between our implementation and other existing tools using the same data is not feasible; the classification tasks and output labels are fundamentally different.

Three types of image-classification tasks were conducted in this study. The first image classification task focused on achieving the previously described Step 1, where a view classifier was used to identify the three cross sections: parasternal long-axis view, parasternal short-axis view at the mitral valve level, and apical four-chamber view. The second task involved estimating positional deviations from the correct probe position using echocardiographic images. This position classification task used a position classifier to categorize 15 types of deviations in the parasternal long-axis view, 19 in the parasternal short-axis view, and 18 in the apical four-chamber view. The third task focused on evaluating the visual quality of the displayed images by categorizing them into four levels–best, acceptable, poor, and bad–using a quality classifier. The position and quality classifiers were designed to achieve Step 2 and were individually trained based on the view classifier. Subsequently, the results predicted by the position and quality classifiers were combined or threshold-processed to enhance the performance of the position classification using quality classification as a weighting factor. The classification results were evaluated using Venn diagrams under two conditions: one where all images, regardless of quality, were defined as targeted images and the other where only high-quality images (best or acceptable) were defined as targeted images. The latter metric was intended to identify images suitable for FoCUS. Finally, Step 3 is discussed based on the results of the position classifier for non-optimal cross sections.

We engaged in an engineering optimization task aimed at the practical implementation of an AI system that assesses TTE image quality based on the FoCUS criteria without requiring supervision from expert sonographers or technicians. While we utilized an existing AI architecture for system development, the novelty of this study lies in the construction of a framework that integrates a unique dataset with distinct evaluation models. In particular, images where the probe position deviated from the optimal cross-sectional plane were collected. Classifiers based on these images were constructed to estimate the probe position and assess diagnostic image quality.

https://doi.org/10.3390/diagnostics16071032

2. Methods

The Ethics Committee of Hamamatsu University School of Medicine (EC HUSM) and the Conflict-of-Interest Management Committee (Approval No. 22-121) approved this study. All methods were carried out in accordance with relevant institutional guidelines and regulations. The study was specifically designed and conducted in accordance with the Declaration of Helsinki. All the participants provided written informed consent. The authors declare that they have no conflicts of interest associated with this study.

2.1. Collection and Annotation of Training and Validation Datasets

In this study, two datasets were obtained: one for training the view classification model, and the other was used as the main dataset to develop the position and quality evaluation models. The models developed in this study were designed to assess whether echocardiographic images acquired during practice sessions by beginner-level FoCUS trainees are appropriate views. Data were collected exclusively from healthy young adults, representing the target population of beginner users of FoCUS. A scanning protocol was developed to systematically acquire various intentionally suboptimal images.

2.1.1. Collection of the View Classification Model Dataset

Between June and August 2023, data were collected from six medical students proficient in TTE, under the supervision of a cardiologist certified by the Japanese Circulation Society. A method was adopted to intentionally capture images that deviated from optimal views, starting from reference standard views.

The dataset included three standard views: the parasternal long-axis (PLAX) view, parasternal short-axis (PSAX) view at the mitral valve level, and apical four-chamber (A4C) view. Each view was categorized into seven to eight scanning sequences. Each sequence began with the optimal view displayed for 5–10 s, followed by systematic probe adjustments to acquire intentionally deviated images (Figures 1–3; Supplementary Table S1). For instance, in the counterclockwise rotation sequence for the PLAX view, the optimal view was shown for 10 s, followed by PLAX_cc1 and PLAX_cc2, each recorded for 5 s. This process was used across all three standard views. Scanning techniques, including rotating, sliding, tilting, and rocking, were not defined by the physical angle of the probe but by the visualization of anatomical structures on the ultrasound images, according to the criteria shown in Supplementary Table S1. In the PLAX view, the images obtained through "sliding lower parasternal" and "tilting posteriorly" were found to be similar in appearance, making it difficult to distinguish them. Therefore, only the "sliding lower parasternal" maneuver was used during data acquisition. Similarly, when multiple scanning methods produced nearly indistinguishable images, only one image was selected for data collection. Consequently, the specific scanning protocols varied slightly across the PLAX, PSAX, and A4C views.

The data acquisition process was conducted using the Aplio verifia™ (Canon Medical Systems, Ohtawara, Japan) with a PST-28BT probe, recording videos at 60 fps. Before data collection, the imaging personnel underwent training in probe manipulation and followed predefined procedures.

The recorded video data were subsequently annotated by four medical students, with the supervision of a cardiologist. Each video was labeled and cross-checked by three separate annotators to ensure the quality of the annotation. View labels were assigned according to the section during data collection, whereas position labels were determined by cross-referencing video timestamps with probe manipulations within each series. Segments with unintended probe displacement or significant image distortion were excluded from the dataset.

https://doi.org/10.3390/diagnostics16071032

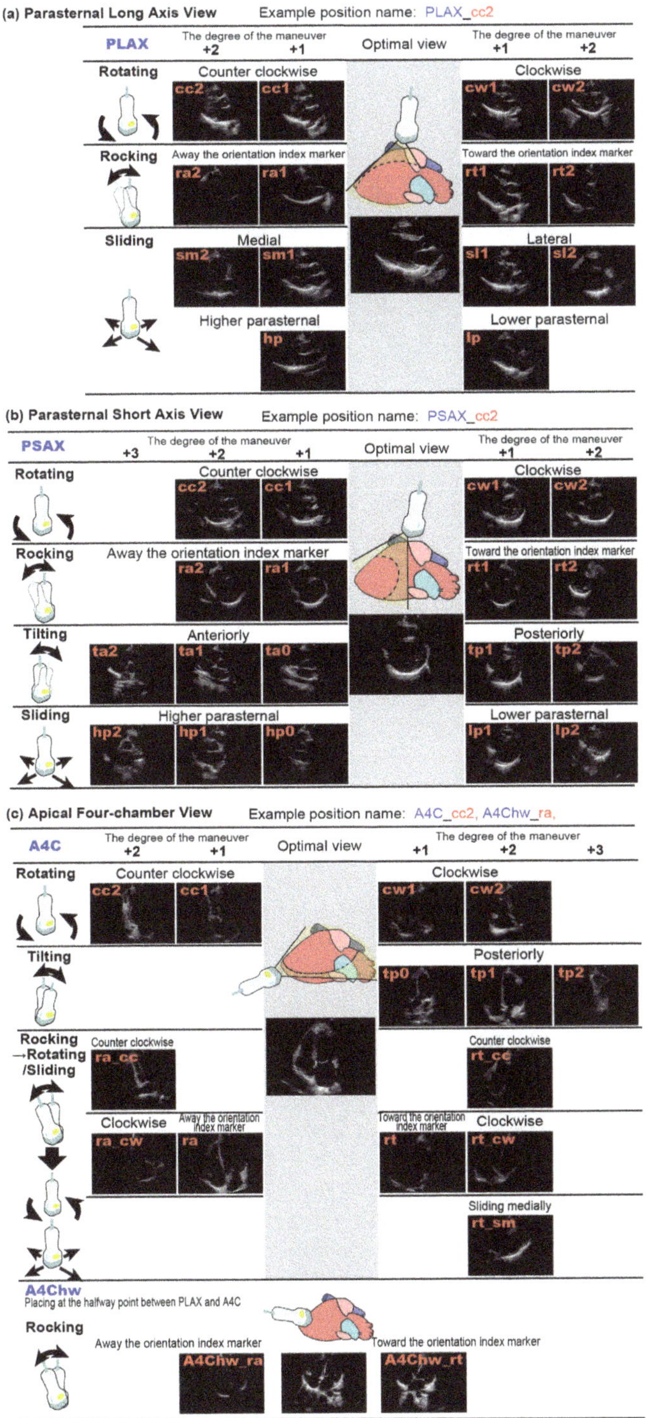

Figure 1. Position Labels for Images Captured in This Study. The optimal cross-sectional view for each position was defined as the starting point, and suboptimal views were obtained by applying

rotating, rocking, tilting, or sliding maneuvers to the transducer. Non-optimal views were labeled by appending an underscore to the name of the optimal view, followed by an abbreviation of the maneuver ("cc" for counterclockwise rotation) and the degree of the maneuver (0, 1, or 2). (**a**) Parasternal long-axis view. (**b**) Parasternal short-axis view at the mitral valve level. (**c**) Apical four-chamber view.

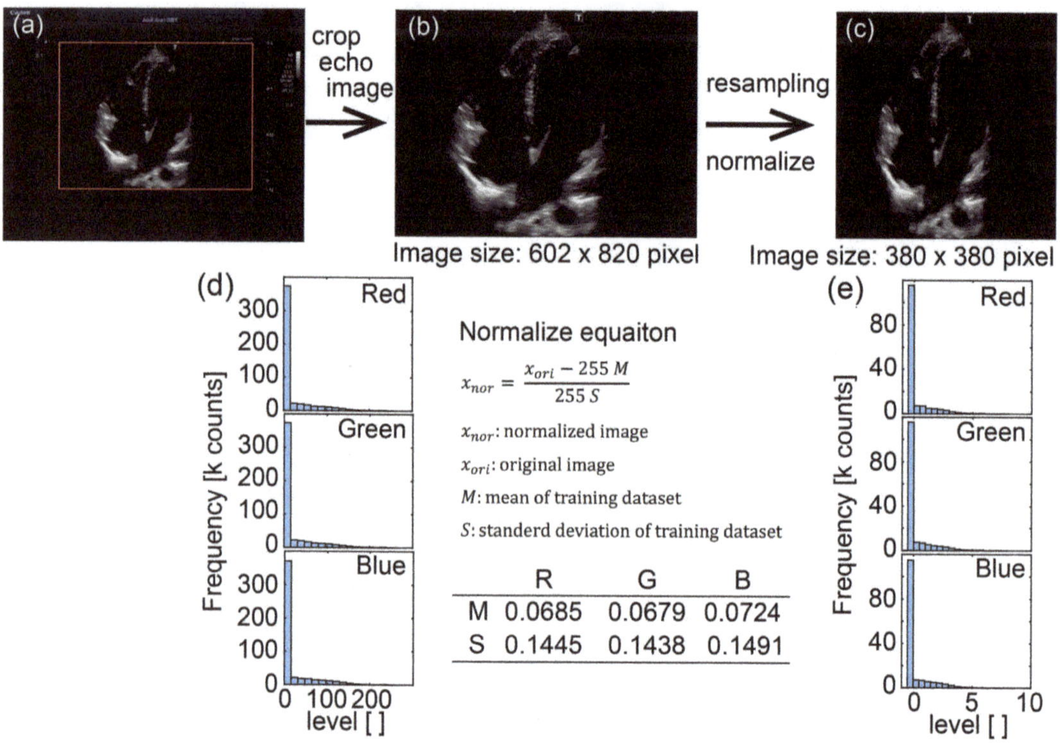

Figure 2. Overview of Preprocessing and Normalization for Images. Preprocessing pipeline for echocardiography images used as inputs. (**a**) Screenshot of the diagnostic ultrasound system. The region of interest (ROI) indicated by the red rectangle is extracted. (**b**) Cropped grayscale echo image from the ROI (image size: 602 × 820 pixels). (**c**) Final CNN input image after resampling to a square format and intensity normalization (image size: 380 × 380 pixels). (**d**,**e**) Pixel-intensity histograms of the cropped image in (**b**) and (**c**), respectively.

2.1.2. Collection of the Main Dataset

Between January and May 2023, the main dataset for the position and quality evaluation models was collected using the same methods and equipment as the pretrained model dataset (Figures 1–3). Data acquisition was conducted using the Aplio verifia™ (Canon Medical Systems) with a PST-28BT probe. The annotation process was conducted by the four medical students who worked on the view classification model dataset.

In addition to the position labels, the main dataset included quality labels categorized into four levels: best, acceptable, poor, and bad. The criteria for quality evaluation were according to the American Society of Echocardiography guidelines [19], with acceptable or higher quality defined as acceptable (Supplementary Table S2). Best: Anatomical structure required for the FoCUS was clearly visible during systole and diastole. Acceptable: At least one systole or diastole showed a partial depiction of the required anatomical structures.

Poor or Bad: Systole and diastole show incomplete or missing depictions of the required anatomical structures.

The main dataset was categorized into two parts: 15 participants were set aside as the test dataset, and the remaining participants were used for training and validation.

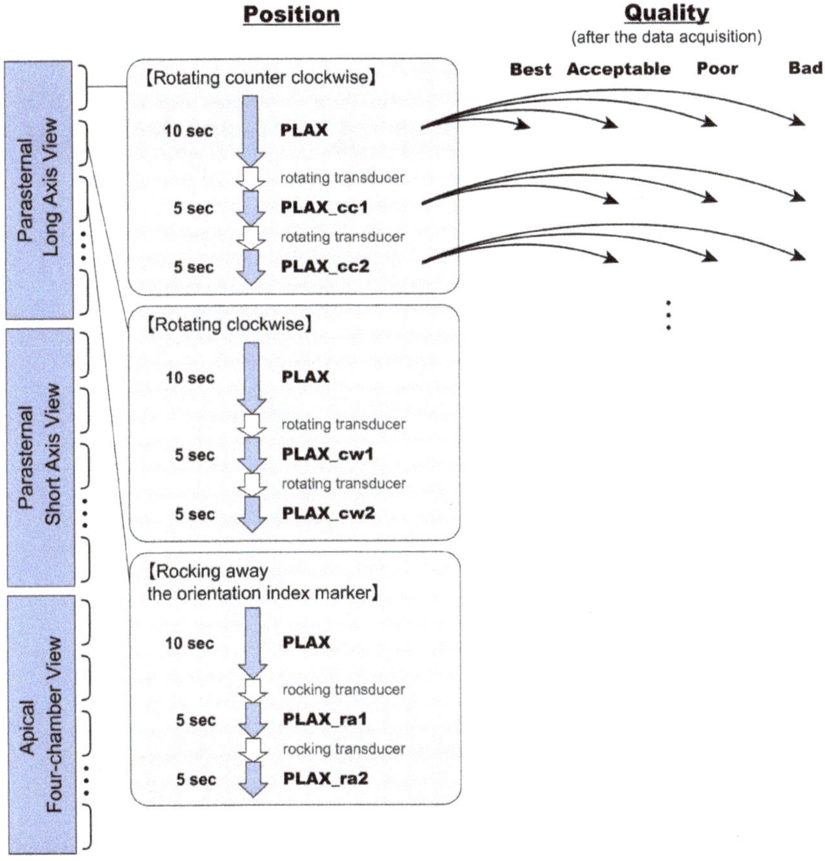

Figure 3. Overview of Data Acquisition and Labeling Methods. The ultrasound data collection was divided into three parts: the parasternal long-axis view, the parasternal short-axis view at the mitral valve level, and the apical four-chamber view. For each standard view, a series of scans was conducted, consisting of seven to eight sequences per part. Each sequence began by continuously capturing the optimal view for 10 s (or 5 s), followed by transducer maneuvers to acquire deviated images. For instance, in the counterclockwise rotating sequence for the parasternal long-axis view part, the PLAX view was recorded for 10 s, followed by a counterclockwise rotation to capture the PLAX_cc1 view for 5 s, and further rotation to capture the PLAX_cc2 view for another 5 s. Similarly, clockwise rotations, rocking toward the transducer marker, and other maneuvers were included in the sequences. The arrows indicate that each image has been assigned a quality label.

2.2. Dataset Processing

For training and inference, standardized preprocessing procedure to the echocardiography frames (Figure 2) was applied. First, we extracted the ROI from the ultrasound system screen by cropping the area indicated by the red rectangle (Figure 2a), resulting in a rectangular image (602 × 820 pixels; Figure 2b). The cropped image data were then randomly undersampled to ensure an equal number of samples per class (Figure 4, Supplementary Table S4).

https://doi.org/10.3390/diagnostics16071032

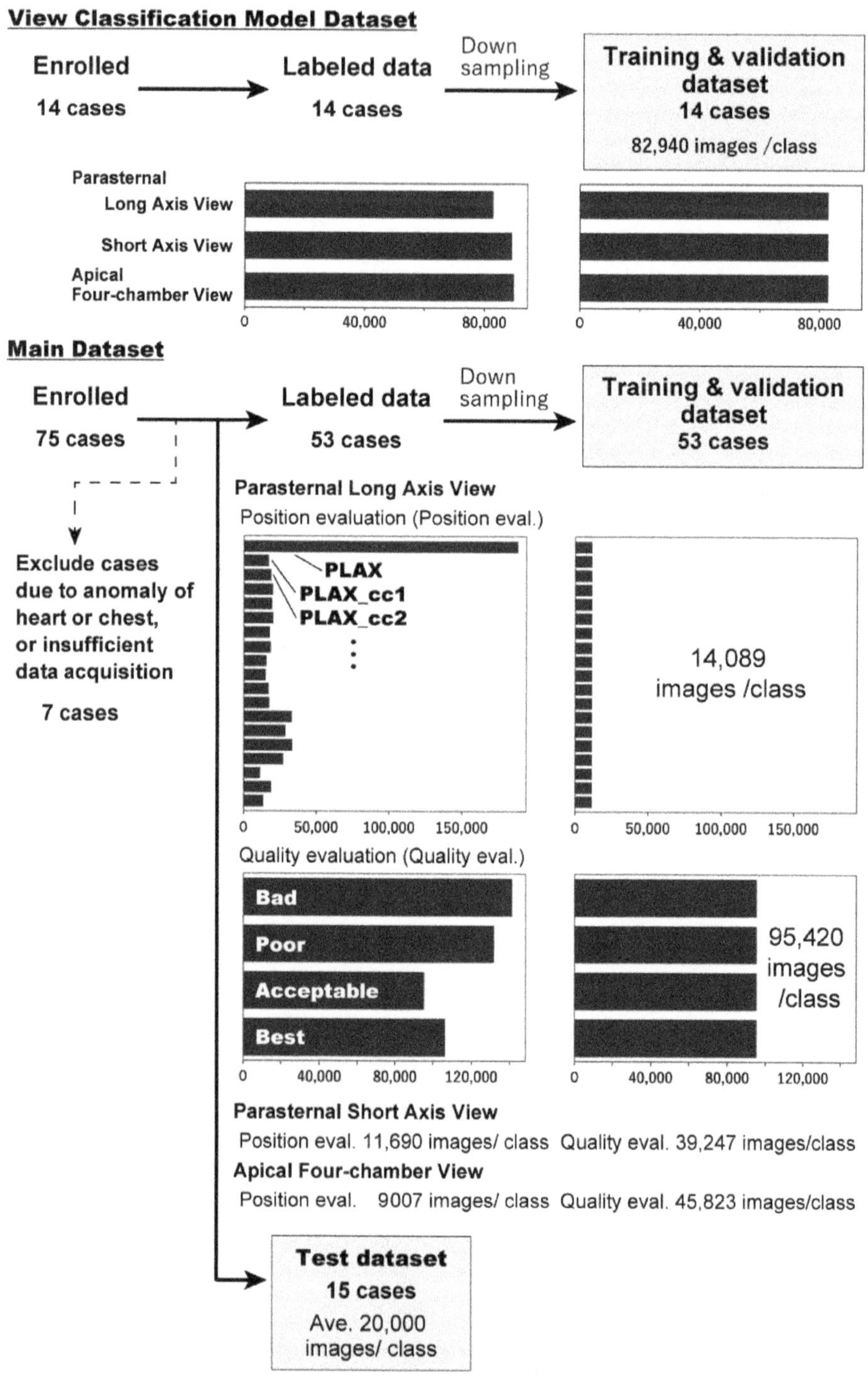

Figure 4. Data Processing Workflow. This figure shows the process of generating training and testing datasets from the collected video data. The arrows indicate the flow of data selected from the enrolled participants.

https://doi.org/10.3390/diagnostics16071032

2.3. Learning Method

2.3.1. Computational Environment

All experiments were conducted using Python version 3.8.10. The following libraries were used: torch v1.13.0, torchvision v0.14.0, pillow v9.3.0, scikit-learn v1.1.3, scikit-image v0.19.3, pandas v1.5.2, numpy v1.23.5, seaborn v0.12.1, cuda v11.7.99, cudnn v8.5.0.96, opencv-python v4.6.0.66, accelerated v0.22.0, timm v0.6.12, and albumentation v1.3.0. The experiments were conducted on two RTX A6000 GPUs (NVIDIA, Santa Clara, CA, USA), each with 48 GB of memory, to ensure sufficient computational resources.

2.3.2. Training View Classification Model

As the first step in this study, a model was developed to classify echocardiographic images into three standard views: the PLAX view, PSAX view at the mitral valve level, and A4C view (the view classification model). The position and quality evaluation models were built using the view classification model as a pretrained backbone.

The view classification model was used as a pretrained model for the position and quality evaluation models for two reasons. First, pretraining on a relatively simple task facilitates adaptation to subsequent tasks. Second, constructing position and quality evaluation models directly from existing pretrained models, such as ImageNet, requires substantial computational resources. The view classification model avoids this issue by improving computational efficiency. Existing image recognition models, such as ImageNet, can be used; however, the domain gap between natural and echocardiographic images poses challenges in achieving high classification accuracy. Additionally, some studies have leveraged open datasets of echocardiographic images; nevertheless [20], there are legal constraints pertaining to their use, limiting their versatility. Therefore, in this study, the view classification model was used as a pretrained model for the position and quality evaluation models.

To obtain a fixed input size compatible with the network, the cropped image was resampled to a square image (380 × 380 pixels; Figure 2c). We then normalized the pixel intensities using the training-dataset statistics to reduce inter-scan intensity variability and stabilize optimization. Mean and standard deviation values were computed from the training dataset. The resulting histograms (Figure 2d,e) show that this procedure shifts and scales the pixel-value distribution toward a near-zero mean with reduced spread, producing standardized inputs for the network.

The following augmentations were used: random rotation within ±20° (Rotate) and random cropping within 90–100% of the original size while maintaining an aspect ratio of 1.36 (RandomResizedCrop). These augmentations were applied to 30% of images in each batch. Following preprocessing, mean values and standard deviations were used to normalize the images.

The images were subsequently input into MobileViTv2_075 [21,22] (Figure 5). The backbone architecture was selected based on a comprehensive comparative evaluation of several lightweight models, including MobileViTv2_075, conducted under identical training and validation conditions. In addition to primary task performance, inference latency, model size, and the propensity for overfitting were systematically assessed (Supplementary Table S3). Vision Transformer (ViT) is a deep learning architecture mainly used for image classification. Unlike convolutional neural networks (CNNs), which perform convolution operations on small local regions of an image (3 × 3 or 5 × 5 pixels) and are, therefore, limited to learning local features, ViTs uses a self-attention mechanism that enables them to process the entire image simultaneously. This enables the model to associate local features directly with the global image context, allowing learning that considers the entire image structure.

https://doi.org/10.3390/diagnostics16071032

ViTs generally require more computational resources than CNNs; nonetheless, MobileViT was developed as a lightweight alternative that maintains similar accuracy with a reduced computational cost. In this study, MobileViTv2_075, an even more efficient version, was used to facilitate real-time image evaluation using smartphones and tablets.

The dataset was divided into training and validation sets using a stratified grouped hold-out approach, with subject identifiers treated as grouping variables. Model training was conducted to ensure independence at the subject level. The split ratio and random seed are specified in the main text. The optimization yielded an ideal batch size of 256 and a learning rate of 0.001. The view classification model was trained for 150 epochs, the position evaluation models for 200 epochs, and the quality evaluation models for 50 epochs. The model corresponding to the epoch with the highest F1-score was retained. The remaining hyperparameters are shown in Tables 1 and 2.

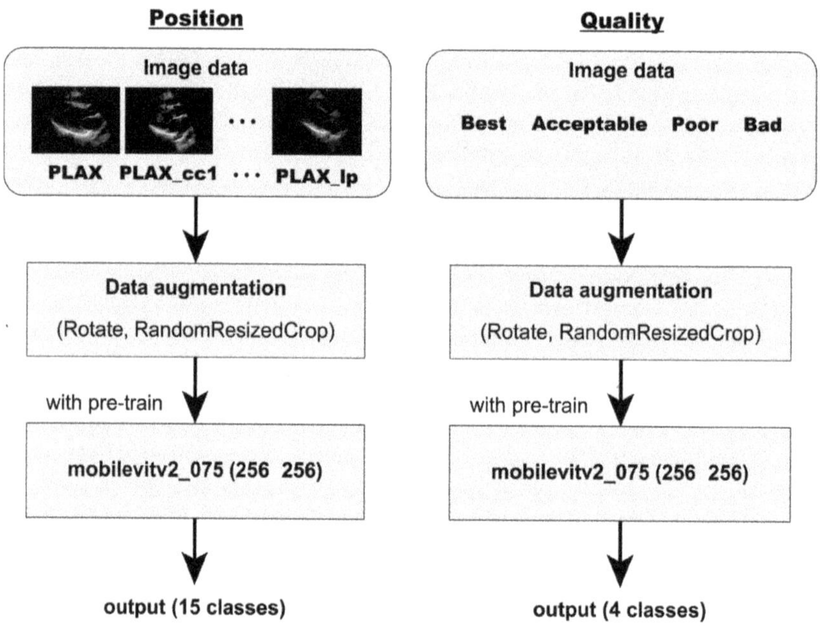

Figure 5. Training Methods for Position and Quality Evaluation Models. The figure shows the training process for the position evaluation model (parasternal long-axis view) and the quality evaluation model. Training images were subjected to data augmentation with a 30% probability and then input into the mobileViTv2_075 model. The model outputs the class probabilities for each evaluation.

Table 1. Overview of Training Methods for Models Developed in This Study.

Model	Optimal View	Total Epoch
Pre-trained	PLAX, PSAX, A4C	150
Position evaluation	PLAX	200
	PSAX	200
	A4C	200
Quality evaluation	PLAX	50
	PSAX	50
	A4C	50

Other training hyperparameters are summarized in Table 2.

https://doi.org/10.3390/diagnostics16071032

Table 2. Hyperparameters Used for Model Training in Position Evaluation Model and Quality Evaluation Model.

Parameter	Value
Batch size	256
Learning rate	0.001
Label smoothing	0.2
Warmup steps	5
Weight decay	0.01

2.3.3. Training Position and Quality Evaluation Models

The position and quality evaluation models for each standard view were trained and validated using a view-classification model. Only data corresponding to each specific view were used for training; for instance, only images from the PLAX view were used for the position evaluation model.

Both models were built using mobileViTv2_075 (input size: 256 × 256 pixels) as the backbone (Figure 5). Identical hyperparameters were used for both the position evaluation model and the quality evaluation model (Table 2). The model corresponding to the epoch with the highest F1-score was retained.

2.4. Evaluation Methods

2.4.1. Evaluation Metrics

The primary evaluation metrics included a recall, a precision, and an F1-score, which showed the harmonic mean of precision and recall for the position evaluation model in determining the correct anatomical planes for each view. Recall, precision, and F1-score of the quality evaluation model for identifying images of acceptable quality (best or acceptable) were measured.

Secondary metrics included a recall, a precision, and an F1-score for the pretrained classification model. Recall, precision, and F1-score for position prediction when combining the position and quality evaluation models were measured. Macro-average recall, precision, and F1-score for identifying images that met positional and quality criteria across all views were calculated.

2.4.2. Test Case Evaluation

The performances of the view classification, position evaluation, and quality evaluation models were assessed using the test data from 15 participants excluded from the training set. The macro-averaged recall, precision, and F1-score were computed for each model.

2.4.3. Combining Position and Quality Evaluation Models

The potential of combining the position and quality evaluation models to improve positional predictions was investigated. The intersections and unions of the positional and quality evaluation results (best or acceptable) were used, while the corresponding recall, precision, and F1-scores were calculated.

2.4.4. Identifying FoCUS-Usable Images

A model for identifying FoCUS-usable images (those meeting the positional and quality criteria) was evaluated. The accuracy of the model was assessed using recall, precision, and F1-scores according to the intersection and union of the position and quality evaluation models.

https://doi.org/10.3390/diagnostics16071032

3. Results

3.1. Participants Characteristics

Table 3 shows the characteristics of the participants whose data were used to train the position, quality, and view classification models. Overall, 281,534 frames from 14 participants (maximum: 22,202 frames; minimum: 1918 frames) were used exclusively for training. An additional 425,781 frames from 15 participants (an average of 28,385 frames per subject) were used for testing.

Table 3. The Characteristics of the Subjects Used in This tudy.

	Training Dataset for View Classification Model	Training Dataset for Position and Quality Evaluation Model	Test Dataset
Number	14	53	15
Age (years) •	22 ± 2.8	23 ± 3.6	22 ± 2.8
Sex			
Male	13 (92.9)	46 (86.8)	13 (86.7)
Female	1 (7.1)	7 (13.2)	2 (13.3)
BMI, kg/m^2 •	22 ± 2.6	21 ± 2.5	22 ± 2.3
Total images	281,534	595,571	425,781

Unless otherwise noted, data are numbers of subjects, with percentages in parentheses. • Data are mean ± standard deviation.

Among the 75 participants initially enrolled in the position and quality evaluation models, seven were excluded owing to factors such as pectus excavatum or insufficient image acquisition. The remaining 1,886,225 frames from 68 participants were grouped as follows: 595,571 frames from 53 participants (average: 11,237 frames per participants after undersampling) were used for training and validation, and 425,781 frames from 15 participants (average: 28,385 frames per subject) were used for testing.

The undersampled data used for training are as follows (Figure 4): view classification model, 82,940 frames per class; position evaluation model, parasternal long-axis, 14,089 frames per class; parasternal short-axis, 11,690 frames per class; apical four-chamber, 9007 frames per class; quality evaluation model, parasternal long-axis, 95,420 frames per class; parasternal short-axis, 39,247 frames per class; and apical four-chamber, 45,823 frames per class.

For testing, the view classification model used 425,781 frames (average: 30,413 frames per participants), while the Position and Quality evaluation models used 141,927 frames (average: 9461 frames per participants). The distribution of quality labels according to their position in the test data is shown in Table 4.

Table 4. Distribution of Positions and Quality.

(I) Parasternal Long-Axis View						
			Quality			
Position		Total	Best	Acceptable	Poor	Bad
Optimal view	PLAX	51,897 (185,140)	28,580 (104,636)	16,655 (52,298)	5431 (25,971)	1231 (2235)
Rotating	PLAX_cc2	9964 (19,338)	0 (0)	66 (0)	4949 (427)	4949 (18,911)
	PLAX_cc1	4641 (16,968)	0 (0)	3087 (0)	777 (12,093)	777 (4875)

https://doi.org/10.3390/diagnostics16071032

Table 4. *Cont.*

(I) Parasternal Long-Axis View						
			Quality			
Position		Total	Best	Acceptable	Poor	Bad
	PLAX_cw1	8306 (18,745)	0 (0)	3634 (0)	2336 (15,258)	2336 (3487)
	PLAX_cw2	9726 (19,741)	0 (0)	0 (0)	4863 (0)	4863 (19,741)
Rocking	PLAX_ra2	10,802 (18,732)	0 (0)	0 (0)	5401 (274)	5401 (18,458)
	PLAX_ra1	8664 (32,771)	0 (868)	3353 (13,900)	4075 (14,416)	1236 (3587)
	PLAX_rt1	8276 (26,316)	0 (29)	932 (2035)	3672 (19,017)	3672 (5235)
	PLAX_rt2	5671 (17,542)	0 (0)	520 (275)	274 (461)	4877 (16,806)
Sliding	PLAX_sm2	10,862 (16,280)	0 (0)	274 (274)	5294 (0)	5294 (16,006)
	PLAX_sm1	4216 (17,108)	0 (274)	4062 (502)	77 (10,593)	77 (5739)
	PLAX_sl1	4675 (21,745)	0 (177)	4151 (768)	524 (14,294)	0 (6506)
	PLAX_sl2	4941 (14,089)	0 (0)	0 (0)	0 (0)	4941 (14,089)
	PLAX_hp	5438 (18,136)	0 (192)	3516 (9120)	1374 (7329)	548 (1495)
	PLAX_lp	6238 (32,060)	0 (0)	3642 (16,248)	1292 (11,654)	1304 (4158)
(II) Parasternal Short-Axis View						
			Quality			
Position		Total	Best	Acceptable	Poor	Bad
Optimal view	PSAX	53,921 (188,329)	46,560 (148,215)	4662 (23,375)	2699 (16,034)	0 (705)
Rotating	PSAX_cc2	5206 (18,924)	0 (0)	0 (0)	0 (1026)	5206 (17,898)
	PSAX_cc1	6092 (17,149)	0 (0)	625 (743)	4800 (11,309)	667 (5097)
	PSAX_cw1	7076 (20,043)	0 (0)	0 (548)	0 (15,995)	7076 (3500)
	PSAX_cw2	1246 (19,475)	0 (0)	345 (0)	901 (0)	0 (19,475)
Rocking	PSAX_ra2	12,481 (17,979)	0 (0)	201 (0)	6916 (0)	5364 (17,979)
	PSAX_ra1	6669 (20,187)	539 (548)	1346 (4194)	4236 (13,010)	548 (2435)

Table 4. *Cont.*

(II) Parasternal Short-Axis View						
			Quality			
Position		**Total**	**Best**	**Acceptable**	**Poor**	**Bad**
	PSAX_rt1	4692 (19,124)	345 (512)	4347 (1297)	0 (14,086)	0 (3229)
	PSAX_rt2	5279 (15,894)	0 (0)	0 (0)	5279 (0)	0 (15,894)
Tilting	PSAX_tp2	17,826 (28,072)	0 (0)	0 (548)	8913 (3135)	8913 (24,389)
	PSAX_tp1	8819 (34,166)	5994 (23,855)	1245 (3386)	1031 (5641)	549 (1284)
	PSAX_ta0	3999 (11,690)	0 (0)	425 (337)	3574 (10,805)	0 (548)
	PSAX_ta1	5228 (19,588)	0 (0)	394 (0)	4834 (4594)	0 (14,994)
	PSAX_ta2	4602 (13,963)	0 (0)	0 (0)	4602 (0)	0 (13,963)
Sliding	PSAX_hp2	4911 (17,973)	0 (0)	0 (0)	0 (0)	4911 (17,973)
	PSAX_hp1	11,240 (17,561)	0 (0)	476 (0)	5382 (3684)	5382 (13,877)
	PSAX_hp0	5467 (15,490)	0 (0)	5467 (1466)	0 (11,088)	0 (2936)
	PSAX_lp1	11,091 (33,854)	8665 (18,710)	2426 (3353)	0 (9322)	0 (2469)
	PSAX_lp2	17,596 (29,512)	0 (0)	0 (0)	8798 (2517)	8798 (26,995)
(III) Apical Four-Chamber View						
			Quality			
Position		**Total**	**Best**	**Acceptable**	**Poor**	**Bad**
Optimal view	A4C	38,733 (123,392)	17,228 (53,867)	10,811 (30,277)	9863 (32,544)	831 (6704)
Rotating	A4C_cc2	5530 (18,571)	0 (0)	0 (0)	5530 (0)	0 (18,571)
	A4C_cc1	4727 (17,418)	0 (0)	3421 (0)	1306 (11,052)	0 (6366)
	A4C_cw1	5530 (19,860)	0 (0)	0 (0)	0 (13,079)	5530 (6781)
	A4C_cw2	5418 (13,103)	0 (0)	0 (0)	0 (0)	5418 (13,103)
Tilting	A4C_tp0	3183 (9007)	0 (0)	2414 (0)	769 (6478)	0 (2529)
	A4C_tp1	4304 (14,638)	0 (0)	1895 (0)	2409 (7864)	0 (6774)
	A4C_tp2	4284 (16,666)	0 (0)	0 (0)	4284 (0)	0 (16,666)

https://doi.org/10.3390/diagnostics16071032

Table 4. *Cont.*

(III) Apical Four-Chamber View						
			Quality			
Position		Total	Best	Acceptable	Poor	Bad
Rocking, followed by Rotating/Sliding	A4C_ra_cc	4965 (17,774)	0 (0)	0 (0)	0 (274)	4965 (17,500)
	A4C_ra	5220 (15,561)	0 (0)	548 (1859)	3153 (12,334)	1519 (1368)
	A4C_ra_cw	4444 (12,964)	0 (0)	0 (0)	0 (0)	4444 (12,964)
	A4C_rt_cc	3695 (12,818)	0 (0)	0 (0)	0 (1113)	3695 (11,705)
	A4C_rt	11,543 (41,674)	343 (822)	2390 (13,687)	6036 (15,262)	2774 (11,903)
	A4C_rt_cw	5250 (15,860)	0 (0)	0 (0)	0 (845)	5250 (15,015)
	A4C_rt_sm	7855 (27,049)	0 (0)	0 (0)	377 (2393)	7478 (24,656)
Placing at the midpoint	A4Chw_rt	4753 (19,342)	0 (0)	274 (0)	4479 (2784)	0 (16,558)
	A4Chw	4097 (13,774)	0 (0)	2314 (0)	1783 (8046)	0 (5728)
	A4Chw_ra	4482 (17,289)	0 (0)	0 (0)	0 (778)	4482 (16,511)

Numbers in parentheses represent the combined number of samples used for training and validation.

3.2. Results of the View Classification Model

Figure 6 shows a confusion matrix for classifying echocardiographic images into three standard views: the parasternal long-axis, parasternal short-axis, and apical four-chamber views. Numbers within each square indicate the number of images in each category.

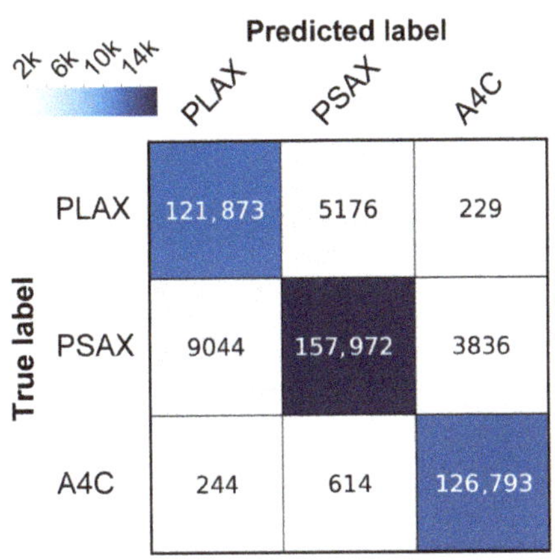

Figure 6. Results of the View Classification Model. The recall, precision, and F1-score of the view classification model for classifying three standard views were 0.959, 0.954, and 0.956, respectively. Among the 425,781 images from the 15 test participants in the dataset, 19,143 images (4.5%) were misclassified as incorrect views.

https://doi.org/10.3390/diagnostics16071032

Overall, 121,873 images with the true label "PLAX" were correctly classified as "PLAX." However, 5176 images with the true label "PLAX" were misclassified as "PSAX", and 229 images with the true label "PLAX" were misclassified as "A4C". Similarly, 9044 images with the true label "PSAX" were misclassified as "PLAX," while 157,972 images with the true label "PSAX" were correctly classified. Additionally, 3836 images with the true label "PSAX" were misclassified as "A4C." For the true label "A4C," 244 images were misclassified as "PLAX," and 614 images with the true label "A4C" were misclassified as "PSAX," while 126,793 images with the true label "A4C" were correctly classified. The rows indicate true labels, and the columns indicate predicted labels. Color intensity corresponds to the number of images, as indicated by the color bars.

Evaluating the view classification model designed to categorize the three standard views achieved an accuracy of 0.955, a recall of 0.959, a precision of 0.954, and an F1-score of 0.956 (Figure 6). Among the 425,781 images in the test dataset (15 participants), 19,143 images (4.5%) were misclassified as incorrect. Given the high F1-score, the view classification model was considered suitable as a pretrained model and was subsequently used to develop the position and quality evaluation models.

3.3. Results of the Position Evaluation Model

The performance of the position evaluation models, which assess whether the probe is correctly positioned to capture optimal views, including the PLAX view, PSAX view, and A4C view, was evaluated using the test dataset (Table 5, Figure 7).

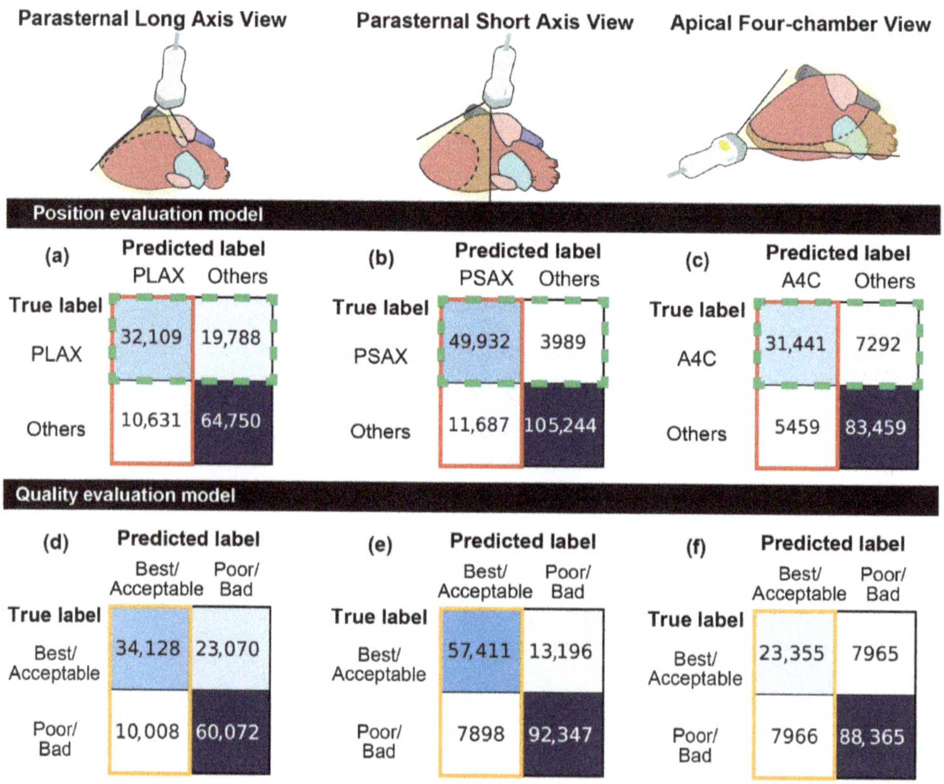

Figure 7. Inference Results of the Position Evaluation Model: The vertical axis represents the actual labels, and the horizontal axis represents the predicted labels. Each cell color indicates the number

of samples. (**a**) Position evaluation model for parasternal long-axis view. (**b**) Position evaluation model for parasternal short-axis view. (**c**) Position evaluation model for apical four-chamber view. Inference Results of the Quality Evaluation Model: The vertical axis represents the actual labels, and the horizontal axis represents the predicted labels. Each cell color indicates the number of samples. (**d**) Quality evaluation model for parasternal long-axis view. (**e**) Quality evaluation model for parasternal short-axis view. (**f**) Quality evaluation model for apical four-chamber view. Three colored outlines (green, red, and orange) enclose parts of the confusion matrix and indicate the combinations corresponding to the regions in the Venn diagram in Figure 8.

Table 5. Performance Metrics of Position Evaluation Models for Standard Views.

View, Optimal View	Accuracy	Recall	Precision	F1-Score
Parasternal Long-Axis View, PLAX	0.761	0.619	0.751	0.679
Parasternal Short-Axis View, PSAX	0.908	0.926	0.810	0.864
Apical Four-chamber View, A4C	0.900	0.812	0.852	0.831

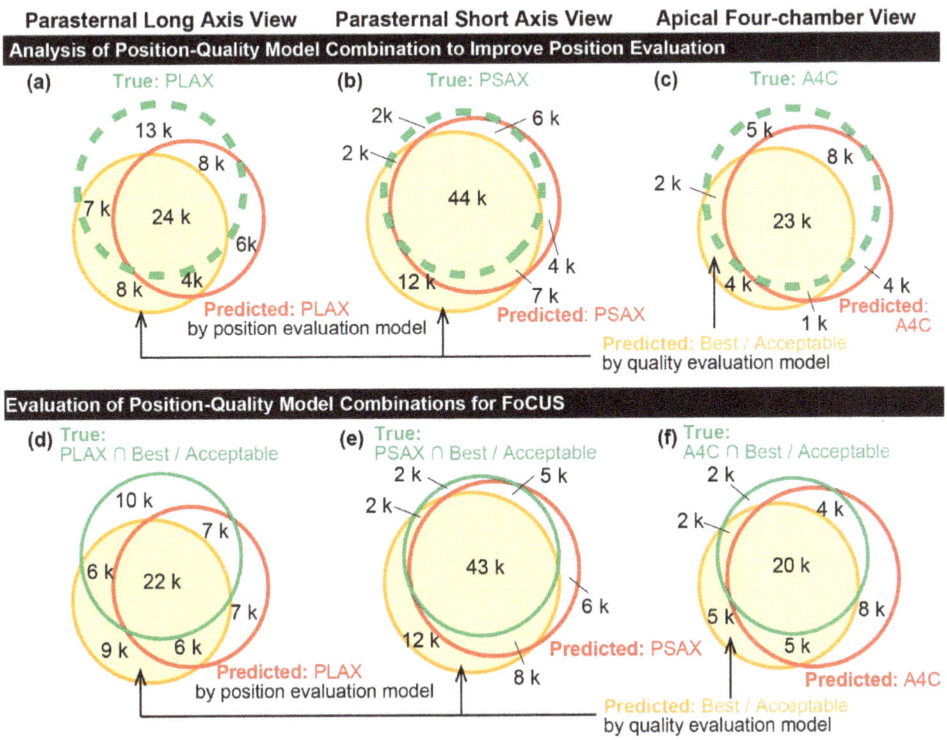

Figure 8. Venn Diagrams of the Predicted Appropriate Sections by Position and Quality Evaluation Models Compared with the Ground Truth: The green dashed line, red solid line, and purple solid line represent the regions of the ground truth labels for appropriate sections, the predicted regions by the position evaluation model, and the predicted regions by the quality evaluation model, respectively. (**a**) Parasternal long-axis view. (**b**) Parasternal short-axis view. (**c**) Apical four-chamber view. Venn Diagrams of the Predicted Appropriate Sections with Best or Acceptable Quality by Position and Quality Evaluation Models Compared with the Ground Truth: The ochre dashed line, red solid line, and purple solid line represent the regions of the ground truth labels for appropriate sections with acceptable quality (best or acceptable), the predicted regions by the position evaluation model, and the predicted regions by the quality evaluation model, respectively. (**d**) Parasternal long-axis view. (**e**) Parasternal short-axis view. (**f**) Apical four-chamber view.

Figure 7 shows the confusion matrices for the position evaluation model applied separately to each standard view: (a) PLAX, (b) PSAX, and (c) A4C views. The matrix compares the number of images correctly classified as "optimal view" versus those classified as "other" positions for each view.

For the PLAX model, 32,109 images with the true label "PLAX" were correctly identified as optimal, while 19,788 were misclassified as "others." Conversely, 10,631 images from other positions were misclassified as "PLAX," and 64,750 were correctly classified as "others." The model achieved an accuracy of 0.761, a recall of 0.619, a precision of 0.751, and an F1-score of 0.679 when evaluating the proper sections. When including optimal and non-optimal sections (full range of positional classes), the model achieved a recall of 0.492, a precision of 0.477, and an F1-score of 0.474 (see Supplementary Figure S1(I) online).

For the PSAX model, 49,932 images with the true label "PSAX" were correctly identified, while 3989 were misclassified as "others." Among the images from other positions, 11,687 were incorrectly classified as "PSAX" and 105,244 were correctly identified as "others." The accuracy, recall, precision, and F1-score for the proper sections were 0.908, 0.926, 0.810, and 0.864, respectively. When including optimal and non-optimal sections, the recall, precision, and F1-score were 0.656, 0.680, and 0.660, respectively (see Supplementary Figure S1(II)).

For the A4C model, 31,441 images with the true label "A4C" were correctly classified, while 7292 were misclassified as "others." Conversely, 5459 images from other positions were incorrectly classified as "A4C," and 83,459 were correctly classified as "others." The model achieved an accuracy of 0.900, a recall of 0.812, a precision of 0.852, and an F1-score of 0.831 for the proper sections. When including both optimal and non-optimal sections, the recall, precision, and F1-score were 0.585, 0.614, and 0.573, respectively (see Supplementary Figure S1(III)).

In all matrices, the rows correspond to the true labels and the columns to the predicted labels. Color intensity represents the number of images, as shown by the color scale above each matrix.

3.4. Results of the Quality Evaluation Model

The performance of the quality evaluation model, which assessed whether key anatomical structures are visible without prominent artifacts, was tested using four quality levels (best, acceptable, poor, and bad) on the test dataset (Table 6, Figure 7).

Table 6. Performance Metrics of Quality Evaluation Models for Standard Views.

View	Accuracy	Recall	Precision	F1-Score
Parasternal Long-Axis View	0.740	0.597	0.773	0.674
Parasternal Short-Axis View	0.877	0.813	0.879	0.845
Apical Four-chamber View	0.875	0.746	0.746	0.746

Figure 7 shows the confusion matrices for the quality evaluation model applied separately to the (d) PLAX, (e) PSAX, and (f) A4C views. In each matrix, true labels are categorized as "Best/Acceptable" or "Poor/Bad," with predictions shown in the same categories.

For the PLAX view, 34,128 images with the true label "Best/Acceptable" were correctly classified, while 23,070 were misclassified as "Poor/Bad." Among the images labeled "Poor/Bad," 10,008 were misclassified as "Best/Acceptable," while 60,072 were correctly classified. The model achieved an accuracy of 0.740, a recall of 0.597, a precision of 0.773, and an F1-score of 0.674 when assessing combined "best" and "acceptable" quality.

For the PSAX view, 57,411 images with the true label "Best/Acceptable" were correctly identified, while 13,196 were misclassified as "Poor/Bad." Among the images labeled "Poor/Bad," 7898 were incorrectly classified as "Best/Acceptable," and 92,347 were correctly identified. The corresponding scores were an accuracy of 0.877, a recall of 0.813, a precision of 0.879, and an F1-score of 0.845.

For the A4C view, 23,355 images with the true label "Best/Acceptable" were correctly classified, while 7965 were misclassified as "Poor/Bad." Among the "Poor/Bad" images, 7966 were incorrectly labeled as "Best/Acceptable," and 88,365 were correctly classified. The corresponding scores were an accuracy of 0.875, while recall, precision, and F1-score were 0.746 across all metrics.

The rows indicate true labels, whereas the columns indicate predicted labels. Color intensity corresponds to the number of images, as shown by the scale bar above each matrix.

3.5. Analysis of Position-Quality Model Combination to Improve Position Evaluation

The combination of the position and quality evaluation models was compared with the standalone position evaluation model (Table 7, Figure 8a–c). The Venn diagrams show the overlap between the ground truth labels (green dashed circles), regions predicted by the position evaluation model (red solid circles), and regions predicted by the quality evaluation model (purple solid circles) for each standard view: (a) PLAX, (b) PSAX, and (c) A4C. The corresponding areas in the confusion matrices in Figure 7 are aligned with the Venn diagrams in Figure 8a–c. The two confusion matrices at the bottom of the figure represent the results of the position evaluation model (left) and quality evaluation model (right).

Table 7. Comparison of Position-Quality Model Combination and Standalone Position Evaluation Model.

Optimal View	Model	Recall	Precision	F1-Score
PLAX	position * ∩ quality evaluation	0.471	0.850	0.606
	position * ∪ quality evaluation	0.757	0.676	0.714
	Standalone position evaluation	0.619	0.751	0.679
PSAX	position ∩ quality evaluation	0.815	0.857	0.835
	position ∪ quality evaluation	0.964	0.687	0.802
	Standalone position evaluation	0.926	0.810	0.864
A4C	position ∩ quality evaluation	0.602	0.946	0.736
	position ∪ quality evaluation	0.869	0.773	0.818
	Standalone position evaluation	0.812	0.852	0.831

* A ∩ B: Intersection of A and B, A ∪ B: Union of A and B.

For the PLAX view, 51,897 images were labeled using PLAX. The position evaluation model predicted 42,740 images as PLAX, while the quality evaluation model predicted 44,136 images as "Best/Acceptable." The intersection of the two models produced 28,722 images, of which 24,419 were correctly labeled as PLAX. Using this intersection, the recall, precision, and F1-score were 0.471, 0.850, and 0.606, respectively. Using the union, 58,154 images were produced, and the corresponding metrics were 0.757, 0.676, and 0.714, representing an improvement in the F1-score compared to the position model alone (0.679).

For the PSAX view, 53,921 images were labeled with PSAX. The position evaluation model predicted 61,619 as PSAX, while the quality evaluation model predicted 65,309 as "Best/Acceptable." The intersection yielded 51,265 images with a recall, a precision, and an

https://doi.org/10.3390/diagnostics16071032

F1-score of 0.815, 0.857, and 0.835, respectively. The union yielded 75,663 images with a recall, precision, and F1-score of 0.964, 0.687, and 0.802, respectively, both of which were lower than the 0.864 achieved by the position model alone.

For the A4C view, 38,733 images were labeled as A4C. The position evaluation model predicted 36,900 as A4C, while the quality evaluation model predicted 31,321 as "Best/Acceptable." This intersection yielded 24,667 images with a recall, a precision, and an F1-score of 0.602, 0.946, and 0.736, respectively. The union yielded 43,554 images with a recall, a precision, and an F1-score of 0.869, 0.773, and 0.818, respectively. However, both approaches showed lower F1-scores than that of the positional model alone (0.831).

Table 7 shows the estimation results. Collectively, these findings indicate that combining the position and quality evaluation models enhanced the F1-score for the PLAX view, especially when using the union approach, while for the PSAX and A4C views, the position model alone outperformed the combined approach.

3.6. Evaluation of Position-Quality Model Combinations for FoCUS

To identify images suitable for FoCUS, position and quality evaluation models were combined to classify images meeting the appropriate position and quality criteria (labeled as either "best" or "acceptable"). The inference results are shown in Table 8 and Figure 8d–f.

Table 8. Comparison of Models for Identifying Images Meeting Both Position and Quality Criteria.

Optimal View	Model	Recall	Precision	F1-Score
PLAX	PLAX * ∩ Appropriate Quality	0.493	0.777	0.603
	PLAX * ∪ Appropriate Quality	0.788	0.613	0.690
PSAX	PSAX ∩ Appropriate Quality	0.843	0.843	0.843
	PSAX ∪ Appropriate Quality	0.971	0.657	0.784
A4C	A4C ∩ Appropriate Quality	0.718	0.816	0.764
	A4C ∪ Appropriate Quality	0.937	0.603	0.734

* A ∩ B: intersection of A and B, A ∪ B: union of A and B.

Figure 8d–f shows the results of combining the position and quality evaluation models to identify the images suitable for FoCUS. The green dashed circles represent the ground truth labels of images classified as appropriate position and "Best/Acceptable" quality, the red solid circles indicate the regions predicted by the position evaluation model, and the purple solid circles indicate the regions predicted by the quality evaluation model.

For the PLAX view, 45,235 images were labeled as PLAX with Best/Acceptable quality. The position evaluation model predicted 42,740 images as PLAX, whereas the quality evaluation model predicted 44,136 images as best or acceptable. The intersection of the two models produced 28,722 images, of which 22,308 were correctly labeled as PLAX with the Best/Acceptable quality. The union resulted in 58,154 images, of which 35,647 were correctly labeled as PLAX with the Best/Acceptable quality. The calculated metrics were as follows: recall 0.493, precision 0.777, and F1-score 0.603 for the intersection, and recall 0.788, precision 0.613, and F1-score 0.690 for the union.

For the PSAX view, 51,222 images were labeled as PSAX with Best/Acceptable quality. The position evaluation model predicted 61,619 images as PSAX, whereas the quality evaluation model predicted 65,309 images as Best/Acceptable. The intersection of the two models produced 51,265 images, of which 43,199 were correctly labeled as PSAX with the Best/Acceptable quality. The union resulted in 75,663 images, of which 49,714 were correctly labeled as PSAX with the Best/Acceptable quality. The calculated metrics are as follows: recall = 0.843, precision = 0.843, and F1-score 0.843 for the intersection, and recall 0.971, precision 0.657, and F1-score 0.784 for the union.

For the A4C view, 28,039 images were labeled as A4C with Best/Acceptable quality. The position evaluation model predicted 36,900 images as A4C, whereas the quality evaluation model predicted 31,321 images as Best/Acceptable. The intersection of the two models yielded 24,667 images, of which 20,134 were correctly labeled as A4C with Best/Acceptable quality. The union resulted in 43,554 images, of which 26,278 were correctly labeled as A4C with Best/Acceptable quality. The calculated metrics were recall 0.718, precision 0.816, and F1-score 0.764 for the intersection, and recall 0.937, precision 0.603, and F1-score 0.734 for the union.

4. Discussion

Securing adequate practice time is essential to improve TTE skills. However, owing to the limited availability of clinical instructors, the demand for educational applications that support independent learning is increasing. The core component of these applications is an AI system capable of evaluating the quality of echocardiographic images and the positioning of the ultrasound probe. In this study, we developed and evaluated an image-assessment AI system using a dataset collected prospectively from healthy volunteers. A two-step framework was used to evaluate the performance of the AI system.

In the first step, a view classification model was developed to classify the images into three standard views: PLAX, PSAX, and A4C. The model achieved a high F1-score of 0.956. In the second step, the system evaluated whether the images depicted optimal cross sections of these three standard views and assessed their quality. The F1 scores of the position evaluation models for each standard view were 0.679 for PLAX, 0.864 for PSAX, and 0.831 for A4C. Conversely, the F1-scores of the quality evaluation model were 0.674 for PLAX, 0.845 for PSAX, and 0.746 for A4C. Inference methods that integrate a position-evaluation model with a quality-evaluation model were explored in this study. Among the three standard echocardiographic views, the union of the position and quality evaluation models yielded a higher F1-score in the PLAX view (0.714) than the position model alone (0.679). Conversely, in the PSAX and A4C views, the F1-score achieved by the position evaluation model alone outperformed the union and intersection combinations of the two models. Notably, the F1-score for the PLAX view using the position evaluation model alone was 0.679, over 0.1 points lower than the scores for the PSAX and A4C views, which were 0.864 and 0.831, respectively. These results indicate that combining the position and quality evaluation models improved the F1-score of the PLAX view, thereby compensating for its relatively lower baseline performance.

The position and quality evaluation models were integrated in this study to evaluate FoCUS usable imaging. For the PLAX view, combining the position and quality evaluation models using a union approach enhanced the F1-score to 0.714. For the PSAX view, the intersection approach produced an F1-score of 0.835, while for the A4C view, the union approach achieved an F1-score of 0.818. The proposed position and quality evaluation models and their integration showed adequate accuracy in assessing the appropriate sections.

The aim of this study was to estimate the probe position from echocardiographic images; however, the accuracy of classifying non-optimal cross-sectional views remained relatively low. One contributing factor was the use of multiclass classification within each standard view, involving 15 to 19 different classes. Grouping similar classes based on the degree of deviation, for instance, by combining PLAX_cc1 and PLAX_cc2, may improve the classification performance for non-optimal sections.

Additionally, the quality evaluation was based on the depiction of anatomical structures, some of which were very small in the images. This unique characteristic of TTE

https://doi.org/10.3390/diagnostics16071032

images, where small anatomical structures significantly influence quality ratings, poses challenges for developing quality evaluation models. The limited training data compared to other studies also restricted the performance. Data augmentation was used to enhance diversity; nonetheless, its impact was insufficient.

The proposed models were designed for use on tablets and smartphones using MobileViTv2_075 for efficient operation in resource-constrained environments. Recent advancements in image recognition, including Vision–Language Models (VLMs) [23,24], may further improve classification accuracy. Among these, Bootstrapping Language–Image Pre-training (BLIP) is a vision–language pre-training framework that supports both understanding and generation tasks [25]. BLIP enhances supervision by generating synthetic captions and filtering out low-quality samples, thereby improving model robustness. Future work may explore the application of VLMs such as BLIP to further improve image assessment performance.

This study makes three major contributions to the literature. First, it is novel because a unique dataset collected from healthy young adults, reflecting practice scenarios among beginner-level trainees, was used for the development of the AI model. Second, the model was trained using a dataset that included a large number of suboptimal images that deviated from optimal cross-sectional views, similar to those beginners often encounter during image acquisition. This deliberate inclusion of non-ideal images has enabled the development of position classifiers capable of evaluating subtle deviations from optimal views, an aspect often overlooked in previous studies. Furthermore, a quality evaluation was conducted based on anatomical visibility to address the unique imaging characteristics of TTE. Third, our proposed framework that integrates position and quality models improved overall performance.

The performances of the models were rigorously evaluated. Position classifiers achieved strong performance for the PSAX and A4C views; however, combining them with quality classifiers further improved the results for views with lower baseline performance, including the PLAX view. The integration of the position and quality models enabled the identification of images suitable for FoCUS, even without the supervision of an expert. These findings highlight the potential of AI-based systems to support independent simulation-based ultrasound training, particularly in environments with limited access to instructors.

Supplementary Materials: The following supporting information can be downloaded at: https://www.mdpi.com/article/10.3390/diagnostics16071032/s1, Table S1: Position evaluation criteria; Table S2: Quality evaluation criteria; Table S3: Inference time, model size, and overfitting-related parameters for each backbone architecture; Table S4: Distribution of positions and quality in the main training dataset; Figure S1: Classification results of the position evaluation model for all sections, including non-optimal sections.

Author Contributions: Conceptualization, S.T. (Sanshiro Togo); methodology, S.T. (Sanshiro Togo), S.T. (Shogo Tsuge), D.I., Y.S., M.H., T.H., Y.K. and K.T.; software, K.S., S.O., D.I. and Y.K.; validation, S.O., D.I. and Y.K.; formal analysis, S.T. (Sanshiro Togo), S.T. (Shogo Tsuge), D.I., Y.S. and S.O.; investigation, S.T. (Sanshiro Togo), S.T. (Shogo Tsuge), D.I., M.H. and T.H.; resources, S.T. (Sanshiro Togo) and K.T.; data curation, D.I., S.O., Y.S. and Y.K.; writing—original draft preparation, D.I.; writing—review and editing, K.T. and S.T. (Sanshiro Togo); visualization, D.I.; supervision, K.N. and T.S.; project administration, S.T. (Sanshiro Togo). All authors have read and agreed to the published version of the manuscript.

Funding: This research received no external funding.

Institutional Review Board Statement: The study was conducted in accordance with the Declaration of Helsinki, and approved by The Ethics Committee of Hamamatsu University School of

Medicine (EC HUSM) and the Conflict-of-Interest Management Committee (Approval No. 22-121, 28 September 2022).

Informed Consent Statement: Written informed consent has been obtained from the subjects to publish this paper.

Data Availability Statement: The datasets generated or analyzed during the current study are held at Hamamatsu University School of Medicine. Access to the data is not publicly available due to institutional and ethical restrictions, but may be granted to qualified researchers upon reasonable request. Data access requests should be submitted to the Hamamatsu University School of Medicine, Next Generation Creative Education Center for Medicine, Engineering, and Informatics (hama-nxcec511@hama-med.ac.jp) and will require prior approval from the committee.

Acknowledgments: We would like to express our heartfelt gratitude to Hiromitsu Kataoka for his technical guidance in TTE and video recording, Nanami Saito and Sayaka Ogawa from HERUS (Hamamatsu ER UltraSound) for their assistance in the video recording of female subjects, and Kenshiro Tajima and Katsuma Horio for conducting the preliminary business investigation into the potential of this study.

Conflicts of Interest: The authors declare no conflicts of interest.

References

1. Neskovic, A.N.; Skinner, H.; Price, S.; Via, G.; De Hert, S.; Stankovic, I.; Galderisi, M.; Donal, E.; Muraru, D.; Sloth, E.; et al. Focus cardiac ultrasound core curriculum and core syllabus of the European Association of Cardiovascular Imaging. *Eur. Heart J. Cardiovasc. Imaging* **2018**, *19*, 475–481. [CrossRef]
2. Andersen, C.A.; Holden, S.; Vela, J.; Rathleff, M.S.; Jensen, M.B. Point-of-care ultrasound in general practice: A systematic review. *Ann. Fam. Med.* **2019**, *17*, 61–69. [CrossRef] [PubMed]
3. Frankel, H.L.; Kirkpatrick, A.W.; Elbarbary, M.; Blaivas, M.; Desai, H.; Evans, D.; Summerfield, D.T.; Slonim, A.; Breitkreutz, R.; Price, S.; et al. Guidelines for the appropriate use of bedside general and cardiac ultrasonography in the evaluation of critically ill patients-part I: General ultrasonography. *Crit. Care Med.* **2015**, *43*, 2479–2502. [CrossRef] [PubMed]
4. Oren-Grinberg, A.; Talmor, D.; Brown, S.M. Focused critical care echocardiography. *Crit. Care Med.* **2013**, *41*, 2618–2626. [CrossRef] [PubMed]
5. Cheng, J.K.; Arntfield, R. Critical care echocardiography: Training, imaging, and indications. *BJA Educ.* **2024**, *24*, 399–408. [CrossRef]
6. Hoppmann, R.A.; Mladenovic, J.; Melniker, L.; Badea, R.; Blaivas, M.; Montorfano, M.; Abuhamad, A.; Noble, V.; Hussain, A.; Prosen, G.; et al. International consensus conference recommendations on ultrasound education for undergraduate medical students. *Ultrasound J.* **2022**, *14*, 31. [CrossRef]
7. Yamada, H.; Ohara, T.; Abe, Y.; Iwano, H.; Onishi, T.; Katabami, K.; Takigiku, K.; Tada, A.; Tanigushi, H.; Mihara, H.; et al. Guidance for performance, utilization, and education of cardiac and lung point-of-care ultrasonography from the Japanese Society of Echocardiography. *J. Echocardiogr.* **2024**, *22*, 113–151. [CrossRef]
8. Kusunose, K.; Haga, A.; Inoue, M.; Fukuda, D.; Yamada, H.; Sata, M. Clinically feasible and accurate view classification of echocardiographic images using deep learning. *Biomolecules* **2020**, *10*, 665. [CrossRef]
9. Zhang, J.; Gajjala, S.; Agrawal, P.; Tison, G.H.; Hallock, L.A.; Beussink-Nelson, L.; Lassen, M.H.; Fan, E.; Aras, M.A.; Jordan, C.; et al. Fully automated echocardiogram interpretation in clinical practice: Feasibility and diagnostic accuracy. *Circulation* **2018**, *138*, 1623–1635. [CrossRef]
10. Madani, A.; Arnaout, R.; Mofrad, M.; Arnaout, R. Fast and accurate view classification of echocardiograms using deep learning. *NPJ Digit. Med.* **2018**, *1*, 6. [CrossRef]
11. Naser, J.A.; Lee, E.; Pislaru, S.V.; Tsaban, G.; Malins, J.G.; Jackson, J.I.; Anisuzzaman, D.M.; Rostami, B.; Lopez-Jimenez, F.; Friedman, P.A.; et al. Artificial intelligence-based classification of echocardiographic views. *Eur. Heart J.-Digit. Health* **2024**, *5*, 260–269. [CrossRef]
12. Gao, Y.; Zhu, Y.; Liu, B.; Hu, Y.; Yu, G.; Guo, Y. Automated recognition of ultrasound cardiac views based on deep learning with graph constraint. *Diagnostics* **2021**, *11*, 1177. [CrossRef] [PubMed]
13. Zhu, Y.; Ma, J.; Zhang, Z.; Zhang, Y.; Zhu, S.; Liu, M.; Zhang, Z.; Wu, C.; Yang, X.; Cheng, J.; et al. Automatic view classification of contrast and non-contrast echocardiography. *Front. Cardiovasc. Med.* **2022**, *9*, 989091. [CrossRef] [PubMed]
14. Ornstein, H.; Adam, D. Classification of Echocardiogram View using A Convolutional Neural Network. *Artif. Intell. Res.* **2021**, *11*, 1. [CrossRef]

https://doi.org/10.3390/diagnostics16071032

15. Zolgharni, M.; Azarmehr, N.; Ye, X.; Howard, J.P.; Lane, E.S.; Labs, R.; Shun-Shin, M.J.; Cole, G.D.; Bidaut, L.; Francis, D.P. Neural architecture search of echocardiography view classifiers. *J. Med. Imaging* **2021**, *8*, 034002.
16. Jansen, G.E.; de Vos, B.D.; Molenaar, M.A.; Schuuring, M.J.; Bouma, B.J.; Išgum, I. Automated echocardiography view classification and quality assessment with recognition of unknown views. *J. Med. Imaging* **2024**, *11*, 054002. [CrossRef]
17. Abdi, A.H.; Luong, C.; Tsang, T.; Jue, J.; Gin, K.; Yeung, D.; Hawley, D.; Rohling, R.; Abolmaesumi, R. Quality assessment of echocardiographic cine using recurrent neural networks: Feasibility on five standard view planes. In *Lecture Notes in Computer Science (Including Subseries Lecture Notes in Artificial Intelligence and Lecture Notes in Bioinformatics)*; LNCS 302–310; Springer: Berlin/Heidelberg, Germany, 2017; Volume 10435.
18. Abdi, A.H.; Luong, C.; Tsang, T.; Allan, G.; Nouranian, S.; Jue, J.; Hawley, D.; Fleming, S.; Gin, K.; Swift, J.; et al. Automatic quality assessment of apical four-chamber echocardiograms using deep convolutional neural networks. In *Proceedings of the Medical Imaging 2017: Image Processing, Orlando, FL, USA, 11–16 February 2017*; SPIE: Bellingham, WA, USA, 2017; Volume 10133.
19. Mitchell, C.; Rahko, P.S.; Blauwet, L.A.; Canaday, B.; Finstuen, J.A.; Foster, M.C.; Horton, K.; Ogunyankin, K.O.; Palma, R.A.; Velazquez, E.J. Guidelines for Performing a Comprehensive Transthoracic Echocardiographic Examination in Adults: Recommendations from the American Society of Echocardiography. *J. Am. Soc. Echocardiogr.* **2019**, *32*, 1–64. [CrossRef]
20. Ouyang, D.; He, B.; Ghorbani, A.; Yuan, N.; Ebinger, J.; Langlotz, C.P.; Heidenreich, P.A.; Harrington, R.A.; Liang, D.H.; Ashley, E.A.; et al. Video-based AI for beat-to-beat assessment of cardiac function. *Nature* **2020**, *580*, 252–256.
21. Mehta, S.; Rastegari, M. MobileViT: Light-weight, General-purpose, and Mobile-friendly Vision Transformer. *arXiv* **2022**. [CrossRef]
22. Mehta, S.; Rastegari, M. Separable Self-attention for Mobile Vision Transformers. *arXiv* **2022**. [CrossRef]
23. Diao, H.; Cui, Y.; Li, X.; Wang, Y.; Lu, H.; Wang, X. Unveiling Encoder-Free Vision-Language Models. *arXiv* **2024**. [CrossRef]
24. Radford, A.; Kim, J.W.; Hallacy, C.; Ramesh, A.; Goh, G.; Agarwal, S.; Sastry, G.; Askell, A.; Mishkin, P.; Clark, J.; et al. Learning Transferable Visual Models from Natural Language Supervision. *arXiv* **2021**. [CrossRef]
25. Li, J.; Li, D.; Xiong, C.; Hoi, S. BLIP: Bootstrapping Language-Image Pre-training for Unified Vision-Language Understanding and Generation. *arXiv* **2022**. [CrossRef]

https://doi.org/10.3390/diagnostics16071032

diagnostics

MDPI

Article

Syncretic Grad-CAM Integrated ViT-CNN Hybrids with Inherent Explainability for Early Thyroid Cancer Diagnosis from Ultrasound

Ahmed Y. Alhafdhi *, Gibrael Abosamra and Abdulrhman M. Alshareef

Department of Information Systems, Faculty of Computing and Information Technology, King Abdulaziz University, Jeddah 21589, Saudi Arabia; gabosamra@kau.edu.sa (G.A.); amralshareef@kau.edu.sa (A.M.A.)
* Correspondence: aalhafdhi@stu.kau.edu.sa

Abstract

Background/Objectives: Accurate detection of thyroid cancer using ultrasound remains a challenge, as malignant nodules can be microscopic and heterogeneous, easily confused with point clusters and borderline-featured tissues. Current studies in deep learning demonstrate good performance with convolutional neural networks (CNNs) and clustering; however, many approaches focus on local tissue and provide limited, non-quantitative interpretation, reducing clinical confidence. This study proposes an integrated framework combining enhanced convolutional feature encoders (DenseNet169 and VGG19) with an enhanced vision transformer (ViT-E) to integrate local feature and global relational context during learning, rather than delayed integration. **Methods:** The proposed framework integrates enhanced convolutional feature encoders (DenseNet169 and VGG19) with an enhanced vision transformer (ViT-E), enabling simultaneous learning of local feature representations and global relational context. This design allows feature fusion during the learning stage instead of delayed integration, aiming to improve diagnostic performance and interpretability in thyroid ultrasound image analysis. **Results:** The best-performing model, ViT-E–DenseNet169, achieved 98.5% accuracy, 98.9% sensitivity, 99.15% specificity, and 97.35% AUC, surpassing the robust basic hybrid model (CNN–XGBoost/ANN) and existing systems. A second contribution is improved interpretability, moving from mere illustration to validation. Gradient-weighted class activation mapping (Grad-CAM) maps demonstrated distinct and clinically understandable concentration patterns across various thyroid cancers: precise intralesional concentration for high-confidence malignancies (PTC = 0.968), edge/interface concentration for capsule risk patterns (PTC = 0.957), and broader-field activation consistent with infiltration concerns (PTC = 0.984), while benign scans showed low and diffuse activation (PTC = 0.002). Spatial audits reinforced this behavior (IoU/PAP: 0.72/91%, 0.65/78%, 0.58/62%). **Conclusions:** The integrated ViT-E–DenseNet169 framework provides highly accurate thyroid cancer detection while offering clinically meaningful interpretability through Grad-CAM-based spatial validation, supporting improved confidence in AI-assisted ultrasound diagnosis.

Keywords: CNN; ViT-E; fusion features; Grad-CAM; XGBoost; ANN; thyroid cancer

1. Introduction

The thyroid gland, an endocrine gland, is located in the front of the neck below the larynx. It is one of the largest endocrine glands, typically weighing between 25 and 30 g. The gland secretes two hormones: triiodothyronine (T3) and thyroxine (T4) [1]. These hormones play crucial roles in regulating metabolism and growth, affecting almost all body tissues. The thyroid gland and its associated hormones affect a wide range of body systems, including the cardiovascular and nervous systems [2]. Common symptoms of thyroid dysfunction include anxiety, impaired cognitive function, menstrual irregularities, rapid heartbeat, muscle pain, weight gain, and elevated cholesterol levels [3]. Some thyroid disorders, including certain types of thyroid cancer, have genetic components. Exposure to ionizing radiation, particularly during childhood, increases the risk of thyroid cancer. Iodine deficiency or excess, as iodine is essential for thyroid hormone production, can lead to thyroid disorders. Iodine deficiency causes conditions such as goiter, whereas excessive iodine intake can lead to hyperthyroidism [4]. Several autoimmune diseases, such as Hashimoto's thyroiditis, are associated with hypothyroidism due to insufficient hormone production by the thyroid gland [5]. Graves' disease also causes hyperthyroidism due to the thyroid gland secreting excess hormones [6]. A thyroid cancer diagnosis requires clinical and imaging evaluations. Several methods exist for diagnosing this cancer, including a thorough neck examination by a specialist to detect any abnormalities, such as masses or nodules [7]. Ultrasound images provide detailed images of the thyroid gland and help evaluate thyroid nodules by determining their size, characteristics, and location [8]. Blood tests measure the concentrations of thyroid hormones (T3 and T4) and thyroid-stimulating hormone (TSH) to assess overall thyroid function [9]. CT and MRI are used to evaluate thyroid cancer and determine its spread to lymph nodes [10]. Manual diagnosis of thyroid cancer has several limitations [11]. In the early stages, the clinical symptoms are mild and similar to those of other conditions, making manual diagnosis difficult [12]. There is no consensus among specialists on how to identify suspicious thyroid nodules or differentiate between aggressive and non-aggressive nodules [13]. AI techniques have proven highly effective for processing manual diagnoses from CT data, particularly in the early stages [14]. CNN models analyze ultrasound images to identify features indicative of thyroid cancer. These networks analyze massive datasets at high speeds and recognize features that are difficult to detect with the naked eye. AI models assist in analyzing and classifying image data and provide a unified interpretation of ultrasound images [15].

This study addresses a long-standing diagnostic challenge in the early detection of thyroid cancer: the thin, gray line on ultrasound images, where human assessment is inadequate and conventional deep learning algorithms are often clinically inaccurate. Although traditional convolutional neural networks can extract local patterns, they are often used with perceptual segmentation. They are used to isolate tissues and margins without grasping the broader anatomical picture necessary for an accurate diagnosis. This creates a significant discrepancy between statistical and clinical validity. We aim to move beyond incremental model optimization and strive for architectural coherence. Our goal is to develop a comprehensive diagnostic system that fundamentally integrates the detail-focused, locally interpretable deep learning with a holistic, relational interpretation, both of which are essential for interpreting the entire ultrasound landscape and making all predictions highly accurate and virtually interpretable.

This paper discusses different methodologies, tools, and results of previous works with the aim of detecting cancerous thyroid nodules.

Sujini et al. [16] presented DL models: a six-layer CNN and a VGGNet-16. A 6-CNN model was developed for efficient end-to-end analysis, and ultrasound images of the thyroid containing malignant and benign cases were used. The combined CNN-VGGNet-16 technique achieved an accuracy of 0.97. Li et al. [17] introduced a transformer fusing the CNet method for automatically segmenting malignant thyroid nodules. The CNet comprises a CNN branch with a Large Kernel Module for precise shape feature extraction and an enhanced transformer branch with an Enhanced Transformer Module for remote-pixel connectivity in ultrasound images. A Multiscale Module was used to combine

https://doi.org/10.3390/diagnostics16070999

features from branches. Zhang et al. [18] presented multi-CNN trained for thyroid disease classification. This study also explored strategies to enhance the diagnostic accuracy of CNNs by combining feature maps of different scales. The multi-CNN outperformed the standard single-channel CNN. It achieved 0.91 accuracy, 0.94 precision, and 0.90 recall for thyroid disease classification. Namdeo et al. [19] presented a model for thyroid disorder diagnosis. First, image and data features were extracted using neighborhood-based PCA. Two classification processes follow: a CNN for image classification and an NN for disease classification, both using features as inputs. The combined results were used to enhance the diagnostic accuracy. Naglah et al. [20] developed a system to extract complex texture patterns using CNN. The system integrates multiple channels for all inputs, merging the collected scans into the DL and utilizing various adjustable diffusion gradient coefficient values. The system achieved accuracies of 0.87 and sensitivities of 0.69. Li et al. [21] presented a CNN-based system for thyroid nodule recognition. It employs an enhanced U-Net segmentation method to isolate the ROI, optimizes the ROIs using image processing, and classifies them as benign or malignant using a CNN-Fusion network. The results showed strong performance, with segmentation and classification values of 0.855 and 0.86, respectively. Zhao et al. [22] trained and tested five different CNNs on thyroid nodules images. The study found that the CNN models had significantly higher diagnostic performance (AUCs ranging from 0.901 to 0.947) for thyroid malignancies. The ensemble model, which combined three of the best-performing CNNs, achieved the highest AUC value, indicating its effectiveness in diagnosing thyroid nodules. El-Hossiny et al. [23] presented a CNN for TC classification. The CNN architecture achieved 94.69% accuracy in thyroid carcinoma classification. Aljameel et al. [24] designed an Explainable-ANN model to classify thyroid nodules and identify predictive factors for malignancy. The SMOTEENN sampling method was applied to address the class imbalance. The Explainable-ANN model achieved an accuracy of 0.82 and an AUC of 0.86. Wu et al. [25] used three CNNs for classifying thyroid ultrasound images. In independent testing, the best-performing DL algorithm achieved AUCs of 0.829 and 0.779, respectively. Zhang et al. [26] studied an InceptionResNetV2-based framework that was developed and evaluated. The framework includes three multichannel models. It outperformed the ML methods, achieving an accuracy rate of 0.971 and a recall of 0.90. Wang et al. [27] presented a CNN for TC image analyzed for clinicopathological factors. Independent risk factors for TC, such as nodule size and BRAF gene mutation, were identified. The CNN achieved an AUC of 0.78. Rho et al. [28] presented a study to evaluate a deep CNN for distinguishing between malignant and benign thyroid nodules. The CNN was trained on ultrasound images of larger nodules ($\geq$10 mm) and tested on smaller nodules (<10 mm). The CNN outperformed radiologists, achieving 0.832 accuracy, 0.383 specificity, and 0.66 in AUC. Vasile et al. [29] developed an ensemble method combining two DL models: one, called 5-CNN, and the other based on a repurposed and optimized VGG-19 architecture, for classifying thyroid ultrasound images. The ensemble CNN-VGG method achieved results surpassing both 5-CNN and VGG-19, with 97.35% accuracy and 95.75% sensitivity.

These studies collectively disclose three interdependent gaps in existing studies. First, most current frameworks are based on convolutional architectures that capture local image features but fail to capture global spatial interactions among ultrasound images. Second, most successful systems are black-box classifiers that offer no interpretability or visual justification for their predictions. Third, a comparatively small range of models attempts to combine feature extraction, global context modeling, and explainable inference into a single, unified architecture.

This study addresses these gaps. The proposed architecture enables local feature learning and long-range contextual reasoning to coexist within a single architecture by combining convolutional processing with vision transformer-based global attention mechanisms. More importantly, an interpretation was provided using Grad-CAM as part of the prediction pipeline. Rather than creating a plain label, the system creates spatially consistent visual explanations to indicate the areas that affect the diagnostic decision. The outcome is not just a classifier but something that can help clinical reasoning another step away; it is more of a digital assistant than a muted statistical engine. These explanations are already included in the updated manuscript to clarify the research gaps underlying the proposed hybrid CNN-ViT architecture.

The key contributions of this study are as follows:

- Development and evaluation of two new concurrent hybrid architectural models, ViT-DenseNet169 and ViT-VGG19. These are not clusters but integrated systems in which the ViT encoder and an optimized convolutional neural network backbone are trained together in continuous dialogue, enabling the simultaneous extraction of local features and their global contextualization.
- Integration of interpretability into core model processes. The Grad-CAM visualization method is integrated into the inference path, making the model appear as a transparent diagnostic partner rather than a black-box classifier. This allows for a graphical review of the model's spatial inference, whether it focuses on the nodule core, invasive margins, or diffuse normal tissue, directly linking its decisions to identifiable acoustic biomarkers and establishing the necessary clinical confidence.
- Development and evaluation of hybrid systems combining XGBoost-CNN and ANN-CNN for the diagnosis of thyroid cancer ultrasound images.
- Development and evaluation of hybrid systems integrating features from multiple CNN and classifying them using XGBoost and ANN were developed and evaluated to support the visual diagnosis of thyroid cancer.

The remainder of this paper is organized as follows. Section 2 describes the materials and methods used, including the dataset, image enhancement, and the architecture of the proposed hybrid systems (Section 2.5 Hybridization of ViT-E with DenseNet169 and VGG19; Section 2.6. Inference and imaging of the attention mechanism). In Section 3, the experimental results are presented and discussed, and the performance of the ViT-CNN hybrid systems is analyzed in the conclusion (Section 3.5). The combination of convolutional and diagnostic accuracy, along with a graphical study of their decision-making processes, is presented (Section 3.6). Grad-CAM was used as an interpretive lens. Section 4 provides a comprehensive discussion and comparison of the performances of all models. Finally, Section 5 concludes the study by summarizing the main findings and their implications for future research.

2. Materials and Methods

Figure 1 illustrates a sophisticated, comprehensive diagnostic system that combines convolutional neural network-based inference with transducers to detect thyroid cancer in ultrasound images. The process begins with specific preprocessing steps (enlargement, median filtering, and Laplace optimization) aimed at minimizing the impact of spotting noise while preserving the diagnostically relevant edges and internal structures. This is a crucial step because anatomically accurate input representation is essential for subsequent data interpretation.

Local morphological markers were simultaneously acquired using the DenseNet169 and VGG19 networks. These convolutional flows encode integrated fine-grained patterns, such as echo gradients, edge irregularities, and internal heterogeneity. Their outputs do not produce independent features; instead, they are integrated with the overall contextual representations learned by an optimized ViT network. Using linear projection and localized

https://doi.org/10.3390/diagnostics16070999

embedding, the ViT branch learns spatial relationship patterns across long distances within the gland, enabling a patch-level contextual interpretation beyond localized areas.

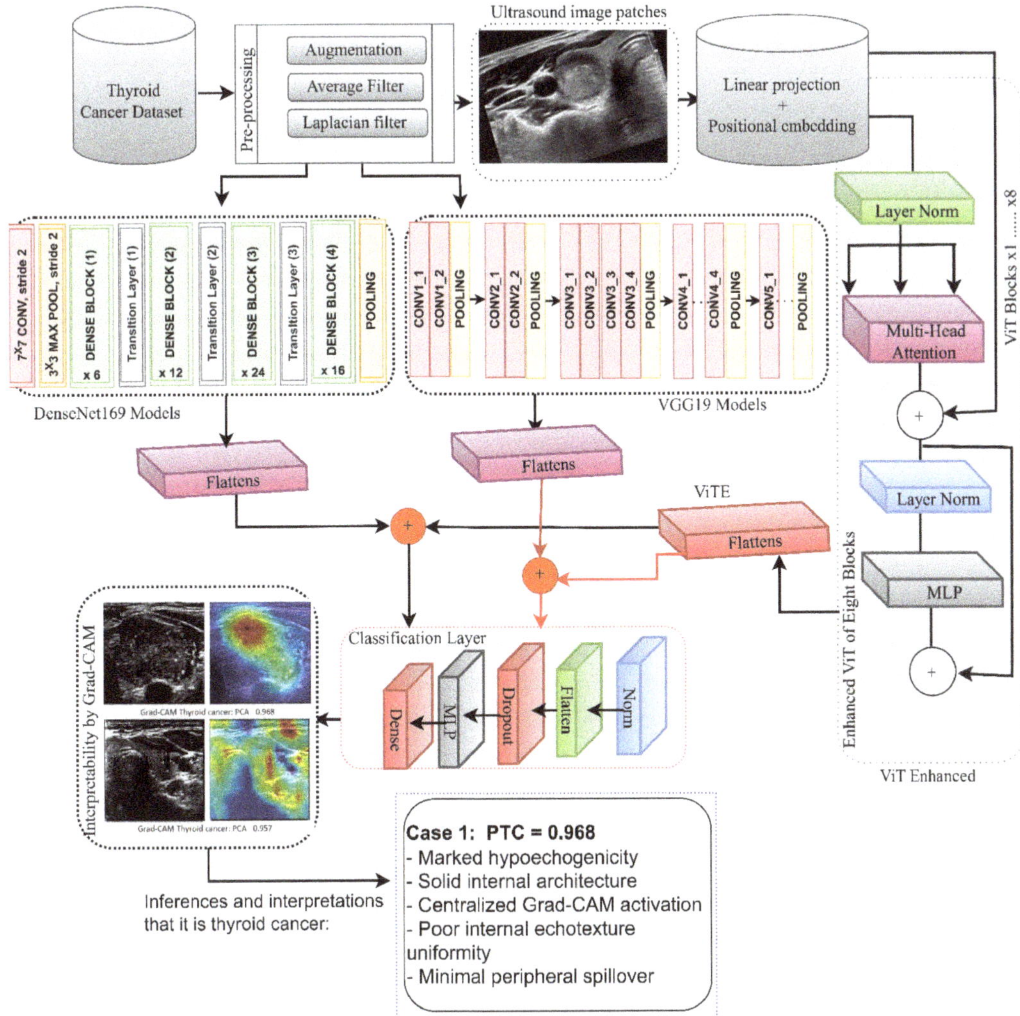

Figure 1. An interpretable end-to-end deep diagnostic architecture integrating dense convolutional encoding and vision transformer reasoning for robust thyroid cancer detection.

Interpretability is directly integrated via Grad-CAM, which operates on the last convolutional feature maps and redraws significance scores on the image. The resulting heat maps indicate the location and cause of the model's presence, which may be focal lesion activation, peripheral capsule focus, or diffuse benign activity. Model attention and known ultrasound biomarkers, such as low-echoic nuclei or invasive edges, were spatially aligned to provide a checklist for clinical inference.

Overall, the figure demonstrates that interpretability is not a supporting visual display layer; rather, it is an inherent feature of the model structure. The framework combines the specificity of DenseNet169/VGG19 features, ViT-based global inference, and Grad-

https://doi.org/10.3390/diagnostics16070999

CAM annotations to enhance thyroid cancer diagnosis, with improved transparency and clinical reliability.

2.1. Description of Dataset

The thyroid cancer ultrasound dataset consisted of 7288 images [30]. This dataset was divided into two categories: TC (4006 images) and normal tissue (3282 images). This dataset was used to train and evaluate the proposed systems for diagnosing thyroid cancer. Figure 2a shows examples of ultrasound images from the thyroid cancer dataset.

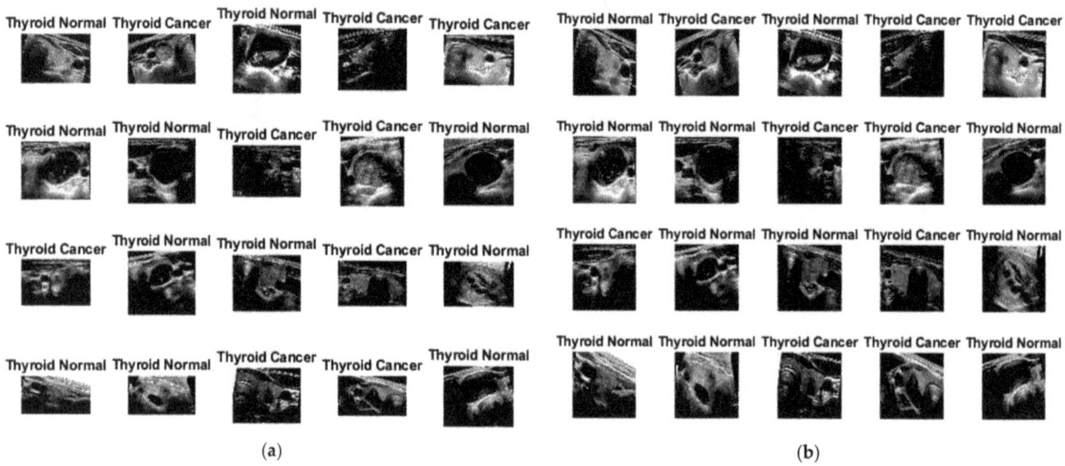

Figure 2. Sample thyroid images of the TC dataset. (**a**) Before enhancement; (**b**) after enhancement.

This study aimed to detect and characterize the early stages of thyroid cancer by applying the XGBoost algorithm and an ANN based on CNN features to ultrasound images from the thyroid cancer dataset. The dataset consisted of 7288 images divided into two categories: thyroid cancer and normal tissue. The dataset was distributed as follows: 4006 images of thyroid cancer tumors and 3282 images of normal thyroid tissue. It should be noted that the dataset is unbalanced between the two categories; therefore, this issue will be addressed. To facilitate training and validation, the dataset was divided into two distinct subsets, as shown in Table 1, with 80% of the data allocated to the training and validation subsets in an 80:20 ratio, respectively. This partitioning strategy enables the algorithms to learn from most of the data while simultaneously validating its performance. Furthermore, a dedicated subset comprising 20% of the dataset was retained for the precise evaluation of the systems.

Table 1. Distribution of thyroid histopathological images across training, validation, and testing sets.

Phase	80%		Testing 20%
Types	Training (64%)	Validation (16%)	
TC	2564	641	801
TN	2100	525	656

After retaining 20% of the 7288 images as a stable test set, we performed five-fold stratified cross-validation on the remaining 80% of the data to select the model and fine-tune the hyperparameters. Within each fold, four folds trained the model, while the remaining fold validated it, and the class ratios were maintained to preserve the TC/TN balance. Data

https://doi.org/10.3390/diagnostics16070999

augmentation (and any fitting steps, such as principal component analysis) was applied only within the training partition of each fold to prevent leakage.

2.2. Augment and Balance the Dataset

Data augmentation is a common method in image classification that increases the size of a dataset by creating new, slightly modified images from the existing dataset. This helps improve the model's performance by providing diverse training samples. In the context of a TC dataset, data augmentation is particularly useful for two main purposes: increasing the number of ultrasound images and balancing the datasets [31]. One way to augment the data is by applying rotation, scaling, and flipping transformations to the original images. For example, the original ultrasound images were rotated by degrees (90°, 180°, etc.) as shown in Figure 3. This simulates the different angles at which the ultrasound images were captured. These transformations create new images similar to the original images but with different features [32]. The images were flipped horizontally and/or vertically. This mimics the orientation of the thyroid gland in different patients. Scaling involves resizing images to slightly different dimensions. This accounts for variations in the image resolution and aspect ratio. The translation shifts the images horizontally and/or vertically [33]. This simulates slight changes in the position during ultrasound scans. The TC dataset is unbalanced; therefore, data augmentation was used to generate additional samples for the minority class (polyp images) [34]. In this study, the TC class images were doubled for each original image, and the normal class images were tripled for each original image. Thus, the training-phase dataset became the TC class 5128 and the normal class 6300.

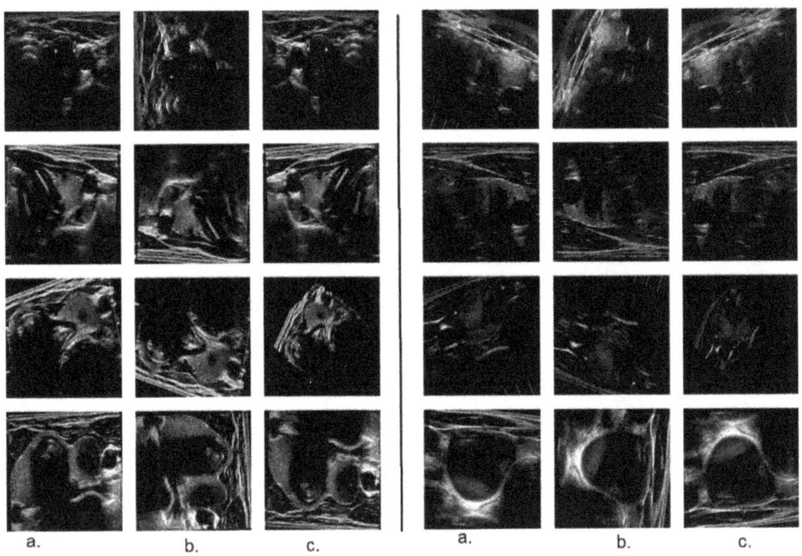

(A) (B)

Figure 3. Representative ultrasound images of thyroid cancer and normal thyroid with corresponding data augmentation transformations. The first three columns display original ultrasound images of thyroid cancer. (**A**) Data augmentation technique: Number of images of a thyroid cancer class. (**a**), alongside augmented variants (**b**,**c**). The following three columns present original ultrasound images of normal thyroid tissue. (**B**) Data augmentation technique: Number of images of a normal thyroid class. (**a**), with their corresponding augmented images (**b**,**c**) generated through data augmentation techniques.

https://doi.org/10.3390/diagnostics16070999

The held-out test set was not used in developing the model. Also, the test set was not subjected to any augmentation, preprocessing, fitting steps, or hyperparameter tuning. Data augmentation operations were confined to the training phase of each cross-validation fold, and the validation and test sets were left unchanged to ensure an unbiased assessment of the model.

2.3. Enhancing Ultrasound Images

Improving the ultrasound images of TC is necessary to ensure an accurate diagnosis in the subsequent stages. High-quality images aid in TC detection. In this study, average and Laplacian filters were employed. Noise appears in sound wave images as grainy spots and random variations in pixel density. An average filter was used to improve image quality by removing artifacts and noise. The use of an average filter with an operating factor of 5 × 5 helps reduce artifacts and noise, thereby improving the image quality of TC ultrasound images [35]. The average filter uses a small matrix, known as a kernel, of size 5 × 5 to process the image. Each pixel in the image was processed by averaging the values of its neighboring pixels. The central pixel is replaced with the average value. After replacing all image pixels with the average of their neighbors, an improved image is obtained [36]. This is because it replaces each pixel with the average of its local neighborhood. To improve the ultrasound images of TC, applying a 5 × 5 average filter enhances the visibility of important structures while reducing noise interference, as shown in Equation (1).

$$A(i,j) = \frac{1}{5x5} \sum_{s=-2}^{s=2} \sum_{t=-2}^{t=2} f(i+s, j+t) \quad (1)$$

In the provided framework, $f(i,j)$ represents the input, $A(i,j)$ indicates the output, and $5x5$ symbolizes the total pixel count.

The Laplacian filter is an image processing technique used to enhance edge details, making it particularly useful for showing boundaries and increasing contrast in ultrasound images of TC. The primary purpose of the Laplacian filter is edge detection. It identifies rapid changes in pixel intensity that correspond to image edges. Edges are abrupt transitions from dark to light or vice versa, and are crucial for delineating structures and boundaries in medical images, such as tumors in ultrasound images of the thyroid. Applying the Laplacian filter to an image computes the second derivative of pixel intensities, thereby emphasizing regions with rapid intensity changes. The negative center pixel in the filter kernel amplifies the difference between neighboring pixel values, emphasizing the edges [37]. This effect increases the contrast along the edges, making them more pronounced. The Laplacian filter effectively enhances the edges of structures in ultrasound images, such as thyroid nodules, for TC diagnosis. This enhancement makes it easier for AI techniques to identify and delineate the affected areas. The increased contrast along the edges improves visualization and aids in diagnosis, as shown in Equation (2).

$$\nabla^2 f(i,j) = \frac{\partial^2 f}{\partial i_2} + \frac{\partial^2 f}{\partial j_2} \quad (2)$$

They used a second-order differential equation. represented the Laplacian operator as $\nabla^2 f$, with indices i and j referring to the spatial coordinates of the pixels.

Finally, the outcomes of the two image filters are combined using the 8-version of the digital Laplacian filter as follows:

$$L(x,y) = \left(\sum_{s=-1}^{1} \sum_{t=-1}^{1} \mathrm{A}(\mathrm{x}+\mathrm{s}, \mathrm{y}+\mathrm{t})\right) - 9A(x,y) \quad (3)$$

https://doi.org/10.3390/diagnostics16070999

Figure 2b shows some TC dataset ultrasound images after applying image enhancement techniques.

2.4. The Improvement of the DenseNet169 and VGG19 Models

The ImageNet weights were strategically used with a pretrained model. Both DenseNet169 and VGG19 early convolutional layers are ImageNet-trained and have universal visual primitives, edges, gradients, and textural discontinuities that work well in ultrasound analysis. Such a head start is critical because most medical imaging datasets are modest. Once the models are initialized, they are fine-tuned on the thyroid ultrasound dataset, and the deeper layers refine their representations to capture the sonographic morphologies of echogenicity change and microcalcification. The heads of the classifiers were removed and replaced with global average pooling, thereby making the networks feature extractors rather than end-to-end classifiers. This mixed-method initialization maintains learning transfer by providing specificity for thyroid cancer detection.

The virtues and burdens of models such as DenseNet169 and VGG19 are their architectural weights. In thyroid ultrasound, their hierarchical structure provides fine-grained resolution of tissue structure, distinguishing between indiscriminate shadowing and unmistakable calcification. However, this richness requires a calculative supremacy that is excessive, even lavish, to burden, so to speak, on the delicate, even diffuse, repertoire of a sonographic image [38]. These canonical structures could not be accepted by us; they were sculpted. In the process of sculpting them, we needed to be particularly careful to avoid deleting unnecessary computational pathways and preserve the layers that produce decisive, clinically salient features. It is a form of surgical pruning guided by the principle that efficiency should never compromise diagnostic acuity in medical imaging.

The choice of DenseNet169 and VGG19 was informed by the fact that these architectures are philosophically in nature and align with the diagnostic requirements of thyroid ultrasound interpretation. Both networks use ImageNet-pretrained weights, which, in early convolutional filters, capture universal visual primitives, including edges, gradients, and texture discontinuities, which are meaningful even when the domain is no longer a natural image but is instead sonographic data. This inheritance is important. Medical imaging data are rarely rich, and a pretrained initialization provides a stable visual vocabulary on which specialized learning can build. However, the two backbones can interpret images differently. VGG19 is structurally clear, with its convolutional hierarchy arranged in an orderly manner; therefore, it is sensitive to changes in boundary continuity and echogenicity. In contrast, denseNet169 with dense connectivity enables the reuse of multiscale features and maintains weak internal textures or microcalcification patterns. The framework achieves this balance by combining these complementary representations with the ViT encoder, providing fine-grained local morphology with global contextual argumentation and an analytical balance comparable to that of practicing radiologists.

The VGG19 structure is notoriously homogeneous, consisting of a series of convolutional and pooling layers. This simplicity renders its optimization an exercise in determining the diminishing returns. Table 2 shows that the early blocks (through Conv3) cannot be negotiated. They build basic gradient maps and edge detectors, which constitute the alphabet of any later interpretation. Nonetheless, the deeper convolutes (Conv4 and Conv5), although theoretically capable of learning high-order abstractions, seem to overparametrize our domain. We discovered that reducing the number of convolutional filters in these later stages by 30–40% does not degrade feature performance in subsequent tasks. Its classification architecture was overlaid with global average pooling by removing the final, fully connected layers. This change shifts the network's mission from direct classifica-

https://doi.org/10.3390/diagnostics16070999

tion to pure and effective feature extraction, producing a lean yet powerful 512-dimensional description [39].

Table 2. Optimized DenseNet169 and VGG19 model structures for thyroid cancer classification.

Model	Component	Original Specification	Optimized Specification	Parameter Reduction	Rationale for Preservation
VGG19	Blocks Conv1–3	64–256 filters, 2–4 conv layers	Unchanged	0%	Extracts foundational edges, textures, and low-level patterns critical for margin and echogenicity assessment.
	Blocks Conv4–5	512 filters, 4 conv layers	384–448 filters, 3 conv layers	~35%	High-level abstraction capacity is partially retained while mitigating over-parameterization for ultrasound's simpler semantic space.
	Classifier	3 Fully Connected Layers	Global Average Pooling only	~90% (of classifier)	Shifts network role to feature extractor, eliminating massive redundant linear transformations.
DenseNet169	Initial Conv & Pool	7 × 7 conv, 64 filters; MaxPool	Unchanged	0%	Critical first-step processing of raw pixel data into initial feature maps.
	Dense Blocks (4)	Growth rate (k) = 32	Growth rate (k) = 24	~25% per layer	Maintains multiscale feature reuse paradigm while curbing computational cost of excessive channel concatenation.
	Transition Layers	1 × 1 conv & 2 × 2 pooling	Unchanged	0%	Essential for feature map compression and downsampling; key to hierarchical structure.
	Classifier	Fully Connected Layer	Global Average Pooling only	~95% (of classifier)	Redirects model output to a dense feature vector suitable for hybrid fusion.

DenseNet169 posed a more complex challenge. Its power lies in its dense connectivity; each layer receives input from all previous layers, enabling rich feature reuse. In this case, pruning does not refer to the elimination of successive stages as much as the scrambling of the growth inside every dense block. We retained the first convolutional layer and the block-to-block transition layers because they regulate the required compression and downsampling of the feature maps. At the scale of the dense blocks, we utilized the fact that the number of output channels per layer (the growth rate) can be reduced by 25. This balances the exponential channel expansion, which is computationally expensive [40].

Most importantly, we retained all skip connections; the integrity of this gradient highway was paramount. Global average pooling replaced the last classification layer. The outcome, as shown in Tables 2 and 3, is a model that retains its typical multiscale feature fusion but in a more efficient and focused manner.

The VGG19 pruning targeted deep Conv4–5 blocks with many filters. Reducing the number of filters by 30–40% and replacing the classifier with a global average grouping resulted in the significant reduction you see—nearly half the parameters were removed while maintaining feature quality. DenseNet169 was more complex. Reducing the growth rate from 32 to 24, pruned each dense block without cutting critical skip links. A 25% reduction in the number of channels per layer translates to an overall parameter reduction of 31.5%, as shown.

https://doi.org/10.3390/diagnostics16070999

Table 3. Computational comparison of original and pruned architectures.

Model	Configuration	Parameters (M)	Inference Time (ms)
VGG19	Original	143.7	156.8
VGG19	Pruned (ours)	81.2	94.3
Reduction		*43.50%*	*39.90%*
DenseNet169	Original	14.3	58.4
DenseNet169	Pruned (ours)	9.8	42.7
Reduction		*31.50%*	*26.90%*

2.5. Principal Component Analysis of Feature Selection

High-dimensional feature representations generated by convolutional neural networks often exhibit significant redundancy, as many channels encode interconnected visual patterns. To create a more concise and statistically stable representation, principal component analysis (PCA) was applied to the extracted convolutional feature vectors before they were used by subsequent classifiers or integrated with contextual features derived from transformers. PCA performs an orthogonal linear transformation that rotates the original feature space into a new coordinate system whose axes correspond to the directions of maximum variance [41].

Formally, given a central feature matrix $F \in R^{n \times d}$, where n represents the number of samples and d denotes the dimensionality of the extracted feature vectors.

PCA calculates the eigenvectors of the covariance matrix of the feature distribution as Equation (4):

$$\Sigma = \frac{1}{n-1} F^{\top} F \tag{4}$$

where $\Sigma \in R^{d \times d}$ represents the covariance matrix that captures the correlations of pairwise features. Analysis of the eigenvalues of this matrix yields a set of orthogonal eigenvectors and their corresponding eigenvalues as Equation (5):

$$\Sigma v_i = \lambda_i v_i \tag{5}$$

where v_i is the eigenvector i, and λ_i its eigenvalue. The eigenvectors define the directions of the principal components, while the eigenvalues define the variance explained by each component. The principal components are then ranked in descending order according to their eigenvalues.

Instead of specifying a fixed dimension for the reduced feature space, the retained components were chosen to preserve the cumulative variance. Specifically, the smallest number of principal components k that achieve in Equation (6):

$$\frac{\sum_{i=1}^{k} \lambda_i}{\sum_{i=1}^{d} \lambda_i} \geq 0.95 \tag{6}$$

Specifically, the transformation preserves the minimum set of key axes needed to retain 95% of the cumulative variance of the original feature distribution.

This variance-preservation strategy compresses the representation while maintaining the dominant statistical structure of the extracted features. The resulting reduced feature vector is non-correlated and dimensionless, minimizing multicollinearity while preserving the information variance captured by convolutional encoders. This compressed representation is then used where dimensionality reduction is required within the framework, ensuring a consistent feature space and preserving variance for subsequent modeling stages.

https://doi.org/10.3390/diagnostics16070999

2.6. Configuration of XGBoost and ANN Classifiers

ANN and XGBoost classifiers were constructed as optimized, non-convolutional baselines to determine the discriminative power of the extracted CNN features alone. Each was set to present a substantial and clear challenge. The search of the validation set showed that the final configuration was a binary logistic objective, a maximum tree depth of 6, a learning rate (eta) of 0.1, and 1000 estimators [42]. It was scaled using subsampling (subsample and colsample by tree at 0.8) and L2 regularization to guarantee generalizability [43]. This forms a highly powerful interaction-conscious model that considers features as tabular data.

ANN: A small, fully connected network trained to take the same feature vectors reduced using PCA [44]. The model was trained using a learning rate of 1×10^{-4} and binary cross-entropy loss. This design offers a rich and spatially agnostic neural benchmark [45]. These models were not selected randomly [46]. Their canonical, discipline-enforced implementations set a strict performance limit on traditional feature-classifier paradigms, thereby enabling the unambiguous quantification of the transformative impact of our ViT-based global contextual fusion [47].

2.7. Hybridizing ViT-E with DenseNet169 and VGG19: A Syncretic Architecture for Thyroid Nodule Parsing

This proposed a dual-pipeline hybrid architecture to meet the clinical need for the localization of morphological details and the global contextualization of anatomy. We implemented two complementary systems: the first was a combination of enhanced ViT-E with DenseNet169, and the second was a combination of ViT-E with VGG19. Both models exploit the special inductive bias of their convolutional networks to extract local features, leverage the ViT-E encoder to capture global relations, and perform the final classification. This design enables comparative analysis and provides a basis for a robust ensemble-based diagnosis [48].

To provide a comprehensive comparative analysis, we developed and evaluated two distinct hybrid architectures in parallel: ViT-E with DenseNet169 and a separate ViT-E with VGG19.

Architecture 1: DenseNet169–ViT-E Hybrid

The model processes an input ultrasound image tensor $X \in \mathbb{R}^{H \times W \times 1}$.

Stream A: Local feature extraction using DenseNet169. A high-dimensional, multiscale feature vector is obtained via the optimized DenseNet169 backbone, which uses dense connectivity to maintain fine-grained textual patterns, as shown in Equation (7):

$$f_{\text{Dense}} = F_{\text{Dense}}(X), f_{\text{Dense}} \in \mathbb{R}^{d_{\text{dense}}}. \tag{7}$$

Stream B: ViT-E Global Context Encoding. Simultaneously, the ViT-E encoder processes the images. It is tiled into NN non-overlapping patches $N = H \times W/p^2$ to which positional encodings are added [49]. The ViT-E encoder (L of multi-head self-attention (MHSA) and feedforward networks) produces a sequence of contextualized token representations. They are summarized (e.g., through global average pooling) into a global feature vector as shown in Equation (8):

$$f_{\text{glob}} = \text{Pool}\left(Z^{(L)}\right), f_{\text{glob}} \in \mathbb{R}^{d_{\text{glob}}}. \tag{8}$$

Fusion and Classification. The local compressed f_{Dense} and the global context f_{glob} are combined with learned, adaptive projection as Equation (9):

$$f_{\text{fused}} = \phi\left(W_{\text{Dense}} f_{\text{Dense}}^{\text{red}} + W_{\text{glob}} f_{\text{glob}} + b_f\right), \tag{9}$$

https://doi.org/10.3390/diagnostics16070999

where ϕ is a ReLU activation, this combination representation is fed through the classification head of the ViT-E, a multilayer perceptron (MLP) to generate logits as Equation (10):

$$o = W_c \cdot \text{Dropout}(\text{LN}(f_{\text{fused}})) + b_c. \tag{10}$$

A softmax activation yields the final predictive probabilities as Equation (11):

$$(y = c \mid X) = \frac{\exp(o_c)}{\sum_{j=1}^{C} \exp(o_j)} \tag{11}$$

The entire framework was developed using a shared-training methodology. The merging layer allows gradients to flow back into the convolutional neural networks, creating features that are not only unique in their own right but also complement the overall narrative provided by the transformer-based model [50]. As a result of this rigorous optimization strategy [51], the model does not simply function as a collection of independent experts but rather becomes a single, comprehensive cognitive framework that not only visualizes the thyroid nodule but also understands the full range of probabilities for its potential transformation into a malignant tumor [52].

The final probability distribution for the two categories (thyroid cancer and normal) was obtained using a softmax function.

Architecture 2: VGG19-ViT-E Hybrid

The second model has a similar scheme, with DenseNet169 being replaced by VGG19.

Stream A: Local feature extraction using VGG19. The underlying VGG19 architecture is optimized to generate a hierarchical feature representation, focusing on edge continuity and local spatial structure, as in Equation (12):

$$f_{\text{VGG}} = F_{\text{VGG}}(X), f_{\text{VGG}} \in \mathbb{R}^{d_{\text{vgg}}}. \tag{12}$$

Stream B and fusion: The global context path (ViT-E) is identical to the path in Structure 1 and shares the same encoder weights. The fusion follows the same pattern, with separate trainable projection weights. Classification is performed in the same way via the ViT-E head [53].

Creating Spatial Detail Enhancement Block layers and headed self-attention in the enhanced ViT:

The ViT-E route involves a radically different analysis. The image X is partitioned into N non-overlapping patches of size $p \times p$ [54].

Each patch x_u is projected in a D-dimensional embedding linearly and flattened as Equation (13):

$$\text{z}_u^{(0)} = \text{E} \cdot \text{Flatten}(\text{x}_u) + \text{b}_e, \text{z}_u^{(0)} \in \mathbb{R}^D \tag{13}$$

where E is a trainable projection matrix and b_e a bias term. Fixed sinuoidal positional encodings were added to this sequence of patch tokens $\text{p}_u \in \mathbb{R}^D$ they are added in order to give to it spatial order as Equation (14) [55]:

$$\text{Z}_u^{(pos)} = \text{z}_u^{(0)} + \text{p}_u \tag{14}$$

Once processed, the $\text{Z}_u^{(pos)}$ string of symbols is passed to the L-layer encoder of the ViT-E converter. The MHSA function represents the core of each L-layer encoder, as shown in Equation (15):

$$\text{Q}_h^l = \text{Z}^{(l-1)}\text{W}_Q^h, \text{K}_h^l = \text{Z}^{(l-1)}\text{W}_K^h, \text{V}_h^l = \text{Z}^{(l-1)}\text{W}_V^h \tag{15}$$

https://doi.org/10.3390/diagnostics16070999

The attention for head h is computed as Equation (16):

$$\text{Attention}_h(Q_h, K_h, V_h) = \text{softmax}\left(\frac{Q_h K_h^\top}{\sqrt{d_k}}\right)V_h \tag{16}$$

The outputs of all heads H are concatenated and projected. This process, followed by layer normalization (LN) and an MLP with residual connections, forms a transformer block as Equation (17):

$$\begin{aligned} Z'^{(l)} &= \text{LN}\left(Z^{(l-1)} + \text{MHSA}(Z^{(l-1)})\right) \\ Z^{(l)} &= \text{LN}\left(Z'^{(l)} + \text{MLP}(Z'^{(l)})\right) \end{aligned} \tag{17}$$

Through successive layers, this builds a representation where features are defined by their contextual associations. The output of the final encoder layer, $Z^{(L)} \in \mathbb{R}^{N\times D}$, is flattened to form a global feature vector as Equation (18):

$$f_{glob} = \text{Flatten}\left(Z^{(L)}\right) \in \mathbb{R}^{N\cdot D} \tag{18}$$

The final fusion step is the culmination of the architecture as we now have two compelling and complementary channel representations; locally, from the information contained in f_{Dense} and with f_{VGG} the aggregate of localized evidence, and globally from the f_{glob} aggregated from across the entire input space. While assuming that the concatenation of these two complementary representations is a simple linear choice for equality, we implement a learnable projection with a dynamic gating process and a non-linear activated representation ϕ(i.e., ReLU) [56].

The trainable matrices W_{red} and W_{glob} allow the network to adaptively calibrate the contribution of each stream. For a diagnosis hinging on internal microcalcifications, the network can learn to weight f_{red} more heavily. For assessing gross extrathyroidal invasion, the global contextual salience in f_{glob} may be prioritized.

The unified representation f_{fused} is then layer-normalized, regularized via dropout, and passed through the ViT-E's classification head—an MLP—to produce logits o for the C target classes.

The ViT-E itself is already a heavy load—86 million parameters. It can be trained on VGG19 with 167 M parameters, achieving an inference time of 218 ms per batch. This is approximately four to five images per second, which is fine with offline analysis. The DenseNet169 hybrid is another that adds only approximately 10 million parameters to the ViT-E base while achieving an inference time of 167 milliseconds as shown in Table 4.

Table 4. Complexity analysis of hybrid architectures versus baseline models.

Model	Parameters (M)	FLOPs (G)	Inference Time (ms)
Pruned VGG19 (standalone)	81.2	11.4	94.3
Pruned DenseNet169 (standalone)	9.8	2.3	42.7
ViT-E (standalone)	86.4	15.8	124.6
VGG19-ViT-E Hybrid	167.6	27.2	218.3
DenseNet169-ViT-E Hybrid	96.2	18.1	167.2

2.8. Inference and Visualization of Attention for Interpretability

The inference process is the most important step in developing a trained model as a diagnostic tool. The proposed ViT-E system, based on DenseNet169-VGG19, uses thyroid ultrasound images and produces two outputs: a classification result and a spatial-attenuation map. This map is essential for interpreting the model, which represents a composite inference. In this model, local feature extraction from DenseNet169 and

https://doi.org/10.3390/diagnostics16070999

VGG19 is combined with the overall contextual analysis of the ViT cipher, resulting in a clinically usable, visual representation. This map indicates the model's readability by visually representing its decision-making logic and highlighting the anatomical regions that contributed most to its predictions [57].

The technical path is tight and does not reflect the model's training architecture. The three main pathways that simultaneously process the input image consist of the DenseNet169 and VGG19 convolutional neural networks and the optimized ViT-E cipher. High-dimensional feature vectors were obtained using DenseNet169 and VGG19. These vectors were merged and reduced using PCA, eliminating redundancies and preserving the most distinctive feature axes [58]. The resulting compressed convolutional feature set was then merged with the ViT-E ciphertext's overall contextual feature vector. This multimodal representation combines localized tissue and shape information with overall contextual knowledge, which serves as the input for the ViT-E classification. The final logarithms were transformed into subsequent probabilities for the TC and NC classes using a softmax function. These probabilities were then translated into a binary clinical diagnosis using an optimized decision threshold to balance sensitivity and specificity in the validation set, which was high owing to the risk of missing malignancy [59].

The Grad-CAM algorithm was used to interpret the reasons for any prediction. This method constructs a discriminative prominence map for each class using gradients of the target class scores from the final convolutional layer of the network. The ViT-E classification header generates a target class score yc (where c is the TC) from the features embedded in our architecture. To illustrate this, we calculated the yc gradient for the Ak feature maps of the final convolutional layers for both DenseNet169 and VGG19 models. These two networks provide spatially conserved feature maps that are essential for positioning.

For each network, the gradient-based significance weight α_k^c for feature map number k is calculated using global mean clustering, as in Equation (19).

$$\alpha_k^c = \frac{1}{Z}\sum_i \sum_j \frac{\partial y^c}{\partial A_{ij}^k} \tag{19}$$

The ReLU activation function retains only the properties that positively influence thyroid cancer prediction. Custom heatmaps for both DenseNet169 and VGG19 were merged, scaled, and subjected to two-linear interpolation to produce a composite prominence map that was accurately overlaid on the original ultrasound image. This latter visualization effectively highlights the pixel regions within the convolutional paths' field of view that are most critical to the ViT-E classifier's decision.

The resulting attention patterns require clinical correlation and are consistent with the known ultrasound biomarkers of malignant tumors. In positive cases, attention is drawn to a narrow, focused halo around the edges of irregular or spinous nodules that visually correspond to invasive growth [60]. The nodule core exhibits a mottled or granular activation pattern, often used to indicate clusters of microcalcifications, which is a hallmark of papillary thyroid carcinoma (PTC). Notably, minor peripheral architectural abnormalities can also be highlighted, suggesting that the model can identify early extrathyroidal expansion. In contrast, the attention characteristic in normal images is diffuse. The heat map shows a smooth, low-density distribution across the homogeneous tissue, with no focal areas. In benign nodules, the focus is usually on their well-defined borders rather than their inward penetration, suggesting a benign morphological assessment [61].

Therefore, the Grad-CAM mechanism can be considered an important tool. This enables the clinician to ensure that the classification used in the system is based on anatomically and pathologically relevant image regions, thereby establishing the necessary confidence in the model. This confidence is further enhanced when the highlighted foci perfectly match the classic malignant features. By prioritizing a small, easily overlooked area that becomes significant upon examination, the model demonstrates a detectability capability that transforms the system into a collaborative diagnostic tool, which appears ambiguous when classifying. The model's synthetic reasoning, based on a highly complex integration of pathways, is highlighted and subjected to expert scrutiny.

3. Results of the Proposed Techniques

3.1. Systems Evaluation Metrics

There are several methods for evaluating the results of AI classifiers, and the confusion matrix is widely considered an important metric. The confusion matrix is a square table that shows the number of images in the test dataset for each category. The rows and columns of the confusion matrix correspond to the actual and expected categories. The diagonal cells of the matrix represent the number of correctly classified samples, which are known as true positives (TPs). The cells below and above the main diagonal indicate incorrectly classified samples, which are further divided into false positives (FPs), false negatives (FNs) and true negative (TNeg). Equations (20)–(25) illustrate the system evaluation metrics, as follows:

$$AUC = \int_0^1 TPR\left(FPR^{-1}(\mathrm{x})\right) dx \tag{20}$$

$$AUC = P\left(\hat{S}(X^{+}) > \hat{S}(X^{-})\right) \tag{21}$$

$$\text{Accuracy} = \frac{\text{TNeg} + \text{TP}}{\text{TNeg} + \text{TP} + \text{FN} + \text{FP}} * 100\% \tag{22}$$

$$\text{Precision} = \frac{\text{TP}}{\text{TP} + \text{FP}} * 100\% \tag{23}$$

$$\text{Sensitivity} = \frac{\text{TP}}{\text{TP} + \text{FN}} * 100\% \tag{24}$$

$$\text{Specificity} = \frac{\text{TNeg}}{\text{TNeg} + \text{FP}} * 100 \tag{25}$$

3.2. Performance Results of Pretrained CNN

This section presents an analysis of the performance results of three CNN models: DenseNet169, AlexNet, and VGG19. These models were trained using the massive ImageNet dataset, which comprises over 1,200,000 images accurately classified into more than 1000 categories. It is important to note that, while the ImageNet dataset is comprehensive, it suffers from limitations in its representation, particularly in the realm of biomedical image datasets. For example, it lacks biomedical data, such as ultrasound images from the TC dataset. The input layers of these models were designed to receive and process images from the TC dataset. Fully connected layers were used to transform these high-level features into feature vectors. Finally, the models classify each feature vector and assign it to its appropriate category, demonstrating their adaptability and usefulness for TC image classification.

Table 5 and Figure 4 in this study provide a comprehensive overview of the performance metrics for the CNN models DenseNet169, AlexNet, and VGG19 when applied to ultrasound image analysis in a TC dataset. The DenseNet169 model achieved excellent results with an AUC of 88.7%, a sensitivity of 88.4%, a specificity of 90.6%, and an accuracy of 90.1%. The AlexNet model demonstrated competitive performance, achieving 87.35% accuracy, 88.05% specificity, 86.15% AUC, 87.65% sensitivity, and 87.5% accuracy. As for the VGG19 model, it achieved strong results with a specificity of 89.85%, an accuracy of

https://doi.org/10.3390/diagnostics16070999

89.6%, an area under the receiver operating characteristic (ROC) curve (AUC) of 87.6%, and a sensitivity of 89.2%.

Table 5. The results of the CNN for analysis for ultrasound to diagnose the TC dataset.

Models	Classes	AUC %	Accuracy %	Precision %	Sensitivity %	Specificity %
DenseNet169	TC	90.6	92	90.2	92.3	88.7
	TN	86.8	87.8	90	88.4	92.5
	Average ratio	**88.7**	**90.1**	**90.1**	**90.35**	**90.6**
AlexNet	TC	86.9	88	89.1	87.9	87.5
	TN	85.4	86.9	85.6	87.4	88.6
	Average ratio	**86.15**	**87.5**	**87.35**	**87.65**	**88.05**
VGG19	TC	88.4	90.8	90.3	90.9	88.4
	TN	86.8	88.1	88.7	87.5	91.3
	Average ratio	**87.6**	**89.6**	**89.5**	**89.2**	**89.85**

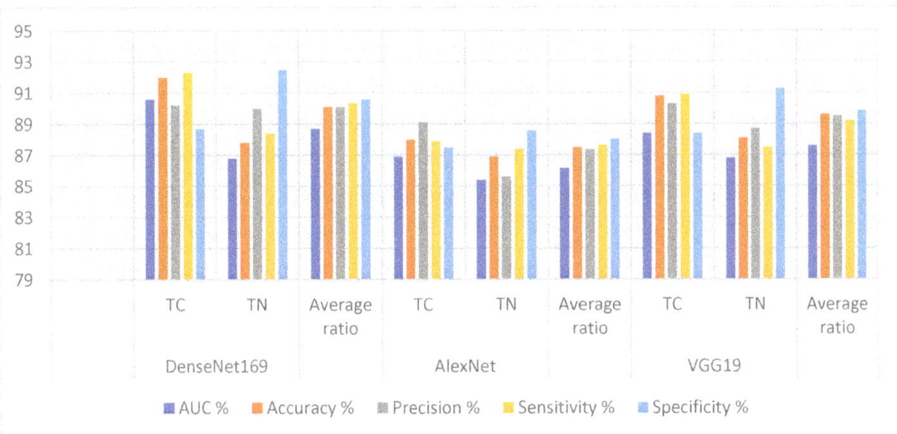

Figure 4. Display the results of the CNN for analysis for ultrasound to diagnose the TC dataset.

3.3. *Results of the Hybrid Method Between CNN with the XGBoost and ANN Networks*

The selection of the XGBoost-CNN and ANN-CNN architectures demonstrated the methodological rigor. Before proposing the shift from hybrid to transformer models, we sought to define the performance limitations of systems enhanced with CNN-ViT. Individual CNNs provided a reasonable foundation, but their accuracy was in the low 90%, which we attribute to their inherent tendency to focus on localized features at the expense of overall anatomical context.

XGBoost and ANN hybrid models were developed to test the hypothesis that more sophisticated classifiers could extract more diagnostic information from these localized features. This approach proved successful, achieving low 90% accuracy, demonstrating the potential to effectively utilize complementary convolutional features. However, despite their power, these models were subject to a localized perspective. They excelled at listing features but failed to integrate them into a unified assessment, as a radiologist would do. This is the fundamental gap that our ViT-CNN hybrid model addresses.

The model can incorporate a vision transformer encoder to capture overall contextual transformations. Distant parts of an image can influence one another, and the model predicts relationships among node boundaries, nuclei, and textures. This is not simply a

https://doi.org/10.3390/diagnostics16070999

replacement of the classifier but a radical shift in the applied thinking methodology by improving the model's features for comprehensive image understanding. This section details the results of combining the XGBoost and ANN models with the CNN models for analyzing ultrasound images of thyroid cancer. Initially, CNN was used for feature extraction, which was subsequently reduced using PCA. XGBoost and ANN networks were used for classification.

Table 6 and Figure 5 summarize the results of combining XGBoost with Dense-Net169, AlexNet, and VGG19 model features for analyzing ultrasound images of the thyroid cancer dataset. The DenseNet169-XGBoost model achieved excellent results, with an AUC of 93.25%, sensitivity of 94.05%, specificity of 96%, and accuracy of 94.2%. The AlexNet-XGBoost model achieved accuracies of 93%, 93%, 92.95%, 93.55%, and 93.1% for quality, AUC, sensitivity, and precision, respectively. The VGG19-XGBoost model performed strongly, achieving 93.6% quality, 93.7% accuracy, 93.4% AUC, and 93.55% sensitivity.

Table 6. The results of XGBoost with CNN features for analysis for ultrasound to diagnose the TC dataset.

Models	Classes	AUC %	Accuracy %	Precision %	Sensitivity %	Specificity %
DenseNet169-XGBoost	TC	93.8	95.7	93.8	95.8	91.8
	TN	92.7	92.3	94.6	92.3	96.2
	Average ratio	**93.25**	**94.2**	**94.2**	**94.05**	**96**
AlexNet-XGBoost	TC	94.1	93.5	93.9	94.2	92.8
	TN	91.8	92.5	92.1	92.9	94.1
	Average ratio	**92.95**	**93.1**	**93**	**93.55**	**93.45**
VGG19-XGBoost	TC	94.2	94.4	94.2	94.3	92.9
	TN	92.6	92.8	93.2	92.8	94.3
	Average ratio	**93.4**	**93.7**	**93.7**	**93.55**	**93.6**

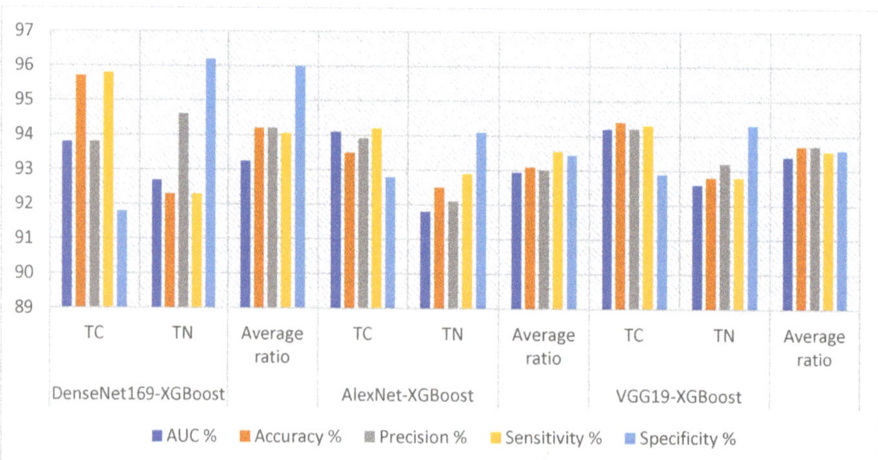

Figure 5. Display the results of hybrid method of CNN and XGBoost for analysis for Ultrasound to diagnose the TC dataset.

Table 7 and Figure 6 summarize the results of the ANN using features from the DenseNet169, AlexNet, and VGG19 models to analyze the ultrasound images in the TC dataset. The DenseNet169-ANN model achieved strong results, with an AUC (92.1%),

https://doi.org/10.3390/diagnostics16070999

sensitivity (93.75%), specificity (93.95%), and accuracy (93.5%). The AlexNet-ANN model yielded a precision (93%), specificity (93.25%), AUC (91.25%), sensitivity (93.45%), and accuracy (93.1%). The VGG19-ANN model produced robust outcomes with a precision (92.6%), accuracy (92.7%), AUC (91.25%), and sensitivity (92.9%).

Table 7. The results of the ANN with CNN features for analysis for ultrasound to diagnose the TC dataset.

Models	Classes	AUC %	Accuracy %	Precision %	Sensitivity %	Specificity %
DenseNet169-ANN	TC	92.3	94	94.2	94.3	93.4
	TN	91.9	93	92.7	93.2	94.5
	Average ratio	**92.1**	**93.5**	**93.45**	**93.75**	**93.95**
AlexNet-ANN	TC	91.8	93.3	94.1	93.4	93.7
	TN	90.7	92.8	91.9	93.5	92.8
	Average ratio	**91.25**	**93.1**	**93**	**93.45**	**93.25**
VGG19-ANN	TC	92.1	93.4	93.3	93.2	92.9
	TN	90.4	91.8	91.9	92.6	93.5
	Average ratio	**91.25**	**92.7**	**92.6**	**92.9**	**93.2**

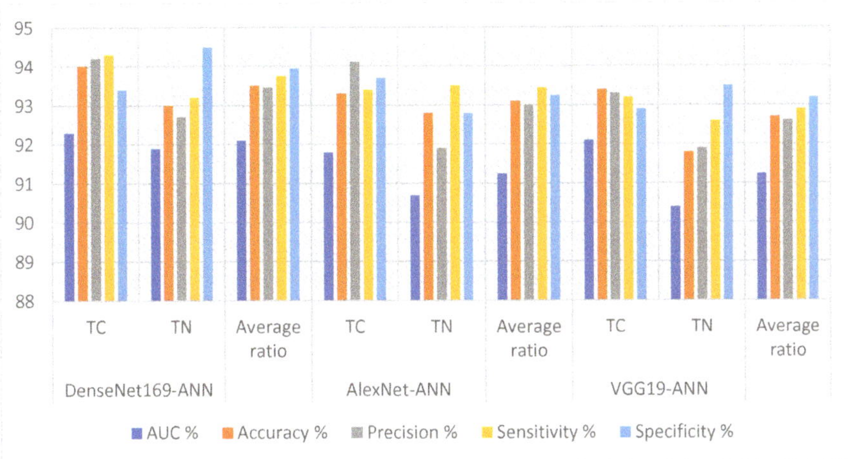

Figure 6. Display results of hybrid method of CNN and ANN for analysis for ultrasound to diagnose the TC dataset.

3.4. Results of the Hybrid Method Between Fusion Features CNN with the XGBoost and ANN Networks

This section summarizes the results of the XGBoost and ANN algorithms for ultrasound image diagnosis using the TC dataset. First, the images were improved, and the required tissues were isolated to extract features. Deep feature extraction from ultrasound images using DenseNet169, AlexNet, and VGG19. The features extracted from the three CNN models were combined to form discriminators from DenseNet169-AlexNet, AlexNet-VGG19, DenseNet169-VGG19, and DenseNet169-AlexNet-VGG19 to form high-level features. PCA was used to handle the high-dimensional features. The selected features were fed to the XGBoost and ANN classifiers for classification.

Table 8 and Figure 7 summarize the results of XGBoost with fusion features from the DenseNet169, AlexNet, and VGG19 models for analyzing ultrasound images in the

https://doi.org/10.3390/diagnostics16070999

TC dataset. The Dense-Net169-AlexNet-XGBoost model achieved reliable outcomes with specificity (96.5%), accuracy (96.4%), sensitivity (96.4%), AUC (95.85%), and precision (96.25%). The AlexNet-VGG19-XGBoost model yielded effective results, with a precision (95.8%), AUC (94%), specificity (96.3%), accuracy (95.8%), and sensitivity (96.05%). The DenseNet169-AlexNet-VGG19-XGBoost model achieved strong performance with accuracy (96.7%), sensitivity (96.65%), AUC (95.4%), specificity (96.8%), and precision (96.65%).

Table 8. The results of hybrid method of fusion features CNN and XGBoost for analysis for ultrasound to diagnose the TC dataset.

Models	Classes	AUC %	Accuracy %	Precision %	Sensitivity %	Specificity %
DenseNet169-AlexNet-XGBoost	TC	95.5	97	96.5	97.2	96.2
	TN class	96.2	95.7	96.3	95.6	96.8
	Average ratio	**95.85**	**96.4**	**96.4**	**96.4**	**96.5**
AlexNet-VGG19-XGBoost	TC	94.2	96	96.4	96.3	96.2
	TN	93.8	95.6	95.1	95.8	96.4
	Average ratio	**94**	**95.8**	**95.8**	**96.05**	**96.3**
DenseNet169-AlexNet-VGG19-XGBoost	TC	95.7	97.1	96.9	97.2	96.5
	TN	95.1	96.2	96.5	96.7	97.1
	Average ratio	**95.4**	**96.7**	**96.65**	**96.95**	**96.8**

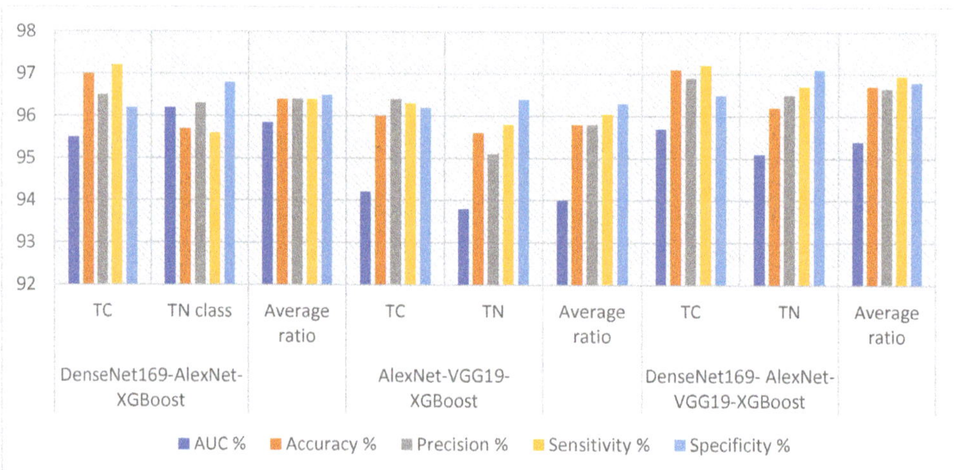

Figure 7. Display results of hybrid method of fusion features CNN and XGBoost for analysis for ultrasound to diagnose the TC dataset.

Figure 8 shows the confusion matrices used to evaluate the performance of the Dense-Net169-AlexNet-XGBoost, AlexNet-VGG19-XGBoost, and DenseNet169-AlexNet-VGG19-XGBoost models. Confusion matrices are valuable visual representations of system evaluation. The Dense-Net169-AlexNet-XGBoost system achieved 97% accuracy for the TC class and 95.7% for the TN class. The AlexNet-VGG19-XGBoost system demonstrated 96% accuracy for the TC class and 95.6% for the TN class. The Dense-Net169-AlexNet-VGG19-XGBoost system showed 97% accuracy for the TC class and 96.2% accuracy for the TN class.

Table 9 and Figure 9 summarize the results of the ANN combined with the DenseNet169, AlexNet, and VGG19 models for analyzing ultrasound images in the TC

https://doi.org/10.3390/diagnostics16070999

dataset. The DenseNet169-AlexNet-ANN model demonstrated excellent performance across specificity (97.1%), sensitivity (96.9%), area under the curve (95.3%), resolution (96.6%), and fine-tuning (96.7%). The AlexNet-VGG19-ANN model achieved good results across the area under the curve (94.25%), fine-tuning (94.9%), resolution (95%), specificity (94.6%), and sensitivity (94.65%). The DenseNet169-AlexNet-VGG19-ANN model achieved strong results across accuracy (97%), fine-tuning (96.95%), area under the curve (95%), sensitivity (96.2%), and specificity (95.95%).

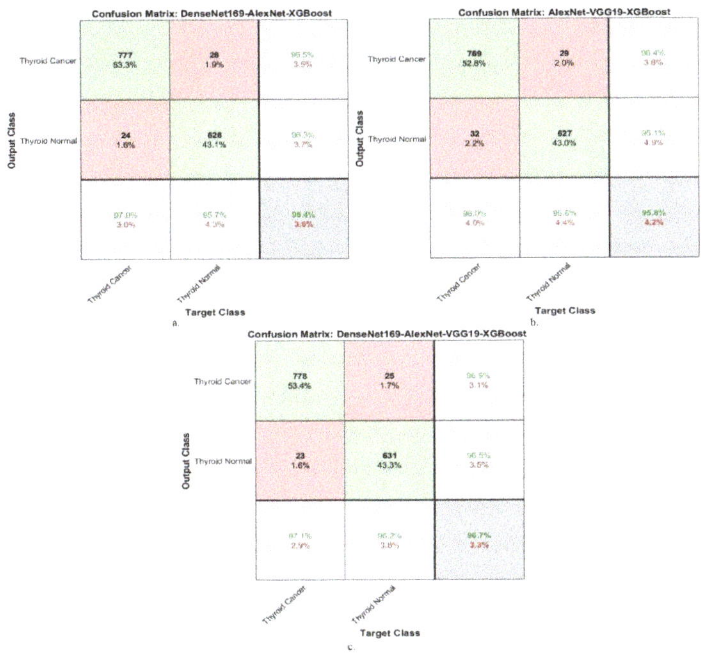

Figure 8. Confusion matrices for display the results of hybrid method of CNN and XGBoost for analysis for ultrasound images to diagnose the TC dataset. (**a**) DenseNet169-AlexNet-XGBoost. (**b**) AlexNet-VGG19-XGBoost. (**c**) DenseNet169-AlexNet-VGG19-XGBoost.

Table 9. The results of hybrid method of fusion features CNN and ANN for analysis for ultrasound to diagnose the TC dataset.

Models	Classes	AUC %	Accuracy %	Precision %	Sensitivity %	Specificity %
DenseNet169-AlexNet-ANN	TC	95.2	97	96.9	97.1	97.3
	TN	95.4	96.2	96.3	96.7	96.9
	Average ratio	**95.3**	**96.6**	**96.7**	**96.9**	**97.1**
AlexNet-VGG19-ANN	TC	94.3	95.3	95.6	95.2	93.9
	TN	94.2	94.7	94.2	94.1	95.3
	Average ratio	**94.25**	**95**	**94.9**	**94.65**	**94.6**
DenseNet169-AlexNet-VGG19-ANN	TC	95.2	97.1	97.4	96.1	96.4
	TN	94.8	96.8	96.5	96.3	95.5
	Average ratio	**95**	**97**	**96.95**	**96.2**	**95.95**

Figure 10 shows the confusion matrices for performance evaluation of the DenseNet169-AlexNet-ANN, AlexNet-VGG19-ANN, and DenseNet169-AlexNet-VGG19-ANNsystems.

https://doi.org/10.3390/diagnostics16070999

Confusion matrices serve as valuable visual representations for evaluating systems. DenseNet169-AlexNet-ANN achieves 97% accuracy for TC and 96.2% for TN. AlexNet-VGG19-ANN achieves 95.3% accuracy for TC and 94.7% for TN. The DenseNet169-AlexNet-VGG19-ANN system achieves 97.1% TC class accuracy and 96.8% TN class accuracy.

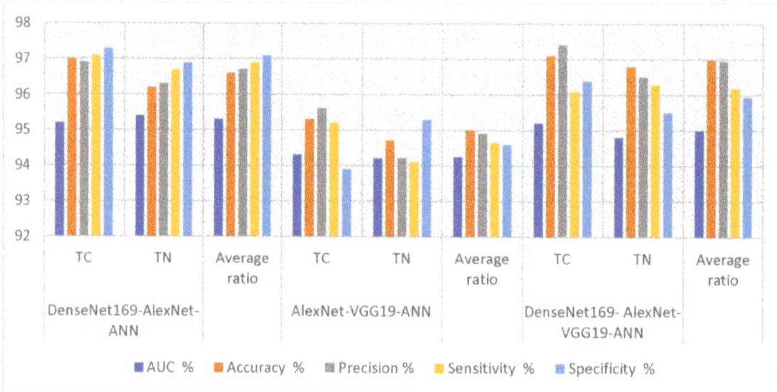

Figure 9. Display results of hybrid method of fusion features CNN and ANN for analysis for ultrasound to diagnose the TC dataset.

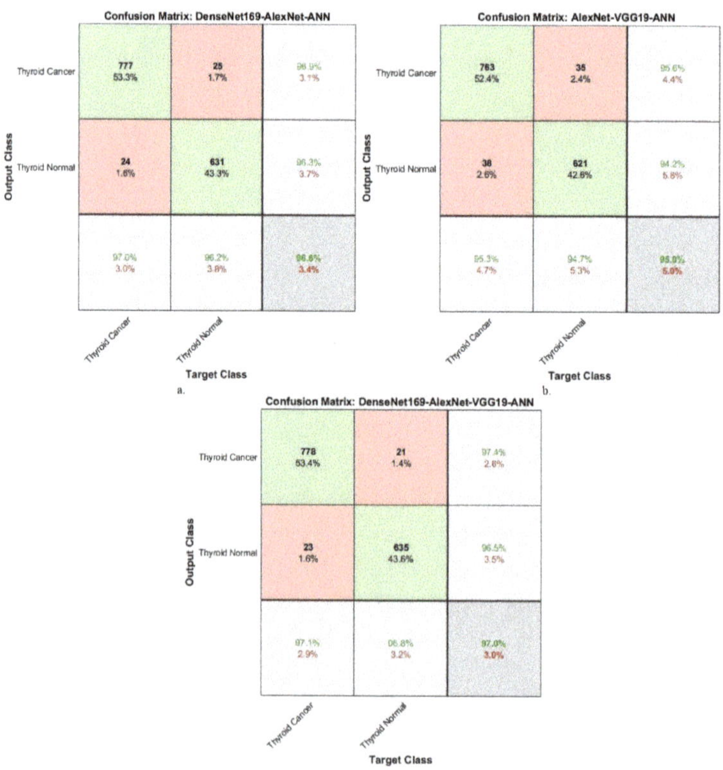

Figure 10. Confusion matrices for display the results of hybrid method of CNN and ANN for analysis for ultrasound images to diagnose the TC dataset. (**a**) DenseNet169-AlexNet-ANN. (**b**) AlexNet-VGG19-ANN. (**c**) DenseNet169-AlexNet-VGG19-ANN.

https://doi.org/10.3390/diagnostics16070999

3.5. Syncretic Fusion and Diagnostic Precision: Evaluating the ViT-CNN Hybrids

The proposed architecture, which combines the comprehensive relational intelligence of the ViT-E network with the local morphological expertise of convolutional networks, significantly improves diagnostic performance. Previous hybrid models integrating convolutional networks with XGBoost classifiers or artificial neural networks achieved low 90% accuracy; however, our ViT-CNN models surpassed this. The results in Table 10 indicate not minor improvements but a qualitative shift in the model's ability to decode ultrasound images of malignant thyroid tumors.

Table 10. Performance metrics of CNN with ViT-E hybrid models for thyroid cancer diagnosis based on data augmentation.

Models	Classes	AUC %	Accuracy %	Precision %	Sensitivity %	Specificity %
ViT-DenseNet169	TC	97.6	98.6	98.6	99.3	98.6
	TN	97.1	98.3	98.3	98.5	99.7
	Average ratio	**97.35**	**98.5**	**98.45**	**98.9**	**99.15**
ViT-VGG19	TC	97.2	98.5	98	99.1	98.2
	TN	96.9	97.6	98.2	97.8	99.3
	Average ratio	**97.05**	**98.1**	**98.1**	**98.45**	**98.75**

Here, we focused on ViT-DenseNet169. Its average metrics show remarkable results: 98.5% accuracy, 99.15% near-perfect specificity, and 98.9% sensitivity. This is a significant achievement in cancer detection. The system's superior ability to identify healthy tissue in its correct location greatly reduces unnecessary patient anxiety and the need for biopsy, owing to its high specificity. However, its sensitivity ensures that malignant nodules are almost never missed. Further examination of the model's performance across categories revealed strong performance in thyroid cancer cases, with sensitivities and accuracies of 99.3% and 98.6%, respectively. This means that when the model is used to identify a cancer case, it is correct almost every time, missing only a small percentage of genuine cancers (false negatives). The ViT-VGG19 model followed with an average accuracy of 98.1% and equally strong sensitivity (98.45%) and specificity (98.75%). The consistent superiority of the ViT-DenseNet169 model suggests that the dense feature reuse at multiple scales can be combined more effectively with the transformer's attentional mechanisms for this task. This performance translates into tangible clinical results, as illustrated by the confusion matrices shown in Figure 11. In the case of ViT-DenseNet169, of the 801 thyroid cancer cases in the test set, only 11 were incorrectly classified as normal. This means that slightly more than 1% of malignant tumors will be missed. However, the false positives in 11 cancer cases were the normal rate for 656 healthy cases. The same trend can be observed with ViT-VGG19, with the margin of error increasing by a small margin (almost negligible) (16 false negatives and 12 false positives). Here, the clinical reality intersects with practical realities. The margin of error in these models was minimal. Not every misclassification indicates system failure. However, it indicates the most ambiguous borderline cases in the dataset—the very images that might make a human expert pause and require further investigation. The models do this not by playing it safe but by making bold and appropriate distinctions in the vast majority of cases.

https://doi.org/10.3390/diagnostics16070999

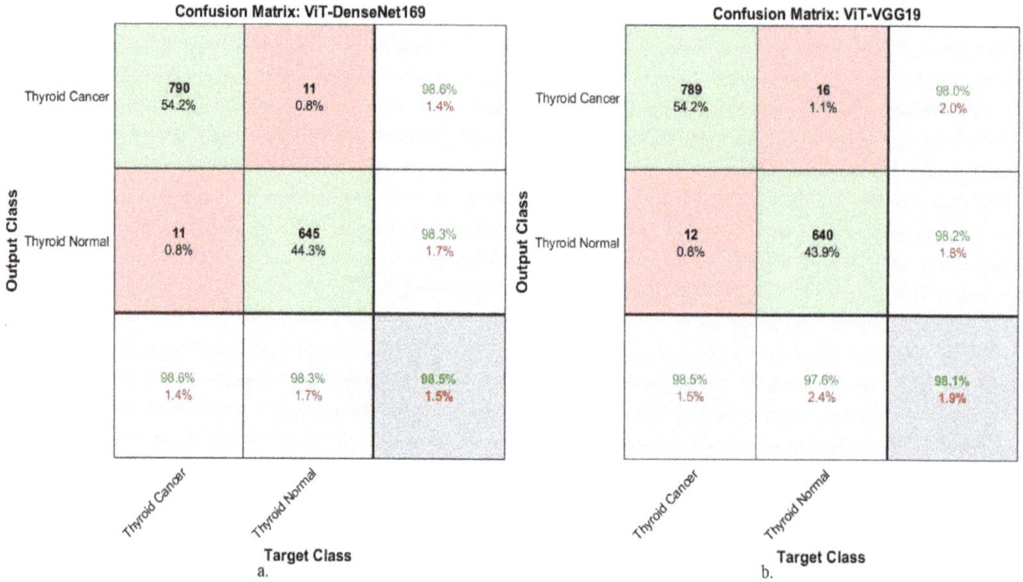

Figure 11. Confusion matrices for the diagnosis of thyroid tumors (**a**) ViT-DenseNet169 and (**b**) ViT-VGG19 hybrid models.

Hybrid ViT-CNN models have improved the accuracy of the two most important metrics—sensitivity and specificity—to over 98.5%, a high level of expert agreement. This means that previous architectures, at least to the extent possible, were limited by their inability to provide a complete synthesis of local pixel-level evidence and a contextual understanding of the image. This is directly addressed in transformer-based data fusion by developing a cognitive model that better reflects radiologists' integrative and interactive thinking. The result is not only a statistically superior model but also a more reliable and clinically understandable one, where its few errors are as significant as the many correct diagnoses.

3.6. Training–Validation Evaluation and the Role of Data Augmentation in Transformer-CNN Hybrid Diagnosis Models

A comparison between Table 11 reveals that the learning process for the hybrid models is informative. Even without additional data, the ViT-DenseNet169 model already achieves consistent diagnostic performance, with an average accuracy of 94.3% and an AUC of 93.6%. The ViTVGG19 model is presented next, with an accuracy of 93.4%. All these findings imply that transformer-CNN fusion can extract meaningful morphological and contextual information of thyroid ultrasound images despite low training variability. However, once augmentation is applied, the models' behavior is significantly altered. ViT-DenseNet169 and ViT-VGG19 achieved performances of 98.5% and 98.1%, respectively. The answer is simple: augmentation introduces networks to realistic probe orientation, tissue presentation, and grayscale texture variations. The models do not simply learn the appearance of malignancy but also its numerous visual variations.

The ViT-DenseNet169 hybrid exhibited a very stable learning behavior, as shown in Figure 12. In the initial three epochs, the validation accuracy increased from 92.5% to 96.6%, whereas the training accuracy was deliberately increased during the first three epochs, which is a good indicator that the model is not merely memorizing. The disparity between training and validation accuracy was narrow at all steps, with a convergence gap of approximately 2%. The loss curves are most revealing: the validation loss decreases in the same direction as the training loss and then levels off at approximately epoch 10, with no indication of creeping upwards. This is a characteristic of a model that generalizes rather than parrots. At epoch 20, the architecture achieves the highest accuracy of 99.5% on training and 97.3% on validation, evidently striking a balance between the transformer's global view and DenseNet169's feature reuse without overfitting.

Table 11. Performance metrics of CNN with ViT-E hybrid models for thyroid cancer diagnosis without data augmentation.

Models	Classes	AUC %	Accuracy %	Precision %	Sensitivity %	Specificity %
ViT-DenseNet169	TC	93.7	94.5	95.1	95.2	94.3
	TN	93.5	94.1	93.3	94.6	95.4
	Average ratio	**93.6**	**94.3**	**94.2**	**94.9**	**94.85**
ViT-VGG19	TC	92.7	93.5	94.5	94.2	93.1
	TN	92.5	93.3	92.2	93.4	94.5
	Average ratio	**92.6**	**93.4**	**93.35**	**93.8**	**93.8**

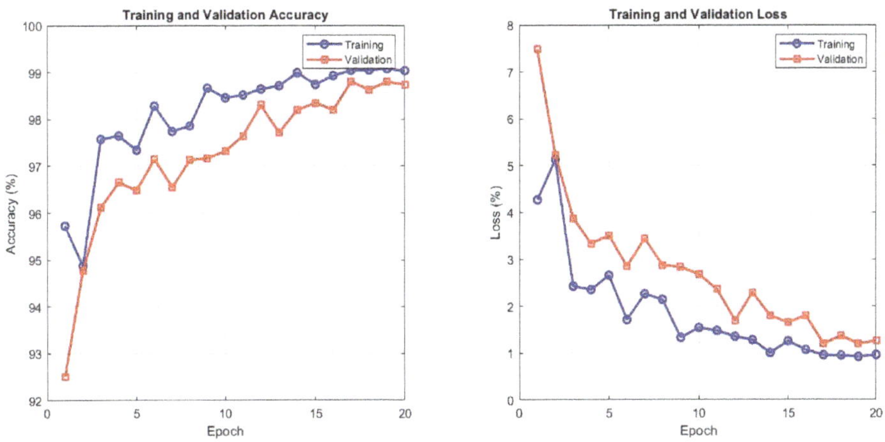

Figure 12. Training and validation accuracy and loss curves of the ViT-DenseNet169 hybrid model.

The ViT-VGG19 hybrid exhibits a slightly different pattern of fast saturation and stability, as shown in Figure 13. The training accuracy stabilized at 98% in epoch 8 and essentially stagnated afterward, whereas the validation accuracy settled at 96.7% in epoch 3 and remained unchanged. The validation loss even goes lower than the training loss after the fifth epoch and stays there, indicating that the representations of the model are transferred to the unknown data with almost perfect accuracy.

Table 12 presents the results that provide clear insights into the behavior of both hybrid architectures with respect to learning and generalization. Training of the ViT-DenseNet169 model indicates excellent stability on the training set, with an AUC of 99.17% and an accuracy of 99.14%, and very close precision, sensitivity, and specificity of 99.1%. These values indicate that the model identifies discriminative patterns in thyroid ultrasound images with high confidence. However, the validation metrics report a less tumultuous tale. The accuracy decreases to 98.74%, with an AUC of 98.85%, and the other indicators remain well balanced in the range of 98.5% to 98.9%. The fact that the training and validation curves are separated by a small margin suggests that the model has acquired generalizable representations rather than memorizing the training samples.

https://doi.org/10.3390/diagnostics16070999

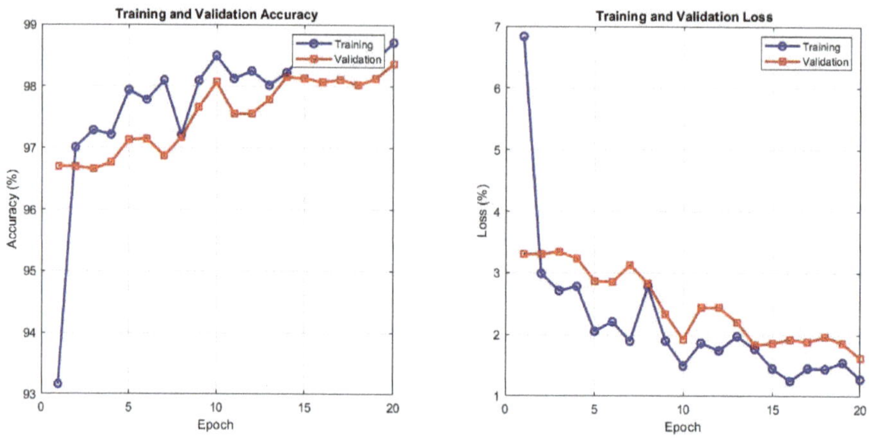

Figure 13. Training and validation accuracy and loss curves of the ViT-VGG19 hybrid model.

The same trend is observed with the ViT-VGG19 structure, but at slightly worse levels. The model achieves 98.50% and 98.43% accuracies and AUCs on the training set, with nearly identical precision, sensitivity, and specificity. The validation performances are also consistent: 98.15% accuracy, 98.10% AUC, and slight variations in the remaining indicators.

Table 12. Comparative training and validation performance of ViT-DenseNet169 and ViT-VGG19 hybrid models for ultrasound-based thyroid cancer diagnosis.

Dataset	Model	AUC %	Accuracy %	Precision %	Sensitivity %	Specificity %
Training	ViT-DenseNet169	99.17	99.14	99.13	99.14	99.1
	ViT-VGG19	98.43	98.5	98.45	98.44	98.47
Validation	ViT-DenseNet169	98.85	98.74	98.9	98.55	98.78
	ViT-VGG19	98.1	98.15	98.2	98.25	98.16

3.7. Mapping the Logic of Doubt: Grad-CAM as an Interpretive Lens for Thyroid Ultrasound

The interpretation of thyroid ultrasound findings lies in a narrow gray area between uncertainty and certainty. Decisions are based on a few millimeters of abnormality, slight echogenicity, and borderline frequencies. Interpretation is not merely a review but a fundamental diagnostic construct. Grad-CAM provides a summary of the deep model's probabilistic outputs for spatial clinical inference by allowing the query of how the deep model arrives at its conclusions. The illustrative Grad-CAM visualization examples of the proposed ViT-DenseNet169 model (shown in Figure 14) support this discussion.

The first sample in Figure 14 (Papillary Thyroid Carcinoma (PTC) = 0.968) shows a well-defined hypoechoic nodule with a central dense activation area. The Grad-CAM heatmap showed a small red center precisely within the lesion border, with minimal extension into the surrounding tissue. This tendency is typical of lesion-based inference and suggests that the model identifies internal structural disturbances rather than the overall tissue. This localized activation is clinically associated with high-risk biomarkers, such as marked hypoechoicity and internal heterogeneity, typically associated with PTC. The confidence score was high and consistent with the visual evidence, confirming the tumor's malignant diagnosis.

https://doi.org/10.3390/diagnostics16070999

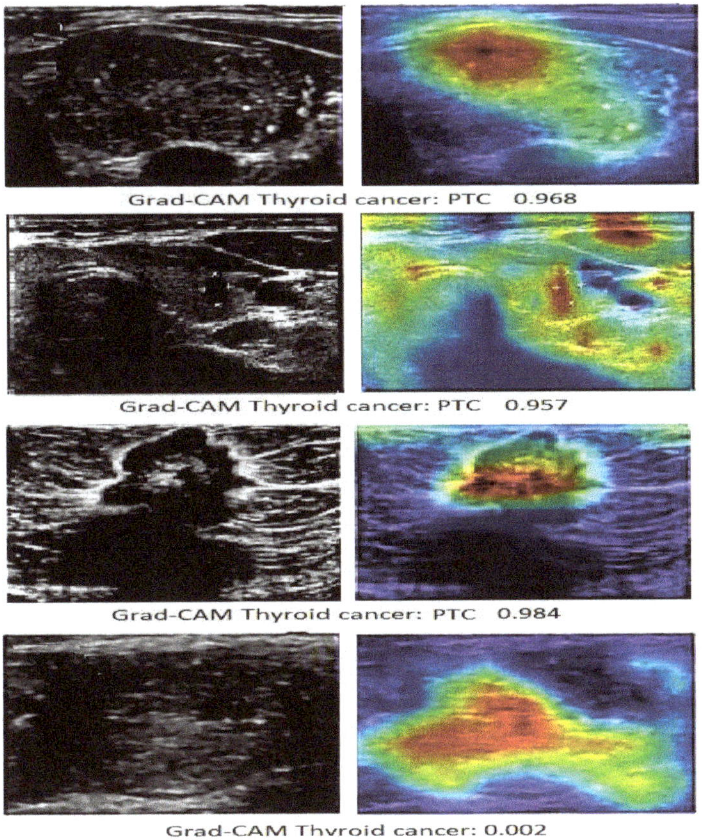

Figure 14. Grad-CAM representations for each ViT-DenseNet169 model, illustrating activation patterns in thyroid ultrasound images, whether the lesion is central, peripheral, infiltrative, or benign.

The second case (PTC = 0.957) shows another, but no less educationally significant pattern of activation. In this case, the Grad-CAM system identified the peripheral border of the nodule, and the second hotspot extended into the intercapsular-visceral interface. This coronal activation pattern indicates that the focus is on the irregularity of the border and the interaction within the capsular tissue, not just on the core of the lesion. This is a concerning aspect for specialists because peripheral focus is often associated with capsular invasion or early extrathyroidal extension. The spatial distribution of attentional signals suggests biologically aggressive behavior rather than simply localized malignancy, despite the high confidence score.

The most extensive and intense malignant cell activation was observed in the third case (PTC = 0.984). The Grad-CAM map extended beyond the central lesion, forming a directional activation field that distorted the expected tissue level. This field effect is not accidental; rather, it demonstrates the model's sensitivity to asymmetric growth and potential secondary effects. These trends are consistent with infiltrative malignancies, in which disease progression extends beyond the visually dominant nodules. The near-definitive model output is explained by the broad, consistent coverage of the activation map, reflecting clinical concerns about extrathyroidal spread.

However, the fourth case (PTC = 0.002) represents only diagnostic adjustment. There was no discrete suspicious nodule on ultrasound, and the Grad-CAM map showed diffuse, low-intensity activation without any focal focus. Attention is focused on the overall tissue profile rather than on the lesions themselves. This non-obsessive orientation is diagnostically significant, indicating that the model does not raise any undue suspicion. Clinically, this behavior supports classifying the case as benign or normal and reinforces confidence in the negative results.

Figure 14 demonstrates that the ViT-DenseNet169 model is not merely an image classification model but also a spatial inference model. Different pathological conditions are associated with central nuclei, peripheral halos, gas fields, and subtle diffusion of attention. Therefore, Grad-CAM acts as a graphical representation of diagnostic ambiguity and confidence, ensuring that the algorithm's decision is linked to known thyroid biomarkers. In this way, deep learning transforms from a purely opaque diagnostic tool into a more transparent clinical partner, capable of simultaneously alerting and reassuring at the point of care.

3.8. Quantifying Interpretability: Spatial Congruence Metrics for Grad-CAM in Thyroid Nodule Analysis

In addition to the qualitative assessment, quantitative measures of spatial concordance between the model focus and the expert's clinical boundaries were provided. This transforms the interpretation from a mere visual impression to a quantifiable diagnostic examination. To establish a dual-focus mask A and isolate the pixels where the model's decision-making logic was concentrated, the threshold for each malignancy was determined using the 90th percentile H of the Grad-CAM heatmap. This mask was then compared with the expert-provided reference manual segmentation mask M.

Two complementary measures were calculated for each participant: The union intersection value (IoU) was used to measure the total overlap between the model, lesion area, and actual lesion, according to Equation 26:

$$\text{IoU} = \frac{|A \cap M|}{|A \cup M|} \tag{26}$$

A value close to 1.0 indicates a high level of concordance, where the model's focus area coincides with the expert's anatomical boundaries. A low IoU value indicates a discrepancy that must be contextualized. This may indicate an error or, as in our case, prioritizing certain malignant features on clinical grounds, such as peripheral tissue invasion rather than nodule size. The percentage of active pixels in the nodule (PAP) is a measure of the model's focus, as defined by Equation (27). It answers a crucial clinical question: Of the areas identified by the model as highly prominent, what percentage actually lies within the pathological zone?

The PAP in the nodule is a measure of model focus, according to Equation (15). It answers a critical clinical question: Of the areas identified by the model as highly prominent, what percentage actually lies within the pathological zone?

$$\text{PAP} = \frac{|A \cap M|}{|A|} \times 100\% \tag{27}$$

Table 13 presents the results of our proposed framework. The case focused on the node center (PTC = 0.968), achieving a high IoU (0.72) and PAP value for active pixels in the node (91), confirming that the model was anchored correctly to the node center with minimal dispersion. The peripheral case (PTC = 0.957) showed significant IoU variability (0.65), but a PAP value of 78%, supporting the 22% of the focus that was not part of the nucleus described at the capsular-visceral boundary. The infiltrating case (PTC = 0.984) showed an IoU variability of 0.58 and PAP of 62. This moderate overlap and large extralesional

https://doi.org/10.3390/diagnostics16070999

profile provide reproducible quantitative evidence for the model in identifying a clinically significant potential infiltration pattern.

Table 13. Quantitative spatial congruence analysis of model attention for representative cases.

Case (PTC Score)	Qualitative Pattern	IoU (90% ile)	PAP (Within Nodule)	Diagnostic Interpretation
First case (0.968)	Central, focused	0.72	91%	High-fidelity localization; attention anchored to nodule core.
Second case (0.957)	Peripheral, rim-like	0.65	78%	Strong overlap with focused margin activation, suggesting capsular assessment.
Third case (0.984)	Infiltrative, extended field	0.58	62%	Moderate overlap with significant extra-nodular attention, indicating potential invasive growth.

This quantitative scrutiny goes beyond simply validating tissue prominence maps; it establishes a diagnostic dialogue between models and clinicians. The high IoU and PAP indicate the model's confidence in localization in classic cases. When associated with aggressive tissue, a low PAP may suggest the model's ability to identify high-risk, subtle features that require more thorough examination. Therefore, these two measures elevate the Grad-CAM visualization to the next level and connect it to the diagnostic reasoning process, providing not only an answer but also a quantitative spatial basis.

4. Performance of the Models in Discussion and Comparison

Diagnosing thyroid cancer using ultrasound presents a significant clinical challenge because of the often indistinct and variable appearances of malignant nodules. This complication necessitates the use of diagnostic tools that not only have high predictive power but also yield clear and interpretable clinical decisions that can be explained to clinicians.

One unresolved issue in previous studies is the lack of correlation between local feature extraction and overall image recognition methods. Most current deep learning models are either local histological classifiers or contextual segments; however, they lack an integrative mechanism for linking subtle morphological details to the overall anatomical context of the glands. This dispersion in perception limits diagnostic accuracy, particularly in early-stage or borderline cases, and negatively affects clinical accuracy by providing predictions without spatially consistent interpretations.

The existing literature review reveals that previous studies have some methodological shortcomings, as shown in Table 14. Although CNN-based models, such as VGGNet and multi-CNN clusters (Sujini et al., Zhang et al., and Vasile et al.), have proven effective for inductive classification, they inherently favor local patterns over long-range relational inference. This limitation sometimes manifests as a trade-off between sensitivity and specificity or as poor performance on small or atypical nodes (Rho et al.). Subsequent developments, such as segmentation-capable hybrid CNN-Transformer models (Li et al.) or fusion networks that combine convolutional neural networks with other classifiers, such as XGBoost or artificial neural networks (Namdeo et al. and Li W. et al.), represent notable improvements. These approaches, along with our core experiments, enhanced the performance through feature integration. However, they typically employ static delayed integration strategies, in which contextual integration occurs sequentially after feature extraction is independent. This cannot be considered a simulation of the dynamic and iterative relationship between pivotal evidence and the radiologist's overall view of the scene. Furthermore, although interpretation techniques have been used in some studies

(Aljameel et al.), they are typically employed as ex-post analyses rather than as design principles, creating a gap between model performance and clinical application.

Table 14. Comparison of the proposed systems with previous works.

Authors	Systems	AUC %	Accuracy %	Precision %	Sensitivity %	Specificity %
Sujini et al. [16]	six-layer CNN and VGGNet-16	94.7	-	-	-	-
Zhang et al. [18]	Multi-CNN	91	-	94	90	-
Naglah et al. [20]	CNN with Texture Patterns	87	-	-	-	97
Li et al. [21]	Eff-Unet + CNN-Fusion	85.5	86	-	-	-
Zhao et al. [22]	CNN Ensemble	94.7	-	-	-	-
El-Hossiny et al. [23]	CNN for Carcinoma Classification	94.69	-	-	-	-
Aljameel et al. [24]	EANN model	86	82	-	-	-
Wu et al. [25]	Three CNNs	82.9	-	97	77.9	-
Zhang et al. [26]	InceptionResNetV2	97.1	-	-	90	-
Wang et al. [27]	CNN with Clinicopathological Factors	78	-	-	-	-
Rho et al. [28]	CNN for Small Nodules	66	83.2	-	89.8	38.3
Vasile et al. [29]	Ensemble CNN Models	-	97.35	95.75	-	-
Our Proposed Systems	ViTE-DenseNet169	**97.35**	**98.5**	**98.45**	**98.9**	**99.15**
	ViTE-VGG19	**97.05**	**98.1**	**98.1**	**98.45**	**98.75**

The importance of this study lies in its attempt to fill these gaps through its original architecture, namely the explicit combination of local convolutional feature extraction and global transformer-based attention in a single, end-to-end trainable model. The ViT-E hybrid models developed, especially ViT-DenseNet169, achieved a new level of performance, as shown in Table 14. These models achieve 98.9% accuracy and 99.15% specificity, which is better than the previous systems, which had an average accuracy of 98.5% across all major measures. This success shows that the sensitivity-specificity trade-off that has been the order of the day in past studies can be overcome by incorporating local and global inference. Although pooled CNNs (Zhao et al. and Vasile et al.) have proven highly accurate, they often exhibit low or nonspecific specificity, a significant demerit for screening tools, as false positives lead to unnecessary surgical intervention. This clinical imperative is specifically met by our more specific model. This is technically achieved by continuously connecting the local morphology structure to the global context during the learning process. The multiscale reuse of DenseNet169 helps detect subtle echogenic gradients and internal abnormalities, whereas the ViT-E encoder enables long-range relational prediction across features, including boundary abnormalities and tissue behavior in the vicinity. The marginally higher score for ViT-DenseNet169 than for ViT-VGG19 indicates that dense connectivity provides a more attention-clustering-congruent feature hierarchy for this task.

In this study, quantitative interpretability is proposed as one of the main validation criteria, alongside the original performance measures.

https://doi.org/10.3390/diagnostics16070999

The Grad-CAM analysis critically demonstrates the model's clinical validity. In cases with a high predicted probability of PTC, the model's attention maps align precisely with established malignant features. A high-confidence malignant case (PTC = 0.968) shows focused central activation, aligning with internal structural disturbance. This direct correlation between high PTC scores, specific spatial attention patterns, and known pathological biomarkers underscores that the model's superior accuracy stems from learning clinically meaningful representations rather than spurious correlations. The further integration of Grad-CAM not only highlights the model's significant areas; it also provides an in-depth analysis of how well the model's attention aligns with clinical evaluation. The quantitative analysis relies on two spatial concordance indicators: IoU and PAP percentage. The proposed model has high diagnostic interest identification accuracy, owing to high IoU and PAP values (e.g., 0.72% and 91% for a central malignant node). Notably, cases with low interconnection (e.g., IoU 0.58 and PAP 62% for the infiltrative pattern) provide distinct diagnostic information, indicating that the model is focused on extranodal extensions, which are signs of invasive growth and a critical risk factor. The model interprets qualitative drawings into a quantitative concordance signal, with its high performance grounded in realistic anatomical evidence. Such spatial precision is not common in previous studies.

This study has implications far beyond incremental gains in accuracy; it addresses the role of artificial intelligence in thyroid cancer screening. Ultrasound diagnosis is an art. Radiologists tend to perceive subtle indicators of unnoticeable echogenic changes, irregular margins, and tiny clusters of calcifications that, alone, are not relevant but gain significance in their spatial context within the gland. The proposed ViT-CNN hybrids follow the same interpretive reasoning, connecting local morphological evidence via convolutional networks with global contextual reasoning via transformer attention. The models obtained have diagnostic accuracy >98%, but the true implication is reliability. High specificity minimizes unnecessary biopsies, whereas high sensitivity minimizes the risk of missed malignancies. More importantly, Grad-CAM visualizations link the algorithm's decisions to clinically recognizable structures. In practice, this will change the system from a black-box classifier to a decision-support partner that can empower clinicians and raise doubts about ambiguous cases that should be subjected to further human investigation.

This framework should be expanded in future studies to include biopsy datasets and longitudinal follow-up to monitor potential trends.

5. Conclusions

This study demonstrates that the main obstacle to using artificial intelligence for thyroid ultrasound imaging lies not only in the classifier's capabilities but also in the discrepancy between how traditional models extract evidence (localized tissue fragments) and how clinicians interpret malignancies (contextual and relational judgments). The proposed hybrid ViT-E models impose a continuous interaction between convolutional morphology and transducer-based global inference. The best model, ViT-E–DenseNet169, achieved 98.5% accuracy, 98.9% sensitivity, 99.15% specificity, and 97.35% AUC, closely followed by ViT-E–VGG19 with 98.1% accuracy and 98.75% specificity—levels exceeding the strongest previous techniques. Grad-CAM analysis clearly demonstrated the model's clinical aspects. When the PTC value reached 0.957, the model highlighted the peripheral borders and the capsule interface, a pattern indicative of invasion. Interpretable accuracy is key: Grand/Grad-CAM heatmaps were qualitatively and quantitatively assessed using the Interrelationship Exchange (IoU) and PAP indices. Typical malignant patterns exhibited anatomically coherent attention with IoU/PAP ratios of 0.72/91% (central), 0.65/78% (peripheral), and 0.58/62% (infiltrative), while benign cases showed diffuse, low-intensity activation. These results suggest that improved performance is associated with coherent spatial inference, supporting safer clinical applications.

Author Contributions: Conceptualization, A.Y.A. and G.A.; methodology, A.Y.A., G.A. and A.M.A.; software, A.Y.A. and G.A.; validation, A.Y.A., G.A. and A.M.A.; formal analysis, A.Y.A., G.A. and A.M.A.; investigation, G.A. and A.Y.A.; resources, A.Y.A., G.A. and A.M.A.; data curation, A.Y.A., G.A. and A.M.A.; writing—original draft preparation A.Y.A.; writing—review and editing, G.A. and A.M.A.; visualization, A.Y.A., G.A. and A.M.A.; supervision, G.A., A.M.A. and A.Y.A.; project administration, G.A. and A.Y.A.; funding acquisition A.Y.A. All authors have read and agreed to the published version of the manuscript.

Funding: This Project was funded by KAU Endowment (WAQF) at King Abdulaziz University, Jeddah.

Institutional Review Board Statement: Not applicable.

Informed Consent Statement: Not applicable.

Data Availability Statement: The datasets used to analyze ultrasound images for early detection of thyroid cancer were obtained from publicly available online repositories and can be accessed via the following link: https://www.kaggle.com/code/garesothmen/thyroid-classification/notebook (accessed on 25 May 2025).

Acknowledgments: This Project was funded by KAU Endowment (WAQF) at King Abdulaziz University, Jeddah. The authors, therefore, acknowledge with thanks WAQF and the Deanship of Scientific Research (DSR) for technical and financial support.

Conflicts of Interest: The authors declare no conflicts of interest.

References

1. Leko, M.B.; Gunjača, I.; Pleić, N.; Zemunik, T. Environmental Factors Affecting Thyroid-Stimulating Hormone and Thyroid Hormone Levels. *Int. J. Mol. Sci.* **2021**, *22*, 6521. [CrossRef]
2. Köhrle, J.; Frädrich, C. Deiodinases control local cellular and systemic thyroid hormone availability. *Free Radic. Biol. Med.* **2022**, *193*, 59–79. [CrossRef]
3. Zhu, Q.; Jiang, G.; Lang, X.; Zhang, J.; Fu, Z.; Zhang, P.; Zhang, X.Y. Prevalence and clinical correlates of thyroid dysfunction in first-episode and drug-naïve major depressive disorder patients with metabolic syndrome. *J. Affect. Disord.* **2023**, *341*, 35–41. [CrossRef]
4. Lee, S.Y.; Pearce, E.N. Assessment and treatment of thyroid disorders in pregnancy and the postpartum period. *Nat. Rev. Endocrinol.* **2022**, *18*, 158–171. [CrossRef] [PubMed]
5. Casto, C.; Pepe, G.; Li Pomi, A.; Corica, D.; Aversa, T.; Wasniewska, M. Hashimoto's Thyroiditis and Graves' Disease in Genetic Syndromes in Pediatric Age. *Genes* **2021**, *12*, 222. [CrossRef] [PubMed]
6. Wu, Y.; Yang, J.; Su, Q.; Gu, H.; Qin, L. Urinary iodine concentration and its associations with thyroid function in pregnant women of Shanghai. *Front. Endocrinol.* **2023**, *14*, 1184747. [CrossRef]
7. Huang, J.; Zhao, J. Quantitative Diagnosis Progress of Ultrasound Imaging Technology in Thyroid Diffuse Diseases. *Diagnostics* **2023**, *13*, 700. [CrossRef]
8. Chan, W.K.; Sun, J.H.; Liou, M.J.; Li, Y.R.; Chou, W.Y.; Liu, F.H.; Peng, S.J. Using Deep Convolutional Neural Networks for Enhanced Ultrasonographic Image Diagnosis of Differentiated Thyroid Cancer. *Biomedicines* **2021**, *9*, 1771. [CrossRef] [PubMed]
9. Capitoli, G.; Piga, I.; L'Imperio, V.; Clerici, F.; Leni, D.; Garancini, M.; Pagni, F. Cytomolecular Classification of Thyroid Nodules Using Fine-Needle Washes Aspiration Biopsies. *Int. J. Mol. Sci.* **2022**, *23*, 4156. [CrossRef]
10. Klain, M.; Zampella, E.; Nappi, C.; Nicolai, E.; Ambrosio, R.; Califaretti, E.; Cuocolo, A. Advances in Functional Imaging of Differentiated Thyroid Cancer. *Cancers* **2021**, *13*, 4748. [CrossRef]
11. Yao, J.; Lei, Z.; Yue, W.; Feng, B.; Li, W.; Ou, D.; Xu, D. DeepThy-Net: A Multimodal Deep Learning Method for Predicting Cervical Lymph Node Metastasis in Papillary Thyroid Cancer. *Adv. Intell. Syst.* **2022**, *4*, 2200100. [CrossRef]
12. Suh, Y.J.; Kwon, M.J.; Noh, H.M.; Lee, H.K.; Ra, Y.J.; Kim, N.Y. Limited Clinical and Diagnostic Utility of Circulating Tumor DNA Detection in Patients with Early-Stage Well-Differentiated Thyroid Cancer: Comparison with Benign Thyroid Nodules and Healthy Individuals. *Healthcare* **2021**, *9*, 386. [CrossRef]
13. Kaliszewski, K.; Ludwig, M.; Ludwig, B.; Mikuła, A.; Greniuk, M.; Rudnicki, J. Update on the Diagnosis and Management of Medullary Thyroid Cancer: What Has Changed in Recent Years? *Cancers* **2022**, *14*, 3643. [CrossRef] [PubMed]

https://doi.org/10.3390/diagnostics16070999

14. Bini, F.; Pica, A.; Azzimonti, L.; Giusti, A.; Ruinelli, L.; Marinozzi, F.; Trimboli, P. Artificial Intelligence in Thyroid Field—A Comprehensive Review. *Cancers* **2021**, *13*, 4740. [CrossRef]
15. Komatsu, M.; Sakai, A.; Dozen, A.; Shozu, K.; Yasutomi, S.; Machino, H.; Asada, K.; Kaneko, S.; Hamamoto, R. Towards Clinical Application of Artificial Intelligence in Ultrasound Imaging. *Biomedicines* **2021**, *9*, 720. [CrossRef] [PubMed]
16. Sujini Ganne, N.; Balakrishna, S. Categorization of Thyroid Cancer Sonography Images Using an Amalgamation of Deep Learning Techniques. In *Soft Computing and Signal Processing*; Springer: Singapore, 2023; pp. 483–491. [CrossRef]
17. Li, G.; Chen, R.; Zhang, J.; Liu, K.; Geng, C.; Lyu, L. Fusing enhanced Transformer and large kernel CNN for malignant thyroid nodule segmentation. *Biomed. Signal Process. Control* **2023**, *83*, 104636. [CrossRef]
18. Zhang, X.; Lee, V.C.; Rong, J.; Lee, J.C.; Song, J.; Liu, F. A multi-channel deep convolutional neural network for multi-classifying thyroid diseases. *Comput. Biol. Med.* **2022**, *148*, 105961. [CrossRef]
19. Namdeo, R.B.; Janardan, G.V. Thyroid Disorder Diagnosis by Optimal Convolutional Neuron based CNN Architecture. *J. Exp. Theor. Artif. Intell.* **2021**, *34*, 871–890. [CrossRef]
20. Naglah, A.; Khalifa, F.; Khaled, R.; Abdel Razek, A.A.K.; Ghazal, M.; Giridharan, G.; El-Baz, A. Novel MRI-Based CAD System for Early Detection of Thyroid Cancer Using Multi-Input CNN. *Sensors* **2021**, *21*, 3878. [CrossRef]
21. Li, W.; Cheng, S.; Qian, K.; Yue, K.; Liu, H. Automatic Recognition and Classification System of Thyroid Nodules in CT Images Based on CNN. *Comput. Intell. Neurosci.* **2021**, *2021*, 5540186. [CrossRef]
22. Zhao, H.B.; Liu, C.; Ye, J.; Chang, L.F.; Xu, Q.; Shi, B.W.; Shi, B.B. A comparison between deep learning convolutional neural networks and radiologists in the differentiation of benign and malignant thyroid nodules on CT images. *Endokrynol. Pol.* **2021**, *72*, 217–225. [CrossRef] [PubMed]
23. El-Hossiny, A.S.; Al-Atabany, W.; Hassan, O.; Soliman, A.M.; Sami, S.A. Classification of Thyroid Carcinoma in Whole Slide Images Using Cascaded CNN. *IEEE Access* **2021**, *9*, 88429–88438. [CrossRef]
24. Aljameel, S.S. A Proactive Explainable Artificial Neural Network Model for the Early Diagnosis of Thyroid Cancer. *Computation* **2022**, *10*, 183. [CrossRef]
25. Wu, G.G.; Lv, W.Z.; Yin, R.; Xu, J.W.; Yan, Y.J.; Chen, R.X.; Dietrich, C.F. Deep Learning Based on ACR TI-RADS Can Improve the Differential Diagnosis of Thyroid Nodules. *Front. Oncol.* **2021**, *11*, 575166. [CrossRef]
26. Zhang, X.; Lee, V.C.; Rong, J.; Liu, F.; Kong, H. Multi-channel convolutional neural network architectures for thyroid cancer detection. *PLoS ONE* **2022**, *17*, e0262128. [CrossRef]
27. Wang, Z.; Qu, L.; Chen, Q.; Zhou, Y.; Duan, H.; Li, B.; Yi, W. Deep learning-based multifeature integration robustly predicts central lymph node metastasis in papillary thyroid cancer. *BMC Cancer* **2023**, *23*, 128. [CrossRef]
28. Rho, M.; Chun, S.H.; Lee, E.; Lee, H.S.; Yoon, J.H.; Park, V.Y.; Kwak, J.Y. Diagnosis of thyroid micronodules on ultrasound using a deep convolutional neural network. *Sci. Rep.* **2023**, *13*, 7231. [CrossRef]
29. Vasile, C.M.; Udriștoiu, A.L.; Ghenea, A.E.; Popescu, M.; Gheonea, C.; Niculescu, C.E.; Alexandru, D.O. Intelligent Diagnosis of Thyroid Ultrasound Imaging Using an Ensemble of Deep Learning Methods. *Medicina* **2021**, *57*, 395. [CrossRef] [PubMed]
30. Thyroid Classification | Kaggle. Available online: https://www.kaggle.com/code/garesothmen/thyroid-classification/notebook (accessed on 10 September 2023).
31. Habchi, Y.; Himeur, Y.; Kheddar, H.; Boukabou, A.; Atalla, S.; Chouchane, A.; Mansoor, W. AI in Thyroid Cancer Diagnosis: Techniques, Trends, and Future Directions. *Systems* **2023**, *11*, 519. [CrossRef]
32. Ahmed, I.A.; Senan, E.M.; Shatnawi, H.S.A.; Alkhraisha, Z.M.; Al-Azzam, M.M.A. Multi-Models of Analyzing Dermoscopy Images for Early Detection of Multi-Class Skin Lesions Based on Fused Features. *Processes* **2023**, *11*, 910. [CrossRef]
33. Gadermayr, M.; Koller, L.; Tschuchnig, M.; Stangassinger, L.M.; Kreutzer, C.; Couillard-Despres, S.; Hittmair, A. MixUp-MIL: Novel Data Augmentation for Multiple Instance Learning and a Study on Thyroid Cancer Diagnosis. In *Medical Image Computing and Computer Assisted Intervention—MICCAI 2023; Lecture Notes in Computer Science (Including Subseries Lecture Notes in Artificial Intelligence and Lecture Notes in Bioinformatics)*; Springer Nature: Cham, Switzerland, 2023; Volume 14225 LNCS, pp. 477–486. [CrossRef]
34. Hamdi, M.; Senan, E.M.; Awaji, B.; Olayah, F.; Jadhav, M.E.; Alalayah, K.M. Analysis of WSI Images by Hybrid Systems with Fusion Features for Early Diagnosis of Cervical Cancer. *Diagnostics* **2023**, *13*, 2538. [CrossRef]
35. Al-Jabbar, M.; Alshahrani, M.; Senan, E.M.; Ahmed, I.A. Analyzing Histological Images Using Hybrid Techniques for Early Detection of Multi-Class Breast Cancer Based on Fusion Features of CNN and Handcrafted. *Diagnostics* **2023**, *13*, 1753. [CrossRef]
36. Bachi, L.; Billeci, L.; Varanini, M. QRS Detection Based on Medical Knowledge and Cascades of Moving Average Filters. *Appl. Sci.* **2021**, *11*, 6995. [CrossRef]
37. Kapoulea, S.; Psychalinos, C.; Elwakil, A.S. FPAA-Based Realization of Filters with Fractional Laplace Operators of Different Orders. *Fractal Fract.* **2021**, *5*, 218. [CrossRef]
38. Alshahrani, M.; Al-Jabbar, M.; Senan, E.M.; Ahmed, I.A.; Mohammed Saif, J.A. Analysis of dermoscopy images of multi-class for early detection of skin lesions by hybrid systems based on integrating features of CNN models. *PLoS ONE* **2024**, *19*, e0298305. [CrossRef]

https://doi.org/10.3390/diagnostics16070999

39. Pino-Ortega, J.; Rojas-Valverde, D.; Gómez-Carmona, C.D.; Rico-González, M. Training Design, Performance Analysis, and Talent Identification—A Systematic Review about the Most Relevant Variables through the Principal Component Analysis in Soccer, Basketball, and Rugby. *Int. J. Environ. Res. Public Health* **2021**, *18*, 2642. [CrossRef]
40. Ghaleb Al-Mekhlafi, Z.; Mohammed Senan, E.; Sulaiman Alshudukhi, J.; Abdulkarem Mohammed, B. Hybrid Techniques for Diagnosing Endoscopy Images for Early Detection of Gastrointestinal Disease Based on Fusion Features. *Int. J. Intell. Syst.* **2023**, *2023*, 8616939. [CrossRef]
41. Abosamra, G.; Oqaibi, H. A signature recognition technique with a powerful verification mechanism based on cnn and pca. *IEEE Access* **2024**, *12*, 40634–40656. [CrossRef]
42. Le, T.T.H.; Oktian, Y.E.; Kim, H. XGBoost for Imbalanced Multiclass Classification-Based Industrial Internet of Things Intrusion Detection Systems. *Sustainability* **2022**, *14*, 8707. [CrossRef]
43. Shamsan, A.; Senan, E.M.; Ahmad Shatnawi, H.S. Predicting of diabetic retinopathy development stages of fundus images using deep learning based on combined features. *PLoS ONE* **2023**, *18*, e0289555. [CrossRef] [PubMed]
44. Bruni, V.; Cardinali, M.L.; Vitulano, D. A Short Review on Minimum Description Length: An Application to Dimension Reduction in PCA. *Entropy* **2022**, *24*, 269. [CrossRef]
45. Almughamisi, N.; Abosamra, G.; Albar, A.; Saleh, M. Anatomy-Guided Hybrid CNN–ViT Model with Neuro-Symbolic Reasoning for Early Diagnosis of Thoracic Diseases Multilabel. *Diagnostics* **2026**, *16*, 159. [CrossRef]
46. Almughamisi, N.; Abosamra, G.; Albar, A.; Saleh, M. Hybrid ConvNeXtV2–ViT Architecture with Ontology-Driven Explainability and Out-of-Distribution Awareness for Transparent Chest X-Ray Diagnosis. *Diagnostics* **2026**, *16*, 294. [CrossRef] [PubMed]
47. Alshari, E.A.; Gawali, B.W. Analysis of Machine Learning Techniques for Sentinel-2A Satellite Images. *J. Electr. Comput. Eng.* **2022**, *2022*, 9092299. [CrossRef]
48. Yu, B.; Yin, P.; Chen, H.; Wang, Y.; Zhao, Y.; Cong, X.; Cong, L. Pyramid multi-loss vision transformer for thyroid cancer classification using cytological smear. *Knowl. Based Syst.* **2023**, *275*, 110721. [CrossRef]
49. Pacal, I.; Ozdemir, B.; Zeynalov, J.; Gasimov, H.; Pacal, N. A novel CNN-ViT-based deep learning model for early skin cancer diagnosis. *Biomed. Signal Process. Control* **2025**, *104*, 107627. [CrossRef]
50. Sun, J.; Wu, B.; Zhao, T.; Gao, L.; Xie, K.; Lin, T.; Sui, J.; Li, X.; Wu, X.; Ni, X. Classification for thyroid nodule using ViT with contrastive learning in ultrasound images. *Comput. Biol. Med.* **2023**, *152*, 106444. [CrossRef]
51. Cece, A.; Agresti, M.; De Falco, N.; Sperlongano, P.; Moccia, G.; Luongo, P.; Parmeggiani, D. Role of Artificial Intelligence in Thyroid Cancer Diagnosis. *J. Clin. Med.* **2025**, *14*, 2422. [CrossRef] [PubMed]
52. Al-Jabbar, M.; Alshahrani, M.; Senan, E.M.; Ahmed, I.A. Multi-Method Diagnosis of Histopathological Images for Early Detection of Breast Cancer Based on Hybrid and Deep Learning. *Mathematics* **2023**, *11*, 1429. [CrossRef]
53. Huang, N.Y.; Liu, C.X. Efficient Tumor Detection and Classification Model Based on ViT in an End-to-End Architecture. *IEEE Access* **2024**, *12*, 106096–106106. [CrossRef]
54. Summia Parveen, H.; Karthik, S.; Sabitha, R. A novel maternal thyroid disease prediction using multi-scale vision transformer architecture with improved linguistic hedges neural-fuzzy classifier. *Technol. Health Care* **2024**, *32*, 4381. [CrossRef]
55. Senan, E.M.; Jadhav, M.E. Diagnosis of dermoscopy images for the detection of skin lesions using SVM and KNN. In *Proceedings of Third International Conference on Sustainable Computing: SUSCOM 2021*; Springer Nature: Singapore, 2022; pp. 125–134.
56. Yoon, H.C.; Lin, L.P. Brain Tumor Classification in MRI: Insights From LIME and Grad-CAM Explainable AI Techniques. *IEEE Access* **2025**, *13*, 154172–154202. [CrossRef]
57. Raghavan, K.; B, S.; v, K. Attention guided grad-CAM: An improved explainable artificial intelligence model for infrared breast cancer detection. *Multimed. Tools Appl.* **2023**, *83*, 57551–57578. [CrossRef]
58. Sankar, S.; Sathyalakshmi, S. Integrating Knowledge-Guided Layers with Fine-Tuned VGG-16 for Improved Thyroid Malignancy Classification: Explainability with Grad-CAM. In *Perspectives on Global Transformation*; Lecture Notes in Electrical Engineering; Springer Nature: Singapore, 2026; Volume 1368 LNEE, pp. 1–20. [CrossRef]
59. Song, D.; Yao, J.; Jiang, Y.; Shi, S.; Cui, C.; Wang, L.; Dong, F. A new xAI framework with feature explainability for tumors decision-making in Ultrasound data: Comparing with Grad-CAM. *Comput. Methods Programs Biomed.* **2023**, *235*, 107527. [CrossRef]
60. Shabrina, N.H.; Gunawan, D.; Ham, M.F.; Harahap, A.S. Papillary Thyroid Cancer Histopathological Image Classification Using Pretrained ConvNeXt Tiny and Grad-CAM Interpretation. In Proceedings of the IEEE Joint International Information Technology and Artificial Intelligence Conference (ITAIC), Chongqing, China, 8–10 December 2023; pp. 1836–1842. [CrossRef]
61. Alshahrani, M.; Al-Jabbar, M.; Senan, E.M.; Ahmed, I.A.; Saif, J.A.M. Hybrid Methods for Fundus Image Analysis for Diagnosis of Diabetic Retinopathy Development Stages Based on Fusion Features. *Diagnostics* **2023**, *13*, 2783. [CrossRef]

https://doi.org/10.3390/diagnostics16070999

diagnostics

MDPI

Article

Enhancing Approaches to Detect Papilloma-Associated Hyperostosis Using a Few-Shot Transfer Learning Framework in Extremely Scarce Radiological Datasets

Pham Huu Duy, Nguyen Minh Trieu and Nguyen Truong Thinh *

Institute of Intelligent and Interactive Technologies, University of Economics Ho Chi Minh City—UEH, Ho Chi Minh City 700000, Vietnam; trieunm@ueh.edu.vn (N.M.T.)
* Correspondence: thinhnt@ueh.edu.vn; Tel.: +84-903-675-673

Abstract

Background/Objectives: The application of deep learning models for rare diseases faces significant difficulties due to severe data scarcity. The detection of focal hyperostosis (PAH) is a crucial radiological sign for the surgical planning of sinonasal inverted papilloma, yet data is often limited. This study introduces and validates a robust methodological framework for building clinically meaningful deep learning models under extremely limited data conditions ($n = 20$). **Methods:** We propose a few-shot learning framework based on the nnU-Net architecture, which integrates an in-domain transfer learning strategy (fine-tuning a pre-trained skull segmentation model) to address data scarcity. To further enhance robustness, a specialized data augmentation technique called "window shifting" is introduced to simulate inter-scanner variability. The entire framework was evaluated using a rigorous 5-fold cross-validation strategy. **Results:** Our proposed framework achieved a stable mean Dice Similarity Coefficient (DSC) of 0.48 ± 0.06. This performance significantly outperformed a baseline model trained from scratch, which failed to converge and yielded a clinically insignificant mean DSC of 0.09 ± 0.02. **Conclusions:** The analysis demonstrates that this methodological approach effectively overcomes instability and overfitting, generating reproducible and valuable predictions suitable for rare data types where large-scale data collection is not feasible.

Keywords: PAH detection; transfer learning; papilloma-associated hyperostosis; *n*-small data; Vietnamese case study

1. Introduction

Sinonasal inverted papilloma (SIP) is a benign epithelial neoplasm characterized by local aggressiveness, a high risk of recurrence of about 13 to 30%, and a significant potential for malignant transformation [1,2]. Effective recurrence control is critically dependent on the complete surgical resection of the tumor's attachment site. On computed tomography (CT), papilloma-associated hyperostosis (PAH) presents as focal bone thickening at the tumor's origin, serving as a crucial imaging marker for this site, with a reported diagnostic accuracy of approximately 89% [2]. However, its recognition is challenging, requiring expert differentiation from non-pathological inflammatory osteitis [3]. This clinical challenge is compounded by the disease's rarity, as SIP accounts for only 0.4–4.7% of all sinonasal tumors. This rarity creates a clear and significant clinical need for an automated, reliable artificial intelligence model to assist in localizing the attachment site [4,5]. Efforts to develop

 https://doi.org/10.3390/diagnostics16020311

artificial intelligence algorithms aim to diagnose diseases in conditions where there is a shortage of expert diagnoses [6,7]. The primary barrier to applying deep learning is severe data scarcity. In this study, a dataset of 20 patients ($n = 20$) was used to train the model; however, the risks of overfitting and poor generalization are significantly high. This is the biggest challenge, so a processing algorithm for scarce data is needed. A recent attempt by McKee et al. [8] applied the state-of-the-art nnU-Net framework on a considerably larger dataset ($n = 58$). Their model achieved a mean Dice similarity coefficient (DSC) of only approximately 0.34, which is insufficient for clinical use.

A critical review of other modern AI paradigms reveals they are equally unsuitable for n-small problems. Although few-shot learning (FSL) has emerged as a promising strategy for handling scarce data, its application to dense 3D segmentation remains computationally unstable compared to 2D tasks. While Transformer-based architectures (e.g., Swin UNETR) have shown promise in medical imaging by capturing global context [9,10], they typically lack the strong inductive biases, such as translation invariance and locality, that CNNs possess [11]. Consequently, recent benchmarks indicate that without sufficient training data to learn these properties, Transformers can underperform for well-configured CNNs in extremely low-data regimes [12,13]. Similarly, while Self-Supervised Learning (SSL) is powerful, its efficacy often relies on large-scale unlabeled pre-training, which poses challenges when the entire available cohort is extremely small [14,15]. Similarly, Generative Data Augmentation using GANs or Diffusion Models is a dead end. These models are notoriously unstable to train on small datasets and often suffer from mode collapse, producing low-fidelity and blurry images that fail to capture the subtle pathological details of PAH, thus acting as noise that harms rather than helps performance [16]. Finally, the most common strategy, out-of-domain transfer learning, is semantically irrelevant, as features for natural images do not map to subtle bone-texture analysis on CT.

This study develops and validates a robust, composite deep learning framework specifically designed to overcome these challenges regarding the dataset and accuracy of the diagnostic process. This strategic approach combines three key components to solve these identified gaps. The nnU-Net architecture is applied as a powerful and self-configuring baseline. Then, the train-from-scratch failure is solved by employing in-domain transfer learning, initializing weights from a pre-trained skull segmentation model. The limitation of standard data augmentation, which typically focuses on geometric transformations but ignores radiometric variations, is solved. A specialized "Window Shifting" technique that simulates the variability in Hounsfield Unit (HU) calibration often observed between different CT scanners is introduced. By perturbing intensity values, this method compels the model to learn robust morphological descriptors of hyperostosis rather than relying on unstable absolute intensity thresholds, thereby enhancing clinical generalizability. The results demonstrate that this synergistic framework can significantly outperform a baseline model trained from scratch, thus providing a viable pathway for building a reliable AI model with small dataset clinical scenarios. The implications of this study offer a path to guide more precise surgical resections and reduce healthcare disparities to support the diagnosis of tumor location.

This paper is organized as follows. The Materials and Methods are presented in the next section, which presents the dataset and annotation, and the proposed segmentation framework, which is the key and main contribution. The experimental design and statistical evaluation are proposed in the next section. Then, the results and discussions are presented in the corresponding section. Finally, the conclusions of this paper are presented.

https://doi.org/10.3390/diagnostics16020311

2. Materials and Methods

2.1. Dataset and Annotation

This retrospective study analyzed 20 patients with histopathology-confirmed sinonasal inverted papilloma (SIP). Data were assembled in collaboration with otolaryngology and radiology teams; only de-identified images were retained for research, and all handling of patient information adhered to ethical principles. The criteria are biopsy-proven SIP and availability of a preoperative non-contrast paranasal sinus CT. Exclusion criteria comprised any prior sinonasal surgery, prior radiation to the sinonasal region, or severe CT artifacts that could preclude reliable assessment. For each subject, the imaging consisted of a 3D axial helical CT volume acquired in routine clinical practice without the use of intravenous contrast. To capture real-world variability, data were collected using multiple scanner platforms. Despite the limited sample size (n = 20), the cohort was carefully curated by the radiological team to encompass a representative spectrum of PAH morphologies, ranging from subtle focal thickening to extensive neo-osteogenesis, while excluding cases with severe artifacts. Typical in-plane resolution was 512 × 512 in native DICOM, and the number of slices and voxel spacing (Δx, Δy, Δz) varied by scanner protocol. After retrieval from the institutional PACS, DICOM headers were de-identified, and volumes were converted to NIfTI-1 using dcm2niix while preserving affine geometry and orientation. Each case was stored as a paired set comprising the CT volume (NIfTI, floating-point, HU) and a binary label mask (NIfTI, unsigned 8-bit) co-registered on the same grid and affine. A versioned index recorded anonymized IDs, file paths, voxel spacing, slice counts, and quality flags to ensure traceability. Curation followed a three-stage workflow as initial screening against eligibility criteria; acquisition–de-identification–conversion, two-tier quality control (QC) comprising automated integrity and subsequent radiologist visual review to confirm adequate sinonasal coverage and to exclude prohibitive artifacts that are shown in Figure 1.

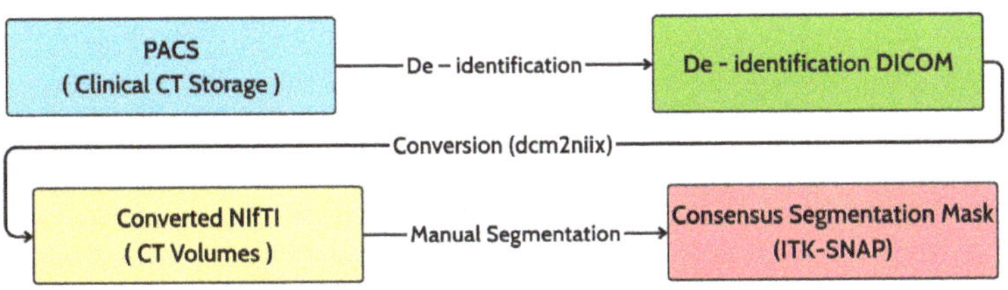

Figure 1. Data processing workflow.

The target is Papilloma-Associated Hyperostosis (PAH), i.e., focal hyperostosis at the presumed tumor attachment site, and explicitly differentiated from diffuse osteitis of chronic rhinosinusitis. Ground truth was established via consensus manual segmentation in ITK-SNAP version 3.8.0 by two board-certified radiologists with over 10 years of experience. To ensure standardized visualization, a constant bone window with the window level set to 500 HU and a window width of 2000 HU was applied. Before formal annotation, the readers conducted a calibration session to harmonize boundary rules (transitional bone, thin/irregular plates, partial-volume effects). For each case, a single consensus mask was produced according to the imaging criteria of Lee DK et al. [2] to ensure reliability and minimize inter-observer variability, where focal bony thickening/neo-osteogenesis localized to the sinus wall indicates the SIP attachment site and must be distinguished from non-focal inflammatory change. Final masks were saved as binary NIfTI volumes co-registered to their corresponding CT, as illustrated in Figure 2.

https://doi.org/10.3390/diagnostics16020311

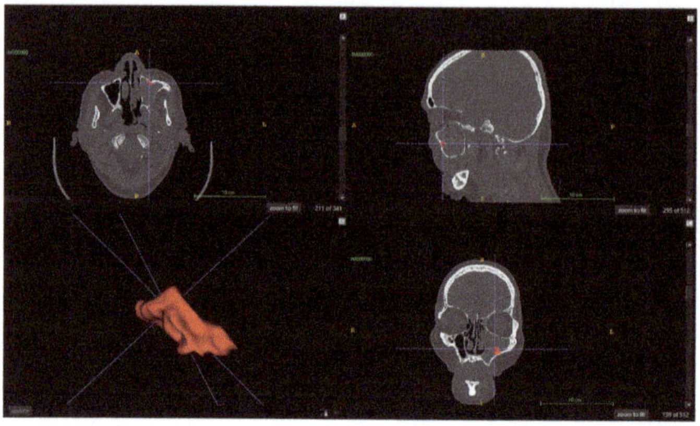

Figure 2. Illustration of the ground truth annotation process using ITK-SNAP ver 3.8.0. (1) The letters denote standard anatomical orientations: A (Anterior), P (Posterior), R (Right), L (Left), S (Superior), and I (Inferior). (2) The blue crosshair lines represent the navigation cursor, indicating the intersection of the axial, sagittal, and coronal planes for 3D triangulation. (3) The red overlay corresponds to the consensus binary mask of the Papilloma-Associated Hyperostosis (PAH), manually delineated by radiologists.

2.2. Data Preprocessing and Augmentation

2.2.1. Preprocessing

Before training, a harmonization step is performed to ensure spatial correspondence between images and labels. Each segmentation mask is resampled to match the spatial grid of its corresponding CT image, including origin, spacing, and orientation, using nearest-neighbor interpolation to preserve binary label integrity. All cases were then processed using the fully automated preprocessing pipeline of nnU-Net. Based on an analysis of dataset properties, nnU-Net automatically configured the core preprocessing steps. Specifically, all volumes were resampled to a target median voxel spacing of $0.6 \times 0.46875 \times 0.46875$ mm. Intensities are subsequently clipped to a Hounsfield Unit (HU) window suitable for bone structures and normalized via z-score standardization, thereby ensuring consistency across patients and scanners.

2.2.2. Input Formulation and Augmentation

Before entering the backbone network, all CT volumes were preprocessed and augmented following the standard nnU-Net pipeline [17], with additional modifications introduced to enhance robustness under extremely small-sample conditions. All CT volumes were first resampled to an isotropic voxel spacing of $0.6 \times 0.46875 \times 0.46875$ mm^3 and clipped to a bone-specific Hounsfield Unit (HU) window $[-1000, 3000]$, suppressing non-relevant soft-tissue signals. Intensity values were then standardized by z-score normalization as in Equation (1):

$$I' = \frac{I - \mu}{\sigma} \tag{1}$$

where I denotes the voxel intensity; μ, σ represent the mean and standard deviation within the corresponding non-zero voxel distribution.

During training, on-the-fly data augmentation is applied to mitigate overfitting and improve generalization. The default nnU-Net transformations included random rotations ($\pm 15°$), scaling (0.9–1.1), elastic deformations, and mirroring along random anatomical planes. In addition to these built-in augmentations, we introduced a custom window-shifting augmentation to simulate inter-scanner variability in bone-window calibration.

https://doi.org/10.3390/diagnostics16020311

After normalization [18,19], the voxel intensities were globally perturbed with a probability $p = 0.3$ as in Equation (2):

$$I'' = I' + \alpha, \alpha \sim U(-0.1, 0.1) \tag{2}$$

where α is a uniformly distributed random shift.

This technique differs fundamentally from conventional density-based enhancements such as random gamma correction or brightness scaling, which typically alter the dynamic range or distribution shape. Instead, Window Shifting applies a linear translational offset to simulate the systematic calibration bias in Hounsfield Units often observed between different scanner manufacturers. This encourages the network to learn morphological rather than absolute intensity cues, thereby improving cross-scanner robustness.

Training is conducted on randomly sampled 3D patches of size 80 × 192 × 160 voxels, ensuring a balanced sampling of lesion-containing and background regions [17]. During inference, overlapping patch predictions were aggregated by weighted averaging and thresholded at 0.5 to reconstruct the final binary segmentation mask. This combination of standardized preprocessing and both default and custom augmentations effectively reduced overfitting and enabled the proposed few-shot framework to learn stable, generalizable representations from only 20 subjects.

2.3. Proposed Segmentation Framework

2.3.1. Overall Design and Rationale

The automated segmentation of papilloma-associated hyperostosis (PAH) on computed tomography (CT) scans poses a formidable challenge because the target lesions are typically small, subtle, and located adjacent to complex bony interfaces, which renders them difficult to distinguish from normal anatomical variations or chronic inflammatory changes [1]. Previous work attempting to tackle this task with the well-established nnU-Net framework on a dataset of 58 patients achieved a mean Dice similarity coefficient (DSC) of approximately 0.34, thereby highlighting the severe barrier imposed by limited sample size [8]. In our setting, this difficulty is even more acute, as only 20 patient scans are available. Attempting to train a high-capacity neural network from scratch under such constraints would be expected to result in severe overfitting and instability, ultimately leading to poor generalization performance on unseen data and unreliable clinical predictions [20].

To mitigate these issues, a principled framework is designed that integrates the standardized nnU-Net v2 pipeline, specifically its 3d_fullres configuration, with an in-domain transfer learning strategy that leverages prior knowledge from a closely related segmentation task. However, recent years have witnessed the emergence of Transformer-based architectures such as UNETR and Swin UNETR, which are theoretically capable of capturing long-range contextual dependencies [21]. The decision to rely on a convolutional neural network (CNN) based U-Net is not simply a matter of tradition but rather a deliberate, evidence-based choice supported by both methodological studies and the specific clinical nature of PAH. When compared under identical conditions within the nnU-Net pipeline, a well-configured CNN-based U-Net remained highly competitive and frequently outperformed Transformer-based alternatives across diverse datasets.

Furthermore, the biological and radiological characteristics of PAH itself favor a CNN-based solution. The lesion is a localized phenomenon, expressed as fine-grained textural and morphological changes confined to the bony sinus wall. Such localized features are precisely the type of patterns that convolutional kernels are optimized to detect, while the primary strength of Transformers, with modeling global contextual relationships, is of secondary importance in this application [21]. In addition, the effectiveness of combining nnU-Net with in-domain transfer learning has been demonstrated in related small-data scenarios. For example, Bareja et al. [22] successfully adapted a model pre-trained on adult

https://doi.org/10.3390/diagnostics16020311

gliomas to segment pediatric medulloblastomas, achieving robust performance across multiple institutions despite limited sample sizes.

Building upon these insights, our solution involves initializing an optimized 3D U-Net with weights derived from a skull segmentation model available in the PYCAD Model Zoo and subsequently fine-tuning it within the nnU-Net v2 framework. This approach is grounded in rigorous empirical evidence, exploits the proven strengths of CNN-based architectures for localized feature extraction, and directly addresses the fundamental challenge of extreme data scarcity by harnessing in-domain transfer learning.

2.3.2. Model Architecture

Building upon the preprocessed dataset described in Section 2.1, the proposed framework employs the three-dimensional U-Net architecture exactly as implemented in the standardized 3d_fullres configuration of nnU-Net. The following description details the key components and design principles of this specific, well-established framework, which our study adopts without modification. This architectural choice is deliberate to design the symmetric encoder–decoder of U-Net, which is uniquely suited to capturing features across multiple spatial scales. The encoder progressively down samples the input to extract high-level contextual information, while the decoder reconstructs spatial resolution. Crucially, skip connections relay high-resolution feature maps from the encoder to their corresponding decoder layers, enabling precise localization of small, subtle targets such as Papilloma-Associated Hyperostosis (PAH) while maintaining awareness of the broader anatomical context. The network follows a six-stage hierarchy, with encoder feature channels increasing to learn abstract representations. Specifically, the encoder utilizes $3 \times 3 \times 3$ convolutional kernels with a starting feature map size of 32 channels, which doubles at each subsequent down-sampling stage (32, 64, 128, 256, 320, 320). The model processes a 3D input patch of size $80 \times 192 \times 160$ voxels, producing a binary segmentation mask of identical dimensions. At each convolutional layer l, the transformation of a 3-D input patch $x \in \mathbb{R}^{C_{in} \times D \times H \times W}$ is defined as (3):

$$y_k^{(l)} = \phi\left(\mathrm{IN}\left(\sum_{c=1}^{C_{\text{in}}} W_{k,c}^{(l)} * x_c + b_k^{(l)}\right)\right) \tag{3}$$

where $W_{k,c}^{(l)} \in \mathbb{R}^{K \times K \times K}$ denotes the 3-D convolution kernel ($K = 3$), $*$ represents the convolution operator, IN $(\cdot)$ is Instance Normalization, $b_k^{(l)}$ is the bias term, and $\phi\,(\cdot)$ is the Leaky ReLU activation function with a negative-slope coefficient $\alpha = 0.01$ as (4).

$$\phi(x) = \begin{cases} x, & x \geq 0 \\ \alpha x, & x < 0 \end{cases} \tag{4}$$

The total number of parameters per layer is $K^3 C_{\text{in}} C_{\text{out}} + C_{\text{out}}$. Each encoder block contains two consecutive $3 \times 3 \times 3$ convolutions, each followed by Instance Normalization and Leaky ReLU activation, expressed as (5).

$$f^{(l)} = \phi\left(\mathrm{IN}\left(\mathrm{Conv}_{3^3}\left(\phi\left(\mathrm{IN}\left(\mathrm{Conv}_{3^3}\left(f^{(l-1)}\right)\right)\right)\right)\right)\right) \tag{5}$$

Between stages, spatial resolution is reduced using strided convolution with a stride of 2, which is a parametric down-sampling operation that replaces max pooling as in Equation (6).

$$f_{\downarrow}^{(l)} = \mathrm{Conv}_{3^3}^{s=2}(f^{(l)}) \tag{6}$$

https://doi.org/10.3390/diagnostics16020311

This approach allows the network to learn adaptive feature compression, which is beneficial for capturing subtle osseous texture variations in CT data. Symmetrically, the decoder pathway progressively up-samples feature maps to reconstruct the full spatial resolution. This is achieved through transposed convolutions with a kernel with a stride of 2, which is shown in (7).

$$g_{\uparrow}^{(l)} = \text{ConvTranspose}_{2^3}^{s=2}\left(g^{(l+1)}\right) \tag{7}$$

The output of this up-sampling step is then concatenated with the corresponding feature map from the encoder via a skip connection (8).

$$\tilde{g}^{(l)} = \text{Concat}\left(g_{\uparrow}^{(l)}, f^{(l)}\right) \tag{8}$$

Each concatenated tensor subsequently passes through a decoder block identical in structure to the encoder block, a process that refines spatial details lost during down-sampling. These skip connections are crucial for preserving fine-grained boundaries, such as cortical bone interfaces and localized PAH. The output head of the network projects the final decoder feature map to the class space using a $1 \times 1 \times 1$ convolution followed by a voxel-wise softmax function (9).

$$p = \text{Softmax}\left(\text{Conv}_{1^3}\left(\tilde{g}^{(0)}\right)\right) \tag{9}$$

This produces a dense probability map $\rho \in [0,1]^{C \times D \times H \times W}$ for C = 2 classes. In the nnU-Net v2 paradigm, auxiliary prediction heads are attached to intermediate decoder stages to enable deep supervision. During training, these auxiliary outputs, denoted as $p^{(s)}$ at scale s, are upsampled to the native resolution and contribute to the total loss, thereby improving gradient propagation across multi-scale representations. An important aspect of this architecture is its receptive field (RF). For a kernel size K = 3 and stage-wise strides $s_j \in (1,2)$. The effective RF after the 1st stage follows the relation (10).

$$R_l = R_{l-1} + (K-1)\prod_{j=1}^{l-1} s_j \tag{10}$$

After five down-sampling operations, the theoretical RF spans the entire $80 \times 192 \times 160$ voxel Three-dimensional patch, enabling the model to integrate long-range sinus-bone context while maintaining high-resolution spatial fidelity through the skip pathways. The design rationale for this architecture is based on several key principles. 3-D kernels are employed because the dataset has nearly isotropic voxel spacing, allowing the network to exploit inter-slice continuity critical for sinonasal morphology. This stands in contrast to conventional 2D U-Net approaches, which process volumes slice-by-slice and consequently fail to capture the volumetric contextual information essential for defining the 3D structure of PAH lesions. Instance Normalization stabilizes feature scaling under small-batch training and reduces dependency on absolute HU values. Strided convolution for down-sampling and transposed convolution for up-sampling preserve learnable spatial mappings, avoiding artifacts from pooling/unpooling. Leaky ReLU ensures non-zero gradients for negative activations, mitigating dead-filter effects in low-contrast osseous regions. Finally, the architecture excludes dropout layers by default to prevent information loss in extremely small-sample settings. A schematic of the proposed 3D U-Net architecture for Papilloma-Associated Hyperostosis segmentation is shown in Figure 3.

https://doi.org/10.3390/diagnostics16020311

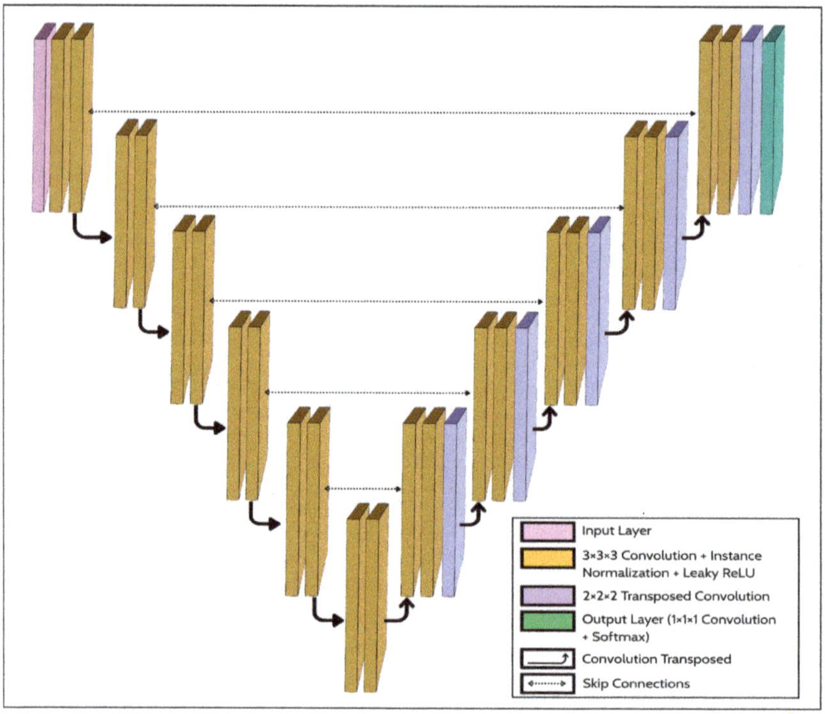

Figure 3. Schematic of the proposed 3D U-Net architecture for Papilloma-Associated Hyperostosis (PAH) segmentation.

Overall, this design is grounded in both empirical and clinical considerations. The use of 3D convolutional kernels leverages the near-isotropic voxel spacing, enabling the model to exploit inter-slice continuity essential for sinonasal morphology. Instance Normalization stabilizes training under small-batch conditions and reduces dependency on absolute HU values. Strided and transposed convolutions preserve learnable spatial mappings, avoiding the artifacts introduced by pooling and unpooling. The Leaky ReLU activation ensures non-zero gradients even in low-contrast osseous regions, minimizing dead-filter effects. Finally, the architecture omits dropout layers to prevent information loss in the small-sample regime. Collectively, these design choices enable the 3D U-Net backbone to achieve a robust balance between spatial precision, contextual awareness, and data efficiency in segmenting Papilloma-Associated Hyperostosis.

2.3.3. Transfer Learning Strategy

Training a deep neural network with millions of parameters from scratch on an extremely small dataset ($n = 20$) is fraught with challenges. With random weight initialization, the model would likely fail to learn meaningful representations and would be highly susceptible to overfitting, where it memorizes the training data instead of learning generalizable features, resulting in poor performance on unseen cases and unreliable validation metrics [9]. To mitigate these risks, a transfer learning strategy is adopted. The fundamental benefit of this approach is that it leverages knowledge from a related, large-scale task to provide the model with a strong initial foundation [2]. Instead of starting from a random state, the model's weights are initialized from a network that has already been trained to understand relevant low-level features. This provides a much better starting point for

https://doi.org/10.3390/diagnostics16020311

optimization, accelerates convergence, and significantly improves the model's ability to learn from limited data. In this study, an in-domain transfer learning approach is employed. This choice is highly strategic because PAH manifests as localized thickening and neo-osteogenesis at bone interfaces, a model already proficient in segmenting complex cranial bone structures, possesses rich shape and texture priors that are far more relevant to this task than features learned from natural image datasets like ImageNet. Then, fine-tuning all layers of this pre-trained model on 20 subjects is performed, allowing the network to adapt its learned knowledge to the specific task of identifying PAH. The fine-tuning process leveraged the robust optimization framework of the nnU-Net model. The optimization settings were conducted using the default nnU-Net v2 optimization settings. Crucially, the fine-tuning utilized the full data augmentation pipeline described in Section 2.2.2, which includes both the standard nnU-Net transformations and our custom window-shifting augmentation designed to enhance cross-scanner robustness. A patient-level 5-fold cross-validation was employed to prevent data leakage and provide a robust performance estimate. The transfer learning process can be expressed as a domain mapping as (11).

$$\mathcal{F}_{\text{pre}} : A \to B \tag{11}$$

where A represents the skull segmentation model pretrained on large-scale cranial CT data, and B represents the PAH segmentation.

The pretrained network $\mathcal{F}_{\text{pre}}$ provides initial parameters θ_A, which are fine-tuned to θ_B through gradient updates on the small PAH dataset (12).

$$\theta_B = \theta_A + \Delta\theta \tag{12}$$

where $\Delta\theta$ denotes the domain adaptation achieved during few-shot fine-tuning. This mapping efficiently transfers anatomical priors, accelerating convergence and improving generalization under extreme data scarcity.

2.3.4. Output Layer and Fully Connected Equivalent

Unlike classification networks ending with a fully connected layer, the 3D U-Net's final convolution acts as its fully connected equivalent, performing dense voxel-level classification [6]. For each voxel v, p_v is calculated as in Equation (13).

$$p_v = \text{softmax}(W \times F_v + b) \tag{13}$$

where F_v is the final feature vector, W and b are learnable parameters.

This produces a 3D probability map $p(x,y,z)$, thresholded at 0.5 to yield the binary segmentation mask. The output thus represents a continuous volumetric probability field, preserving spatial structure and enabling detailed delineation of hyperostotic regions. This is a more informative outcome than a single global prediction. Clinically, these voxel-wise outputs highlight bony attachment zones of sinonasal inverted papilloma, providing quantitative and visual guidance for preoperative planning. The framework integrates preprocessing, transfer learning from a pre-trained skull model, and data augmentation with window shifting into a fine-tuned 3D U-Net with deep supervision, optimized by Dice combined with cross-entropy loss, yielding robust segmentation results on a scarce SIP CT dataset (n = 20). The framework begins with a preprocessing stage in which sinonasal CT volumes are de-identified, resampled, and intensity-normalized to ensure spatial and radiometric consistency. The pre-trained nnU-Net model is originally trained for skull segmentation and provides the encoder–decoder weight initialization used for in-domain transfer learning. During fine-tuning, all network layers are updated under a composite loss function combining Soft Dice and Cross-Entropy terms, while deep supervision is applied through auxiliary decoder outputs to stabilize gradient propagation. The pipeline also integrates a specialized window-shifting augmentation, which introduces controlled intensity perturbations to improve cross-scanner generalization. The final fine-tuned model produces voxel-wise probability maps highlighting focal hyperostosis regions corresponding to tumor attachment sites on sinonasal CT slices. This integrated strategy combines transfer learning and targeting augmentation to enable robust segmentation performance even under extreme data scarcity (n = 20). Figure 4 shows the proposed pipeline for hyperostosis segmentation.

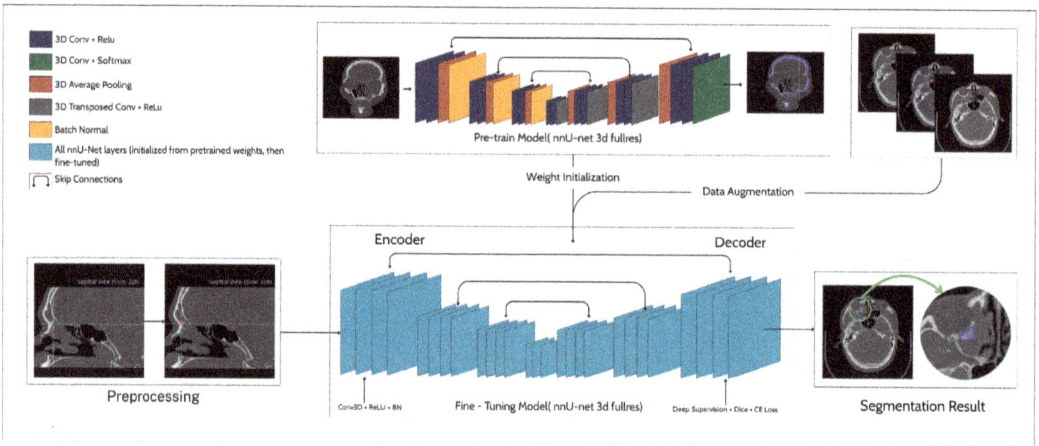

Figure 4. Proposed pipeline for hyperostosis segmentation.

2.4. Training and Optimization

2.4.1. Loss Function

The model is optimized using the standard composite loss function based on the nnU-Net model. This protocol combines the soft dice loss and the voxel-wise Cross-Entropy (CE) loss, a formulation designed to balance region-level overlap accuracy with local voxel-wise classification stability. The total loss is expressed as (14):

$$\mathcal{L}_{\text{total}} = \lambda_{\text{Dice}}\mathcal{L}_{\text{Dice}} + \lambda_{\text{CE}}\mathcal{L}_{\text{CE}} \tag{14}$$

where λ_{Dice} and λ_{CE} denote the weighting coefficients.

The Soft Dice loss directly optimizes spatial overlap between prediction and ground truth, effectively mitigating the severe class imbalance inherent in PAH segmentation. It is defined as (15):

$$\mathcal{L}_{\text{Dice}} = 1 - \frac{2\sum_{i=1}^{N} p_i g_i + \epsilon}{\sum_{i=1}^{N} (p_i^2 + g_i^2) + \epsilon} \tag{15}$$

where p_i and g_i represent the predicted probability and ground-truth label of a voxel i, respectively, N is the total number of voxels, and $\epsilon = 10^{-6}$ ensures numerical stability. Complementarily, the Cross-Entropy loss provides smooth per-voxel gradient feedback, stabilizing the training process and improving convergence. It is formulated as (16).

$$\mathcal{L}_{\text{CE}} = -\frac{1}{N}\sum_{i=1}^{N} [g_i \log(p_i) + (1 - g_i)\log(1 - p_i)] \tag{16}$$

https://doi.org/10.3390/diagnostics16020311

This combined formulation enhances both global shape agreement and local boundary precision, leading to more reliable segmentation results. Training was performed on 3D patches of 80 × 192 × 160 voxels with a batch size of 2, allowing efficient GPU utilization while preserving volumetric context. To further stabilize gradient flow, the framework employs deep supervision by attaching auxiliary outputs to intermediate decoder stages. The multi-scale loss is computed as (17):

$$\mathcal{L}_{\mathrm{DS}} = \sum_{s=1}^{S} \alpha_s \mathcal{L}_{\mathrm{total}}^{(s)} \tag{17}$$

where S denotes the number of supervised scales and α_s are decreasing weights for deeper layers.

The entire fine-tuning process followed the nnU-Net optimization protocol, using stochastic gradient descent (SGD) with Nesterov momentum (μ = 0.99) and a polynomial learning rate decay defined as Equation (18):

$$\eta_t = \eta_0 (1 - \frac{t}{T_{\mathrm{max}}})^{0.9} \tag{18}$$

where η_t is the learning rate at iteration t, η_0 is the initial learning rate, and T_{max} represents the total number of training iterations.

2.4.2. Optimization Setup

The training, optimization, and inference process strictly followed the standardized protocol of nnU-Net with the transfer learning for weight initialization. Given the large memory footprint of volumetric CT data and the limited number of annotated cases, training was performed on 3D patches of modified size. This configuration was selected as an empirically verified trade-off between GPU memory efficiency and contextual coverage, ensuring that each patch fully encapsulated the sinonasal anatomy relevant to Papilloma-Associated Hyperostosis. A stochastic gradient descent (SGD) with Nesterov momentum to accelerate convergence and stabilize training dynamics is used, with the update rule at iteration t can be expressed as (19):

$$\begin{aligned} v_{t+1} &= \mu v_t - \eta_t \nabla_\theta \mathcal{L}(\theta_t + \mu v_t) \\ \theta_{t+1} &= \theta_t + v_{t+1} \end{aligned} \tag{19}$$

where θ_t denotes the trainable parameters at iteration t, v_t is the velocity term, $\mu = 0.99$ represents the Nesterov momentum coefficient, and η_t is the learning rate. And the initial learning rate is chosen as (20) to maintain smooth convergence:

$$\eta_t = \eta_0 (1 - \frac{t}{T})^{p} \tag{20}$$

where T denotes the total number of training iterations and $p = 0.9$ controls the rate of decay.

To ensure stable gradient propagation across the deep architecture, deep supervision was employed by attaching auxiliary segmentation heads to intermediate decoder stages. Each auxiliary output $\hat{Y}^{(s)}$ at scale s contributed to the overall multi-scale loss function as (21):

$$\mathcal{L}_{\mathrm{total}} = \sum_{s=1}^{S} w_s \mathcal{L}(\hat{Y}^{(s)}, Y) \tag{21}$$

https://doi.org/10.3390/diagnostics16020311

where Y is the ground-truth mask, $\mathcal{L}$ denotes the composite Dice–Cross-Entropy loss, and w_s are scale-specific weighting coefficients that decrease exponentially with layer depth ($w_s = 1/2^{s-1}$).

This hierarchical supervision framework ensures that both shallow and deep layers receive informative gradient signals, mitigating vanishing gradients and promoting balanced feature learning across multiple resolutions. Weight initialization followed the transfer learning paradigm to make sure that all layers were fine-tuned end-to-end without freezing, allowing full adaptation to the PAH domain while retaining low-level structural priors related to craniofacial bone morphology. Regularization is implicitly achieved through a combination of data augmentation, deep supervision, and transfer learning, with no dropout layers applied, consistent with nnU-Net's design for small-batch stability. Gradient clipping at a maximum L2-norm of 12 was employed to prevent exploding gradients. Each training run was repeated across five cross-validation folds, ensuring independence between training and validation subsets to eliminate data leakage. The model with the best mean Dice score on the validation set of each fold was retained for inference. During testing, overlapping patch predictions are aggregated via Gaussian-weighted averaging, and a probability threshold is applied to produce the final binary segmentation mask. This approach yielded smooth probability transitions between patches and minimized boundary artifacts.

2.4.3. Validation Strategy

Given the extremely limited dataset size (n = 20), the design of the validation protocol was critical to obtain statistically reliable and unbiased estimates of model performance. Following best practices in medical research [23], a five-fold cross-validation scheme is adopted to maximize data utilization while maintaining strict separation between training and validation samples. The dataset was randomly partitioned into five non-overlapping folds at the patient level, ensuring that no slices or volumes from the same subject appeared in both training and validation subsets, thereby eliminating the risk of data leakage [24]. This process was repeated five times so that every patient contributed exactly once to the validation set. To address the potential instability inherent in small validation splits (4 cases per fold), we implemented a stratified sampling strategy based on lesion volume. This ensures that each fold contains a representative distribution of both subtle (hard-to-segment) and extensive (easier-to-segment) PAH cases. By balancing the difficulty across folds, we minimized the risk of having a specific fold dominated by outliers, thereby stabilizing the error rate and ensuring a more reliable performance estimate. Each fold was trained independently from scratch using the fine-tuned pre-trained weights as initialization. The validation subset was not used for any form of model selection, early stopping, or hyperparameter tuning beyond the automatically configured nnU-Net pipeline, thereby preserving the integrity of the evaluation. Within each training session, the best-performing model checkpoint was selected based on the lowest validation loss and highest Dice Similarity Coefficient (DSC) observed during training. The primary evaluation metric was the Dice Similarity Coefficient (DSC), which measures the spatial overlap between the predicted segmentation (P) and the ground truth mask (G) as (22).

$$\mathrm{DSC} = \frac{2 \mid P \cap G \mid}{\mid P \mid + \mid G \mid} \tag{22}$$

To complement the quantitative analysis, qualitative visual inspection was performed for each validation case by two radiologists to assess spatial alignment between predicted PAH regions and the manually annotated ground truth. To assess the robustness of the framework, the variance across folds was explicitly analyzed. A low standard deviation in

https://doi.org/10.3390/diagnostics16020311

DSC (<0.06) indicated consistent performance across different validation splits, suggesting stable learning behavior and minimal dependence on specific training subsets. This multi-fold evaluation thus provides a statistically sound and generalizable estimate of the model's true performance under conditions of extreme data scarcity. Given the extremely small sample size, the choice of validation method is critical to avoid biased performance estimates. A 5-fold cross-validation strategy is employed, which is a standard and robust method widely used in medical imaging research, including similar studies with small datasets [8,10]. This approach provides a good balance between bias and variance: the dataset was split into 5 folds, with each fold using 16 cases for training and 4 for validation. Using a validation set of 4 cases offers a more stable performance estimate during training compared to a single-case validation, reducing the impact of outliers. The final performance was calculated by averaging the Dice Similarity Coefficient (DSC) across all 5 folds. The DSC was chosen as the primary metric as it is the gold standard for assessing spatial overlap in medical segmentation tasks.

3. Results

All experiments are conducted on an NVIDIA RTX 4090 GPU with mixed precision (AMP) enabled for computational efficiency. The rigorous 5-fold cross-validation yielded stable and clinically relevant segmentation results. Our proposed framework, leveraging in-domain transfer learning and specialized data augmentation, achieved a mean Dice Similarity Coefficient (DSC) of 0.48 ± 0.06 across the 5 validation folds. This level of performance indicates a substantial spatial overlap between the model's predictions and the expert-defined ground truth. A crucial component of our study was the ablation experiment comparing this approach to a baseline model. As shown in Table 1, the model trained from scratch on the same 5-fold cross-validation splits failed to converge to a meaningful solution, resulting in a poor and clinically insignificant mean DSC of 0.09 ± 0.02. Furthermore, we compared our results with existing state-of-the-art benchmarks for this specific pathology. Our proposed framework (DSC 0.48) demonstrates a substantial improvement over the standard nnU-Net implementation reported by McKee et al. [8], which achieved a DSC of only 0.34 despite utilizing a dataset nearly three times larger ($n = 58$). This indicates that standard SOTA methods struggle with the subtle features of PAH, whereas our proposed Transfer Learning and Window Shifting strategy effectively captures these nuances. This stark contrast underscores the infeasibility of standard deep learning approaches in this data-scarce regime and provides strong evidence for the necessity of our proposed methodological framework. To rigorously validate this improvement, a paired t-test was conducted comparing the fold-wise Dice scores of the proposed framework against the baseline. The analysis yielded a p-value of $p < 0.001$, confirming that the performance gain is statistically significant. Furthermore, we calculated the 95% Confidence Interval (CI) for the proposed model's mean DSC, resulting in a range of (0.427, 0.533). This narrow interval, relative to the mean, further reinforces the stability and reliability of the reported results despite the small sample size. And the detailed performance of the proposed framework across each validation fold is presented in Table 2. The individual Dice scores, ranging from 0.39 to 0.54, demonstrate consistent performance across different data splits.

Table 1. Segmentation performance of different models.

	Baseline Model	Proposed Framework
DSC	0.09 ± 0.02	0.48 ± 0.06

https://doi.org/10.3390/diagnostics16020311

Table 2. Proposed Framework Performance per Fold.

	Dice Score
Fold 0	0.5401
Fold 1	0.3928
Fold 2	0.5274
Fold 3	0.4339
Fold 4	0.5134

To better understand these quantitative results, a qualitative analysis of the segmentation outputs is performed. For this purpose, a custom software tool is developed to visualize the base CT images with overlays for both ground truth and AI-predicted segmentations. It also included features for quantitative assessment, such as real-time Dice score calculation, and functions to automatically navigate to slices with the largest segmentations or maximal prediction discrepancies, facilitating a targeted analysis of model performance. This detailed visual inspection revealed that while the mean DSC is moderate, the model is capable of achieving excellent results on a subset of the data. For instance, Figure 5 showcases a representative example of the model's upper-bound performance (DSC = 0.8950). This case exemplifies a scenario where the radiological signs of hyperostosis are distinct, allowing the model to perform an accurate delineation. The predicted segmentation mask (blue) demonstrates a high degree of concordance with the ground truth outline (red), capturing the lesion's morphology and extent with high precision. Presenting this successful case is intended to illustrate the model's learned capabilities and its potential clinical utility when faced with non-ambiguous pathology.

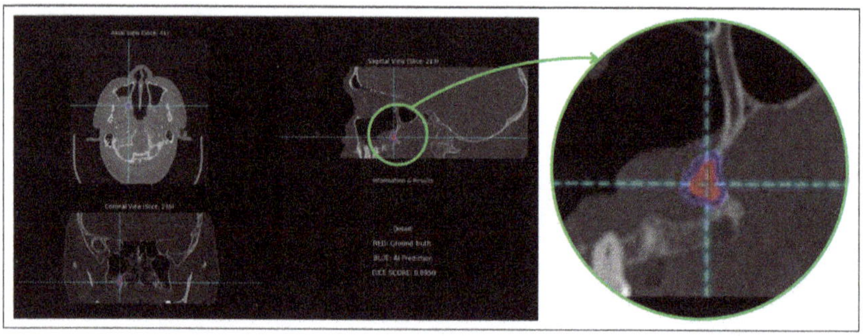

Figure 5. Qualitative comparison showing high segmentation accuracy between the model's prediction (blue overlay) and the ground truth (red overlay), achieving a Dice score of 0.8950.

A critical examination of cases with lower DSC scores revealed two primary modes of failure, as illustrated in Figure 6. In some instances, the model's lack of specificity led to significant over-segmentation errors. A severe example is illustrated in Figure 6a, where the AI Prediction (blue) vastly overestimates the lesion's extent. While the ground truth (red) is localized to the left ethmoid sinus and nasal septum, the model's prediction incorrectly expands to include large areas of normal anatomy, encompassing bilateral ethmoid and sphenoid sinuses. This error, resulting in a low DSC of 0.1266, suggests that while the model attempts to identify bone thickening, its specificity is severely challenged in anatomically complex regions, where it struggles to distinguish pathological hyperostosis from benign physiological sclerosis. False negatives are predominantly observed in cases with subtle or small lesions. Figure 6b provides a clear example, where the ground truth (red) indicates a distinct lesion, but the AI prediction (blue) is almost non-existent, resulting in a very

https://doi.org/10.3390/diagnostics16020311

low Dice score of 0.1108. This highlights a significant limitation in the model's sensitivity, indicating a detection threshold below which subtle lesions are completely missed.

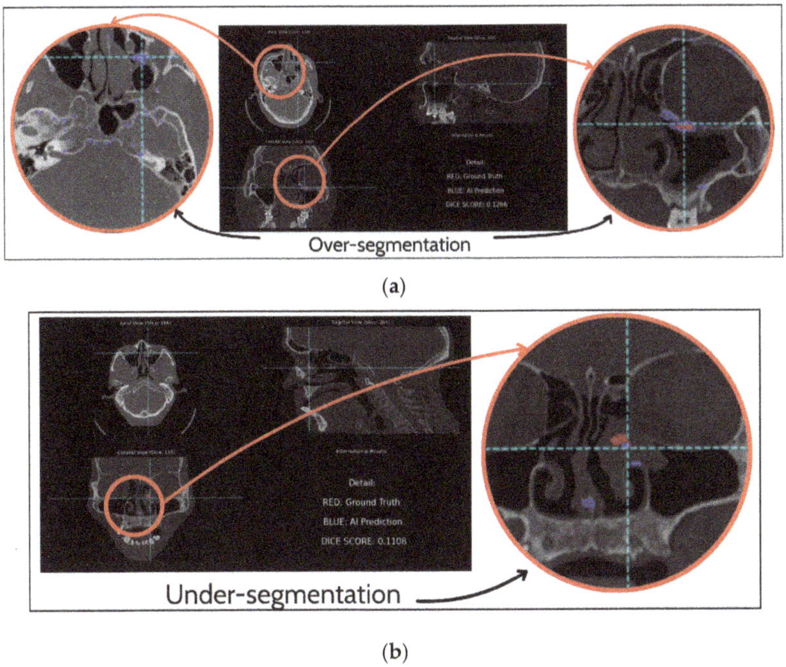

Figure 6. Examples of segmentation failure modes. (**a**) Over-segmentation; (**b**) under-segmentation.

4. Discussion

In this study, a methodological framework is proposed to address the critical challenge of developing deep learning models from extremely scarce medical imaging data. Our approach successfully developed a model for the automated segmentation of papilloma-associated hyperostosis, a key radiological sign for surgical planning in sinonasal inverted papilloma, with a dataset of only 20 patients. By integrating in-domain transfer learning, specialized data augmentation, and a robust cross-validation strategy, this framework achieved stable and relevant segmentation performance, specifically DSC of 0.48 ± 0.06, demonstrating a viable approach for building algorithms to support clinical diagnosis. The critical importance of the proposed strategic approach is underscored by the results of the ablation study. The baseline model, an nnU-Net trained from scratch, failed to converge to a meaningful solution, yielding a clinically insignificant mean DSC of 0.09 ± 0.02. This outcome is consistent with established principles that high-capacity neural networks are prone to severe overfitting and instability when trained on very small datasets. In stark contrast, a proposed framework produced stable and valuable segmentation results, providing strong evidence that the synergistic combination of pre-trained weight initialization and clinically informed data augmentation is essential for overcoming the challenges of extreme data scarcity.

While a mean DSC of 0.48 is generally considered moderate in large-organ segmentation, its interpretation requires nuance in the context of PAH. Mathematically, the Dice coefficient heavily penalizes boundary errors for small structures; a deviation of merely a few voxels can reduce the score significantly even when the lesion is correctly localized [25]. Clinically, the primary value of this AI model is not pixel-perfect delineation, but rather the

https://doi.org/10.3390/diagnostics16020311

precise localization of the tumor attachment site to guide surgical drilling. Visual inspection confirms that despite the moderate DSC, the model consistently identifies the correct focal hyperostosis region in successful cases, providing actionable confidence for preoperative planning. Regarding the comparison with McKee et al. [8], one might hypothesize that our superior performance with 14% higher despite less data. However, our cohort exhibits significant heterogeneity, acquired from multiple scanner platforms (GE, Siemens) with varying slice thicknesses. Therefore, the performance gain is driven by methodological innovations, namely the use of in-domain transfer learning, which provides stronger shape priors than training from scratch and window shifting. This consistency, achieved through the synergistic combination of in-domain transfer learning and stratified cross-validation, provides strong evidence that this methodology can generate reproducible and reliable predictions, even when facing extreme data scarcity. The implications of this work, however, extend far beyond the specific task of PAH segmentation. The data scarcity problem remains one of the most significant barriers to the application of AI in thousands of rare diseases, where collecting large-scale datasets is impractical or impossible. Our study provides a detailed and reproducible blueprint for researchers facing these challenges. It demonstrates that by strategically combining powerful, self-configuring architectures like nnU-Net with in-domain transfer learning and extensively informed data augmentation, it is possible to build valuable and reproducible AI models without relying on big data. Future research should prioritize the validation of this framework on larger, multi-center datasets to rigorously evaluate its generalizability and robustness. Such studies will be crucial in determining the real-world clinical performance of the model.

Finally, regarding the choice of methodology, we deliberately prioritized in-domain transfer learning over other few-shot or augmentation strategies proposed in the recent literature. While metric-based few-shot learning methods (e.g., MAML, Prototypical Networks) show promise for 2D images, applying them to high-dimensional 3D volumetric data remains computationally unstable and susceptible to overfitting due to the exponential increase in feature space complexity [12]. Similarly, geometric interpolation techniques like Mixup or Cutmix were excluded because they generate non-physical artifacts (e.g., "ghost" bone structures) that violate anatomical plausibility. In contrast, our "Window Shifting" approach simulates a real-world physical phenomenon, specifically radiometric calibration bias, thereby preserving the structural integrity essential for clinical interpretation.

5. Conclusions

In conclusion, this study proposes a methodological framework that enables the development of a model for automatic Papilloma-Associated Hyperostosis identification from extremely scarce medical imaging data. A novel algorithm proposed combining specialized transfer learning, extensive data augmentation, and a rigorous validation strategy, which not only provides a potential solution for the segmentation of hyperostosis in sinonasal inverted papilloma but, more importantly, offers a proven roadmap for the wider research community facing the ubiquitous challenge of the data scarcity problem. Future work is expected to combine this method with larger data to increase model accuracy and apply it to support clinical diagnosis.

Author Contributions: Conceptualization, N.T.T.; Methodology, N.M.T. and N.T.T.; Software, P.H.D. and N.M.T.; Validation, P.H.D. and N.M.T.; Formal analysis, P.H.D. and N.M.T.; Investigation, P.H.D. and N.M.T.; Resources, N.M.T.; Data curation, P.H.D. and N.M.T.; Writing—original draft, P.H.D. and N.M.T.; Writing—review and editing, N.T.T.; Visualization, P.H.D.; Supervision, N.T.T.; Project administration, N.T.T.; Funding acquisition, N.T.T. All authors have read and agreed to the published version of the manuscript.

https://doi.org/10.3390/diagnostics16020311

Funding: This research was funded by the University of Economics Ho Chi Minh City-UEH, Vietnam.

Institutional Review Board Statement: The study has been reviewed and approved by the Ethics Committee in Biomedical Research of the Ear–Nose–Throat Hospital (In Vietnamese), Ho Chi Minh City, Vietnam (reference number 68/GCN-BVTMH, dated 2 October 2025).

Informed Consent Statement: Informed consent for participation was obtained from all subjects involved in the study.

Data Availability Statement: The original contributions presented in this study are included in the article. Further inquiries can be directed to the corresponding author.

Conflicts of Interest: The authors declare no conflicts of interest.

References

1. Thapa, N. Diagnosis and treatment of sinonasal inverted papilloma. *Nepal. J. ENT Head Neck Surg.* **2010**, *1*, 30–33. [CrossRef]
2. Lee, D.K.; Chung, S.; Dhong, H.-J.; Kim, H.; Kim, H.-J.; Bok, K. Focal hyperostosis on CT of sinonasal inverted papilloma as a predictor of tumor origin. *Am. J. Neuroradiol.* **2007**, *28*, 618–621. [PubMed]
3. Marino, M.J.; Riley, C.A.; Patel, A.S.; Pou, J.D.; Kessler, R.H.; McCoul, E.D. (Eds.) Paranasal sinus opacification-to-pneumatization ratio applied as a rapid and validated clinician assessment. In *International Forum of Allergy & Rhinology*; Wiley Online Library: Hoboken, NJ, USA, 2017.
4. Taciuc, I.-A.; Dumitru, M.; Vrinceanu, D.; Gherghe, M.; Manole, F.; Marinescu, A.; Serboiu, C.; Neagos, A.; Costache, A. Applications and challenges of neural networks in otolaryngology. *Biomed. Rep.* **2024**, *20*, 92. [CrossRef] [PubMed]
5. Demir, E.; Uğurlu, B.N.; Uğurlu, G.A.; Aydoğdu, G. Artificial intelligence in otorhinolaryngology: Current trends and application areas. *Eur. Arch. Oto-Rhino-Laryngol.* **2025**, *282*, 2697–2707. [CrossRef] [PubMed]
6. Minh Trieu, N.; Truong Thinh, N. The anthropometric measurement of nasal landmark locations by digital 2D photogrammetry using the convolutional neural network. *Diagnostics* **2023**, *13*, 891. [CrossRef] [PubMed]
7. Toan, N.K.; Tuan, H.N.A.; Thinh, N.T. Non-Neural 3D Nasal Reconstruction: A Sparse Landmark Algorithmic Approach for Medical Applications. *Comput. Model. Eng. Sci. (CMES)* **2025**, *143*, 1273. [CrossRef]
8. McKee, S.P.; Liang, X.; Yao, W.C.; Anderson, B.; Ahmad, J.G.; Allen, D.Z.; Hasan, S.; Chua, A.J.; Mokashi, C.; Islam, S. Predicting sinonasal inverted papilloma attachment using machine learning: Current lessons and future directions. *Am. J. Otolaryngol.* **2025**, *46*, 104549. [CrossRef] [PubMed]
9. Shamshad, F.; Khan, S.; Zamir, S.W.; Khan, M.H.; Hayat, M.; Khan, F.S.; Fu, H. Transformers in medical imaging: A survey. *Med. Image Anal.* **2023**, *88*, 102802. [CrossRef] [PubMed]
10. Tang, Y.; Yang, D.; Li, W.; Roth, H.R.; Landman, B.; Xu, D.; Nath, V.; Hatamizadeh, A. (Eds.) Self-supervised pre-training of swin transformers for 3d medical image analysis. In Proceedings of the IEEE/CVF Conference on Computer Vision and Pattern Recognition, New Orleans, LA, USA, 19–24 June 2022.
11. Trieu, N.M.; Thinh, N.T. A Novel Method in Wood Identification Based on Anatomical Image Using Hybrid Model. *Comput. Syst. Sci. Eng.* **2023**, *47*, 2381–2396. [CrossRef]
12. Dissanayake, T.; George, Y.; Mahapatra, D.; Sridharan, S.; Fookes, C.; Ge, Z. Few-Shot Learning for Medical Image Segmentation: A Review and Comparative Study. *ACM Comput. Surv.* **2025**, *58*, 1–36. [CrossRef]
13. Dosovitskiy, A. An image is worth 16 × 16 words: Transformers for image recognition at scale. *arXiv* **2020**, arXiv:201011929.
14. Azizi, S.; Culp, L.; Freyberg, J.; Mustafa, B.; Baur, S.; Kornblith, S.; Chen, T.; MacWilliams, P.; Mahdavi, S.S.; Wulczyn, E.; et al. Robust and efficient medical imaging with self-supervision. *arXiv* **2022**, arXiv:2205.09723. [CrossRef]
15. Espis, A.; Marzi, C.; Diciotti, S. Comparative analysis of supervised and self-supervised learning with small and imbalanced medical imaging datasets. *Sci. Rep.* **2025**, *15*, 32345. [CrossRef] [PubMed]
16. Shin, H.-C.; Tenenholtz, N.A.; Rogers, J.K.; Schwarz, C.G.; Senjem, M.L.; Gunter, J.L.; Andriole, K.P.; Michalski, M. Medical image synthesis for data augmentation and anonymization using generative adversarial networks. In *Proceedings of the International Workshop on Simulation and Synthesis in Medical Imaging*; Springer: Berlin/Heidelberg, Germany, 2018.
17. Isensee, F.; Jäger, P.F.; Full, P.M.; Vollmuth, P.; Maier-Hein, K.H. (Eds.) nnU-Net for brain tumor segmentation. In *Proceedings of the International MICCAI Brainlesion Workshop*; Springer: Berlin/Heidelberg, Germany, 2020.
18. Alshardan, A.; Alruwais, N.; Alqahtani, H.; Alshuhail, A.; Almukadi, W.S.; Sayed, A. Leveraging transfer learning-driven convolutional neural network-based semantic segmentation model for medical image analysis using MRI images. *Sci. Rep.* **2024**, *14*, 30549. [CrossRef] [PubMed]
19. Sun, L.; Li, C.; Ding, X.; Huang, Y.; Chen, Z.; Wang, G.; Yu, Y.; Paisley, J. Few-shot medical image segmentation using a global correlation network with discriminative embedding. *Comput. Biol. Med.* **2022**, *140*, 105067. [CrossRef] [PubMed]

https://doi.org/10.3390/diagnostics16020311

20. Varma, S.; Simon, R. Bias in error estimation when using cross-validation for model selection. *BMC Bioinform.* **2006**, *7*, 91. [CrossRef] [PubMed]
21. Chen, J.; Lu, Y.; Yu, Q.T. Transformers make strong encoders for medical image segmentation. *arXiv* **2021**, arXiv:2102.04306. [CrossRef]
22. Bareja, R.; Ismail, M.; Martin, D.; Nayate, A.; Yadav, I.; Labbad, M.; Dullur, P.; Garg, S.; Tamrazi, B.; Salloum, R. nnU-Net–based Segmentation of Tumor Subcompartments in Pediatric Medulloblastoma Using Multiparametric MRI: A Multi-institutional Study. *Radiol. Artif. Intell.* **2024**, *6*, e230115. [CrossRef] [PubMed]
23. Usmani, U.A.; Happonen, A.; Watada, J. (Eds.) Enhancing medical diagnosis through deep learning and machine learning approaches in image analysis. In *Proceedings of the Intelligent Systems Conference*; Springer: Berlin/Heidelberg, Germany, 2023.
24. Lee, H.-T.; Cheon, H.-R.; Lee, S.-H.; Shim, M.; Hwang, H.-J. Risk of data leakage in estimating the diagnostic performance of a deep-learning-based computer-aided system for psychiatric disorders. *Sci. Rep.* **2023**, *13*, 16633. [CrossRef] [PubMed]
25. Taha, A.A.; Hanbury, A. Metrics for evaluating 3D medical image segmentation: Analysis, selection, and tool. *BMC Med. Imaging* **2015**, *15*, 29. [CrossRef] [PubMed]

https://doi.org/10.3390/diagnostics16020311

diagnostics

MDPI

Article

Convolutional Neural Network-Based Approach for Cobb Angle Measurement Using Mask R-CNN

Marcos Villar García [1,*], José-Benito Bouza-Rodríguez [1,2] and Alberto Comesaña-Campos [1,2,*]

[1] Department of Design in Engineering, University of Vigo, 36208 Vigo, Spain; jbouza@uvigo.gal
[2] Design, Expert Systems and Artificial Intelligent Solutions Group (DESAINS), Galicia Sur Health Research Institute (IIS Galicia Sur), SERGAS-UVIGO, 36213 Vigo, Spain
* Correspondence: marcos.villar.garcia@uvigo.gal (M.V.G.); acomesana@uvigo.gal (A.C.-C.)

Abstract: Background: Scoliosis is a disorder characterized by an abnormal spinal curvature, which can lead to negative effects on patients, affecting their quality of life. Given its progressive nature, the classification of the scoliosis severity requires an accurate diagnosis and effective monitoring. The Cobb angle measurement method has been widely considered as the gold standard for a scoliosis assessment. Commonly, an expert assesses scoliosis severity manually by identifying the most tilted vertebrae of the spine. However, this method requires time, effort, and presents limitations in measurement accuracy, such as the intra- and inter-observer variability. Artificial intelligence provides more objective tools that are less sensitive to manual intervention aiming to transform the diagnosis of scoliosis. **Objectives:** The objective of this study was to address three key research questions regarding automated Cobb angle quantification: "Where is the spine in this radiograph?", "What is its exact shape?", and "Is the proposed method accurate?". We propose the use of Mask R-CNN architecture for spine detection and segmentation in response to the first two questions, and a set of algorithms to tackle the third. **Methods:** The network's detection and segmentation performance was evaluated through various metrics. An automated workflow for Cobb angle quantification and severity classification was developed. Finally, statistical methods provided the agreement between manual and automated measurements. **Results:** A high segmentation accuracy was achieved, highlighting the following: mIoU of 0.8012, and a mean precision of 0.9145. MAE was 2.96° ± 2.60° demonstrating a high agreement. **Conclusions:** The results obtained in this study demonstrate the potential of the proposed automated approach in clinical scenarios, which provides experts with a clear visualization of each stage in the scoliosis assessment by overlaying the results onto the X-ray image.

Keywords: scoliosis; Cobb angle measurement; Mask R-CNN; vertebrae detection and segmentation

1. Introduction

Scoliosis is a condition of lateral deviation of the spine with a curvature greater than 10 degrees [1,2]. Additionally, the vertebrae may twist around their axis, presenting torsion that add complexity. As shown in Figure 1, the human vertebral column is divided into five regions, which are structured from bottom to top as follows: coccyx, sacrum, lumbar vertebrae, thoracic vertebrae, and cervical vertebrae. Of the total number of vertebrae, 5 of them (L1–L5) belong to lumbar region, and 12 (T1–T12) are located in the thoracic region. Scoliosis assessment is performed by measuring the spinal curvature in degrees. Different authors use the following classification to evaluate the severity of the scoliosis:

mild scoliosis is considered at 10–20 degrees; moderate between 20 and 40 degrees; and severe above 40 degrees [2,3]. Other studies have used different threshold values to define curvature degrees. In the treatment of scoliosis there are several levels of severity and ways to deal with it. An early and precise diagnosis allows selection and planning the most suitable treatment. In an initial evaluation, doctors rely on clinical methods with no patient radiation exposure. Observational tests to detect suspect scoliosis include assessment of trunk rotation, the Adams forward bend test, and spine measurement instruments [4]. Through a physical examination, the specialist looks for asymmetries in the patient's back. When clinical examination suggests an abnormal curvature of the spine, anteroposterior X-ray images, considered the gold standard, are used for scoliosis assessment [5].

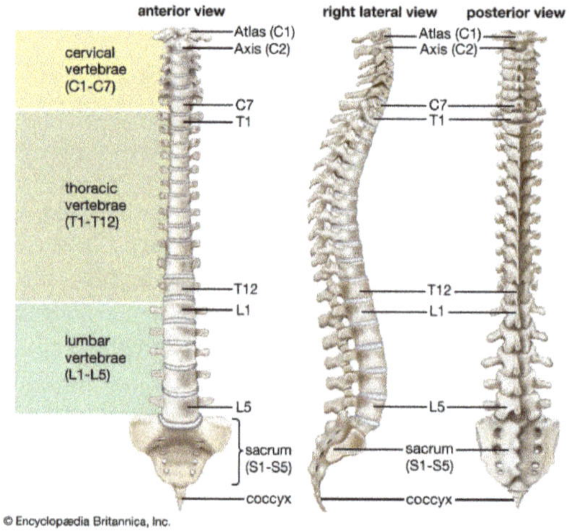

Figure 1. The vertebral column [6].

Radiography experts measure the Cobb angle in the X-ray image by using instruments such as a goniometer. They find the most tilted vertebrae and quantify the angle of the spinal curvature directly onto the radiograph. This manual method requires time and effort. Manual measurements present a large intra- and inter-observer variability due to the wide range of the scoliotic curvature among patients, measurement errors, and low image quality [7]. Moreover, the correct definition of anatomical landmarks, such as end vertebrae and tilted vertebrae, is essential to guarantee precise measurements and to reduce subjective errors associated with their manual identification [8,9]. Manual measurements depend on expert experience and judgment. These limitations can negatively impact the accuracy of measurements, opening opportunities for the development of methods that ensure the correct diagnosis. This research introduces an automated approach based on a convolutional neural network to reduce the intra- and inter-observer variability in the scoliosis severity assessment through Cobb angle measurement. To this end, we propose the use of Mask R-CNN [10] for object instance segmentation in X-ray images. This method predicts a binary mask for each candidate object. Due to this particular capability, we employ Mask R-CNN for single-instance segmentation of the spine, generating a mask that replicates the scoliotic curvature. This mask is then processed using an algorithm that extracts the midline of the scoliotic curvature by connecting the midpoints derived from the spine's boundary. The algorithm further identifies anatomical landmarks, calculates the Cobb angle, and classifies the severity of the spinal curvature. Additionally, it provides a clear visualization of each stage in the measurement process, which can improve the interpretability of the results.

The article is divided into six sections. In Section 2, a review of related works is provided. Section 3 describes the materials and methods used in the study. Section 4 details the experimental analysis and results. Section 5 discusses the study and its scope. Finally, Section 6 presents the conclusions of the study and future work.

2. Related Works

Precision in spinal curvature assessment is required to guarantee precise diagnosis. Several parameters directly impact accuracy and repeatability of measurements such as manual intervention, observer expertise, image quality, and assessment criteria. Variability in inter- and intra-observer assessment led to the development of more objective tools, less sensitive to manual intervention, to improve reliability scoliosis evaluation. The Cobb angle method, which quantifies the degree of spinal curvature, has been considered the reference standard for its assessment [3,11–15]. Wiliński et al. [5] demonstrated that the Cobb angle was the most reliable method through a comparison of measurements performed by inexperienced users employing the Cobb, Centroid, and Ferguson techniques.

In 2009, Vrtovec et al. [16] highlighted the advantages of using image modern processing techniques, which help experts in improving diagnostic and therapeutic planning for spinal conditions. Jin et al. [17] addressed challenges such as the identification accuracy and computational time in computer vision and deep learning. The use of techniques to improve the quality of medical images has increased in recent years, aiming to facilitate the Cobb angle measurement. Despite the good results, there are several limitations in the field. Landmark and segmentation approaches are the most used for Cobb angle quantification [14]. In this study, we have reviewed both vertebral landmark prediction [3,4,8,9,11–13,15,18–28] and vertebrae segmentation [2,14,29–38]. However, those approaches that identify vertebrae individually are not sufficiently accurate [39]. According to Gstoettner et al. in 2007 [8], the selection of the superior and the inferior end vertebrae is a difficult task and represents the main sources of error. The effect of applying filters to improve endplate identification in X-ray images with low-resolution was studied by Anitha et al. [9] in 2014. In 2017, the relationship between anatomical landmarks and the Cobb angle was addressed using structured support vector regression (SSVR) by Sun et al. [18]. Wu et al. [19] proposed a deep learning approach that combined the CNNs' capabilities for feature extraction and statistical methods. These authors also studied the effect of occlusion in Cobb angle estimation using anteroposterior and lateral views of the spine. In 2019, Chen et al. [21] used two networks separately to identify anatomical landmarks and for Cobb angle measurement. Yang et al. [22] identified vertebrae from the thoracic and lumbar region as critical using two convolutional architectures, Faster-RCNN and ResNet 101. An order that facilitates learning, first by detecting vertebrae, followed by locating landmarks, was proposed by Khanal et al. [11]. Cerqueiro et al. [24] explored automated estimation of the Cobb angle using image processing and active contour algorithms to identify the spine contour. However, the snake model can lose features in regions with poorly defined boundaries, particularly when vertebral bodies overlap each other, as noted by Jin et al. [17] and Hoblidar and Prabhu [29]. Given the high performance of both segmentation and landmark estimation, several authors addressed Cobb angle quantification by combining these deep learning techniques. However, these deep learning techniques are commonly used in combination. In 2021, Fu et al. [12] proposed an architecture aiming to combine segmentation with landmark estimation. A study aiming to evaluate variations in the measurements performed by different observers through the application of Cobb, Ferguson, Diab, and Centroid methods was conducted by Thalengala et al. [3], who emphasized the importance of evaluating the inclination of vertebral bodies. In 2022, Huang et al. [13] proposed the use of artificial neural networks and oscillograms that represent endplate inclinations to calculate the Cobb angle. Sun et al. [25] addressed the Cobb angle measurement in X-ray images through two sequential operations: vertebrae segmentation and the application of crop and zoom to locate the corners of each vertebra. In 2023, Han et al. [4] proposed a machine learning algorithm that quantifies scoliosis by detecting maximum pixel values, and fitting a spline curve through anchor points to draw the spinal curvature. The CNN method proposed by Maeda et al. [26] consisted of thoracic and lumbar region identification, and the detection of the four corners of each vertebra as anatomical landmarks. The points identified as the most tilted vertebrae were used to measure the Cobb angle. Qiu et al. [27] addressed noise complex computational operations by introducing a novel deep learning approach which generates spine region, centerline, and the boundary as multiple sources of morphological information. Suri et al. [28] reported a neural network architecture that enables Cobb angle measurements even when the presence of hardware obstructs parts of the spine. In 2024, Chui et al. [15] employed a feedforward neural network (FNN) architecture for Cobb angle prediction and spinal curvature progression analysis. Using a landmark detection algorithm, they identified the center and the edge points of each vertebra to calculate the angles. Maharasi et al. [40] proposed the use of the U-Net architecture to evaluate the scoliosis stage through the detection of several anatomical landmarks. Most automated methods use the superior and inferior endplate regions of each vertebra, commonly named as vertebral plates, to calculate the Cobb angle based on the most tilted vertebrae, known as the end vertebrae. However, the identification of these end vertebral plates remains the main source of error. To address this challenge, Hoblidar and Prabhu [29] proposed an automatic vertebrae segmentation system using image processing techniques to ensure the identification of the most tilted vertebrae at the superior and inferior ends of the spinal curvature. In 2019, Horng et al. [2] applied the original U-Net architecture with two of its variants, Residual U-Net and Dense U-Net, to improve vertebrae segmentation performance on anteroposterior (AP) spinal X-ray images. Nevertheless, selecting an appropriate novel and training configuration still requires human knowledge, as noted by Vuola et al. [31]. These authors compared two widely used segmentation approaches, U-Net and Mask R-CNN, in a nuclei segmentation task to evaluate their strengths and limitations. While Mask R-CNN is better at detecting objects by predicting bounding boxes, U-Net is more precise at segmentation tasks. Pan et al. [41] employed two Mask R-CNN approaches: one trained for spine segmentation and the other designed for vertebrae segmentation. Convolutional neural networks are the most used architectures for object detection and classification in images, according to Alharbi et al. [32]. In 2020, these authors used a pretrained ResNet50, to detect vertebrae in spine X-rays. In 2021, Zhang et al. [33] included sacral vertebrae in their study, providing an efficient and accurate solution for whole-spine vertebra segmentation. In 2022, Caesarendra et al. [14] introduced the concept of intervertebral displacements for Cobb angle quantification using a convolutional neural network. Zhao et al. [34] improved the U-Net architecture by generating a binary segmentation map of each vertebra and identifying the most tilted candidates to calculate the Cobb angle. In 2023, Wong et al. [35] developed an AI-based algorithm that combined two convolutional neural networks: the first for spinal column segmentation, and the second for vertebral bodies' segmentation and bounding box extraction. The tilted angles of these boxes were used to compute the Cobb angle. In 2024, Low et al. [36] developed a deep learning model using a two-stage approach; an attention-based deep neuronal network to segment and identify individual vertebrae, followed by polynomial curve fitting to the vertebral centroids for Cobb angle calculation. Rahmaniar et al. [38] proposed a convolutional neural network-based approach that integrates vertebrae segmentation and hourglass modules to enhance landmark localization by detecting five coordinates per vertebra (center and four corners), which are used to identify the most tilted vertebrae.

In this section, different studies related to this article have been reviewed. Image feature extraction and segmentation capabilities of convolutional neural networks have been demonstrated. Image processing techniques and combinations of networks are widely employed in the state of the art, aiming to calculate the Cobb angle and severity scoliosis. Key features are detected using novel strategies based on the location of the anatomical landmarks and vertebral segmentation with the objective of detecting the most tilted vertebrae, which remains a challenge and is the basis for Cobb angle quantification. Precision in spine detection or vertebrae segmentation is essential in scoliosis assessment. Mask R-CNN incorporates the prediction of a binary mask for each region of interest (RoI); in our case, a single binary mask is generated for the spine. This particular characteristic is relevant in our decision. Our motivation is focused on the use of Mask R-CNN as a single-instance segmentation approach to predict the mask of the spine and use the spinal boundary as the basis for our scoliosis assessment approach. However, the reviewed studies do not utilize the additional information provided by the Mask R-CNN network in the form of a mask to draw the midline of the spine and, through the analysis of its curvature, identify key anatomical landmarks that enable a simplified calculation of the Cobb angle based on the most tilted vertebrae and the assessment of the scoliosis severity. We offer to the users the visualization and recording of the workflow, displaying the main results of each step, as support for improving the interpretability of the scoliosis assessment procedure, to understand the prediction and collect data for further comparison. In the following section, we describe the experimental setup of this study.

3. Materials and Methods

In this study, the performance of the Mask R-CNN method [10] was analyzed for object detection and segmentation. Our approach was designed with the aim of identifying the spine as a single instance within anteroposterior X-ray images. To this end, we used a network that generates the mask corresponding to the spine. Various metrics were employed to validate the network's reliability and robustness in spine segmentation. Considering the segmentation accuracy, different algorithms were developed to identify anatomical landmarks, enabling the Cobb angle quantification and severity classification. Additionally, we developed a visual interface where users can follow the main steps of the process and view the results at each stage by uploading the X-ray images.

3.1. Materials

The collection of spinal images was obtained from the public AASCE MICCAI 2019 dataset [42], which contains 98 raw anteroposterior (AP) spine X-ray images with different dimensions, ensuring a standardized source of data. Mask R-CNN was implemented for object detection and segmentation in a virtual environment using Python 3.12.7, TensorFlow 2.16.2, and Keras 3.7.0. The model was trained locally on a GeForce RTX 4080 Laptop GPU (12 GB) with an Intel i9-14900HX processor (Intel Corporation, Santa Clara, California, USA). Statistical analyses were conducted using the Python libraries NumPy, Pandas, and Pingouin. The algorithms were developed in the JupyterLab environment, generating multiple outputs, including Cobb angle calculation, scoliosis severity classification, spine landmarks, and clear representations of the results on the input X-ray image.

3.2. Methods

The method proposed in this study provides a systematic approach that enables the evaluation of spinal curvature severity based on vertebral alignment. In response to the questions "Where is the spine in this radiograph?" and "What is its exact shape?", Mask R-CNN addresses both: the bounding box localizes the spine within the image (where), and the segmentation mask defines its exact shape (what). The third question, "Is the proposed method accurate?", is answered by evaluating the error between the measurements performed by the two observers (ground truth) and those obtained using our approach.

To answer these questions, we developed a four-stage workflow based on the raw dataset: (1) a preprocessing stage, including image rescaling and annotation; (2) network processing stage; (3) Cobb angle and severity classification processing stage; and (4) post-processing stage for error evaluation. The workflow design is illustrated in Figure 2.

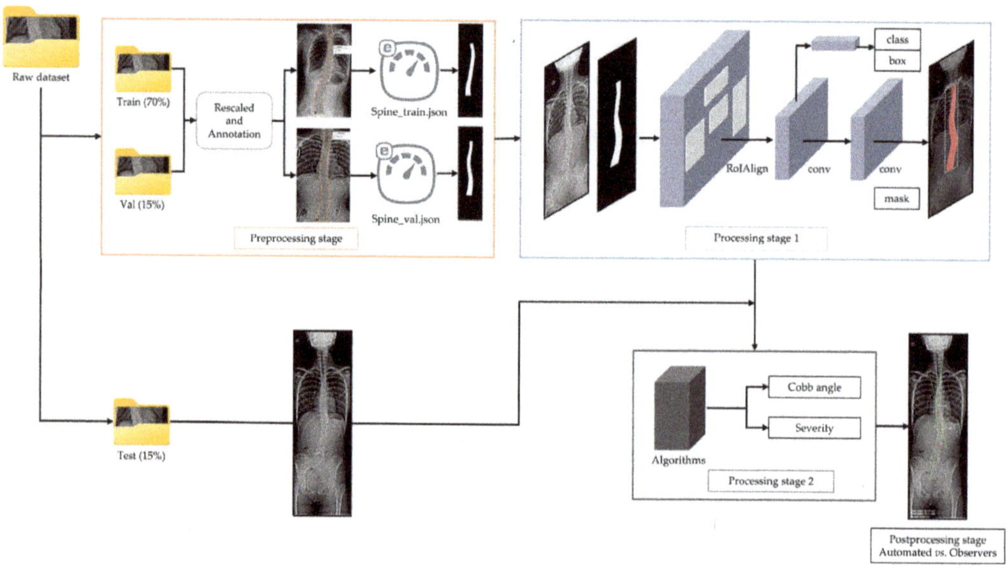

Figure 2. Overview of proposed workflow. First, the raw images (training and validation subsets) are rescaled to a height of 2000 pixels, and the spine contour (the label) is annotated during the preprocessing stage as part of dataset construction process. In the processing stage 1, the network learns to identify the spine based on the annotated binary mask. Processing stage 2 involves Cobb angle quantification and scoliosis severity assessment. At this stage, the input from the test subset is automatically rescaled to a height of 2000 pixels. Finally, in the postprocessing stage, automated measurements are compared with those provided by the human observers.

In the following subsections, a detailed explanation of each stage comprising the proposed workflow is provided.

3.2.1. Preprocessing Stage

Before the preprocessing stage, 70% of the raw images were assigned to the training set, with the remaining 30% equally divided between validation and testing sets. This proportion was selected to ensure a sufficient number of images for model learning. The images in each subset were distributed randomly to minimize selection bias.

The network was trained using only the spine class label plus background. To this end, X-ray images from both the training and validation subsets were employed. Accordingly, the output of the model was limited to the spine localization and mask generation, while the degree of curvature was calculated during the inference stage using a set of algorithms, thereby enabling Cobb angle quantification and scoliosis severity assessment. The original spine X-ray images corresponding to training and validation subsets were rescaled to a height of 2000 pixels. Manual annotation of these images was performed using VGG Image

Annotator (VIA) software, version 2.0.12. This software allows exporting annotations to a JSON file with the coordinates of the spine boundary. During this stage, we annotated the contour of the spine using points aiming to detect the spine and mask generation. The boundary of the spine comprised the ground truth (GT) of our dataset and was used to train the model. The objective of this labelling phase was to prepare the training and validation dataset. The spinal contour was identified and annotated in each image. We used VIA to place points along the boundary of the spine in each radiograph. The coordinates of these points that define the contour of the spine in the training and validation dataset were stored in a JSON file. These annotations were used to replace each X-ray by its binary mask. The pixels of the spine were marked in white, and the rest are marked in black. This ensured the availability of the training labels required by the Mask R-CNN and also enabled the model to distinguish the region corresponding to the spine. This stage allowed the network to learn the shape of the spine and its spatial arrangement during the training process. Mask R-CNN requires a binary mask as the training label associated with the X-ray image from which it was generated.

We used skimage.draw.polygon function to convert these coordinates into a binary mask suitable for the network. Although the original radiographs were greyscale (single-channel), the Mask R-CNN architecture expects input in RGB format (three channels); therefore, a conversion from greyscale to RGB was performed. In this operation, the anatomical information is preserved while an image visual identical and compatible with the model is created. The JSON file, which stores the manually annotated coordinates of the spinal contour, is used to generate the binary masks that serve as training labels. These labels were generated once for the entire dataset. A dataset is specifically created for identify the region of the spine, without considering the severity of the idiopathic condition, or even whether it is present. The network does not aim to discriminate or classify possible cases of scoliosis, only to identify the spinal region. For this reason, this process is mandatory, as it constitutes the training and validation dataset for the network. However, once this dataset has been annotated, the process does not need to be repeated to analyze new radiographs for testing or future data. In such cases, only the input radiographs must be rescaled.

While the raw images were specifically used to train the model for spine detection and segmentation, we also analyzed the degree of spinal curvature to extract additional information from the dataset that may be relevant for future work and necessary for error evaluation on the test subset. We used the severity classification provided by Horng et al. [2]. The raw dataset was distributed as follows: 52% of the cases were classified as moderate, 28% as mild, 13% as severe, and 7% as spinal curve (i.e., absence of scoliosis). The authors noted the dataset presents an imbalanced distribution based on the values of the Cobb angle. Although this imbalance may introduce a morphological bias, favoring more frequently curvatures, this study does not constitute a multiclass annotation problem.

3.2.2. Processing Stage 1: Network Training

In this stage, the convolutional neural network is training for spine detection and segmentation task, using the labels annotated during the previous stage. As mentioned, Mask R-CNN requires a binary mask for training. We used the skimage.draw.polygon function to convert the coordinates of the spine boundary stored in the JSON file into a binary mask suitable for the network. Although the original radiographs were greyscale (single-channel), the Mask R-CNN architecture expects input in RGB format (three channels); therefore, a conversion from greyscale to RGB was performed.

A brief explanation of the model pipeline is provided as follows: The Region Proposal Network (RPN), a core component of the network, generates candidate regions of interest (RoIs) with a high probability of containing the spine. Then, it predicts anchors with a probability score indicating the presence of the spine (binary classification: spine/background). Finally, the top proposal's RoIs are selected and passed through fine classification, bounding box regression, and mask segmentation.

The training subset was used to fit the model, while the validation subset was employed in the optimization process. The test subset was reserved for the next stage. We optimized hyperparameters through multiple trials using Optuna, with training monitored by callbacks such as EarlyStopping and CSVLogger. The objective of this stage was to minimize validation loss and identify the best epoch for spine detection and segmentation.

To this end, we evaluated the performance of the Mask R-CNN network through a comprehensive analysis of the training process. We implemented the Mask R-CNN architecture using the GitHub repositories referenced in [43,44], specifically employing the implementation based on TensorFlow 2.0. As mentioned in the Materials subsection, the dataset comprised 98 anteroposterior spine X-ray images, split into 70% for training, 15% for validation, and 15% for testing. We employed transfer learning by applying COCO pre-trained weights and fine-tuning the model for spine segmentation in X-ray images. Preliminary experiments guided a hyperparameter optimization phase, where various fine-tuning strategies were explored. Multiple training sessions were conducted, monitoring training and validation loss curves to refine model performance. No data augmentation techniques were applied during training. The final model was trained for 300 epochs, demonstrating consistent performance throughout the training process, as illustrated in Figure 3.

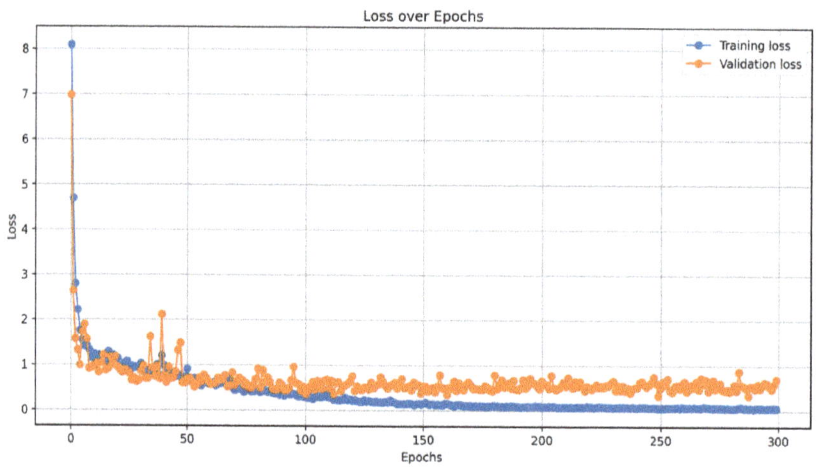

Figure 3. Training and validation loss function over epochs.

As shown in the previous figure, readers may observe that the validation loss function seems to reach a stable performance trend after approximately 50 epochs. In this study, the authors decided to extend and monitor the training process to achieve the best segmentation results and define the most appropriate epoch. Figure 4 presents the evolution of the segmentation metrics, which supports the decision to extend the monitoring beyond 50 epochs, up to 300 epochs.

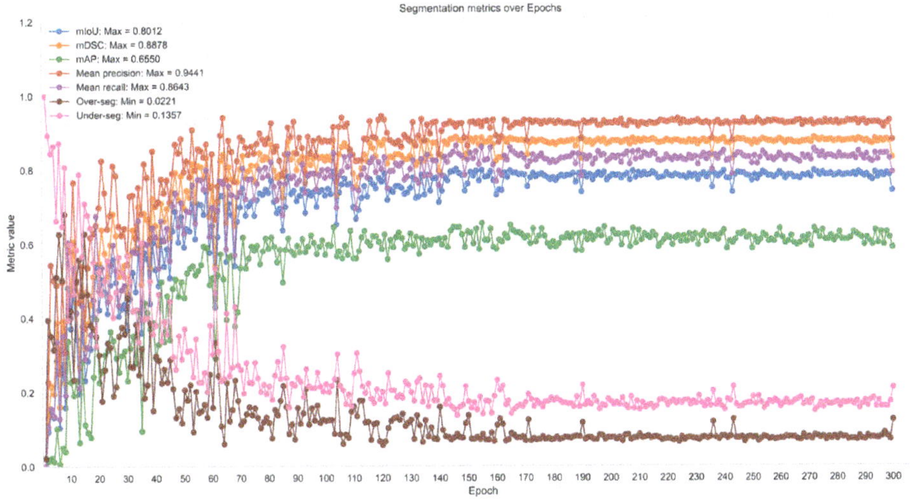

Figure 4. Evolution of segmentation metrics over 300 epochs.

The segmentation model was evaluated using complementary metrics: intersection over union (IoU), also referred to as the Jaccard index, evaluates the overlap between the predicted mask and the ground truth mask; dice similarity coefficient (DSC) reflects the global similarity between the segmented regions and the reference; average precision (AP) serves as an overall performance indicator in terms of detection and precision of segmented regions; precision indicates how many of the positives predicted by the model were actually correct; and recall quantifies how many true positives were successfully detected by the model. Finally, over-segmentation occurs when the model predicts a region larger than the actual object, including areas that do not belong to the target, and under-segmentation happens when the predicted region is smaller, missing relevant parts of the target.

The best epoch was determined on the validation loss function, reaching a minimum value of 0.3472 at epoch 287. Additionally, a comparative analysis was performed using different detection confidence thresholds, ranging from 0.70 to 0.90 in increments of 0.05, to identify the optimal segmentation accuracy in terms of mean IoU (mIoU) and mean AP (mAP). As a result, the highest mIoU metric was achieved at epoch 146, indicating the most accurate segmentation adjustment, while epoch 155 provided a slightly higher value in overall detection accuracy (mAP). Given that our approach focuses on extracting the spine contour from the mask generated by the Mask R-CNN network, we consider the mIoU metric as a key indicator of the agreement between the manually annotated mask (ground truth) and the predicted one. The analysis also revealed that a minimum detection confidence value of 0.85 provided the best segmentation result. To demonstrate network's reliability and robustness, even with a limited dataset, we assessed the discrepancies between the predicted masks and the ground truth, which consisted of the annotated images in VGG Annotator. We employed intersection over union (IoU) to evaluate segmentation accuracy. Figure 5 illustrates the method used to calculate the mean IoU by processing the original image, the ground truth mask previously identified and annotated, the predicted mask, and the corresponding IoU values of several X-ray images.

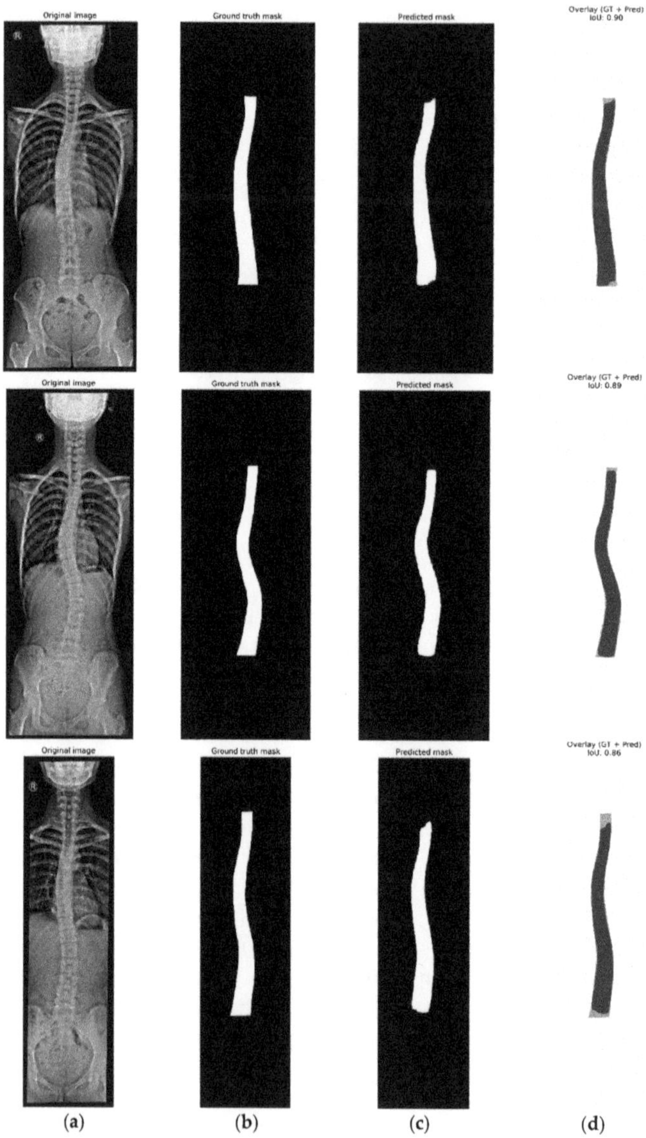

Figure 5. Images for (**a**) original image, (**b**) ground truth mask, (**c**) predicted mask, and (**d**) IoU.

3.2.3. Processing Stage 2: Cobb Angle Quantification and Scoliosis Severity Assessment

Once the mask was generated, we applied various algorithms developed to identify key anatomical landmarks along the spine boundary. The model calculated the position of the most tilted vertebrae, which allowed the quantification of the Cobb angle and the classification of scoliosis severity directly on the original image. Figure 6 shows a depiction of this stage, supported by the representation using the weights from the best epoch identified in the previous stage. The input AP X-ray image, corresponding to the test subset, was automatically rescaled to a height of 2000 pixels.

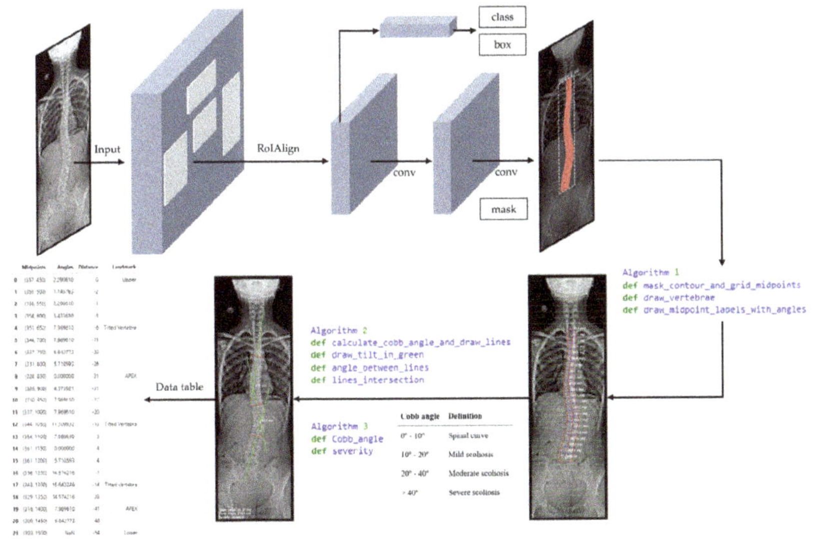

Figure 6. Detailed inference workflow for scoliosis assessment.

The postprocessing stage requires the visualization of the results. To this end, we developed an interface in JupyterLab. Figure 7 presents a visualization of the results obtained during the inference workflow within the JupyterLab environment. This workflow operates as follows:

- Image acquisition is visualized in Figure 7a. The process begins with image acquisition through a user interface that allows the upload of an anteroposterior (AP) full spine X-ray image.
- Figure 7b shows the spine segmentation step using Mask R-CNN. Before inference, the input image is rescaled to a height of 2000 pixels. Once the model is loaded, the inference is initiated on the input image. We defined the confidence value during training. Only masks with a confidence score above this threshold are considered as detected regions. The mask with the largest segmented area is selected as the spinal region, assuming it corresponds to the spine. The Cobb angle quantification is highly dependent on the precision of the generated mask. The more accurate the mask, the more precise the assessment. To this end, the prediction of the model was optimized during training, and the best epoch was used for inference, aiming to ensure the highest possible accuracy in the generation of the mask.
- In Figure 7c, contour extraction and midpoint detection are illustrated. Once the segmentation mask is obtained, the spinal contour is extracted using OpenCV's cv2.findContours() function, which is employed for contour detection in binary images. This contour with the largest segmented area is overlaid onto the image. Our approach is based on the capacity of the algorithms developed to build the spinal curvature within the contour extracted. To this end, we proposed the use of a grid, only considering the horizontal lines drawn in the image, to establish points that will be used to connect and build the spinal curvature. The interface provides a widget to define the grid interval, offering flexibility during the adjustment process. The objective of this method is to ensure that these horizontal lines fall within the lumbar and thoracic vertebrae with the highest possible accuracy. This technique facilitates an approximation of individual vertebrae segmentation. Each vertebra is segmented, without the need to train the model to detect each vertebra individually. The optimal

grid interval value, predefined to 50 pixels, was defined through experimental testing using the widget and observing that the estimation of the midline spine curvature replicates the contour shape. At each grid line, two intersection points are detected where the line crosses the spinal contour. The algorithm calculates the mean distance between these two points, defining the midpoint on the image. These midpoints are the key reference for spinal curve construction through their connection. The extracted midpoints approximate the spinal curve, with the first and last point identified as upper and lower, limiting the length of the spine. A spline interpolation is applied to refine the connections between midpoints, smoothing the spinal path. Then, the algorithm draws a line perpendicular to the tangent of the curve at each midpoint, excluding the upper and lower points. The angle between these perpendicular lines with respect to the horizontal represents the vertebrae inclination at each midpoint and is computed and annotated on the image. The algorithm is designed to draw these tilted lines within the contour as a simplified representation of the vertebrae. This approach just described, which enables the spinal curvature estimation, has been designed as a proof of concept for the proposed methodology. It clearly depends on prior segmentation and assumptions such as local vertebral symmetry. However, these elements are part of an improvement process, and its optimization further strengthens the results of this initial proposal.

- Figure 7d depicts spinal curve estimation and Cobb angle calculation, based on the analysis of the curvature. The algorithm swipes the curvature and identifies the key anatomical landmarks following their definition: tilted vertebrae are defined as the vertebrae with the greatest inclination angle. They represent the greatest spinal deviation. Apex points correspond to the locations where the spinal curvature reaches its maximum deformation; that is, the points with the greatest lateral displacement relative to a vertical reference line. In our study, this reference is defined as the vertical line drawn from the upper reference point. It is important to note that when the algorithm addresses complex curvatures, two apex points are detected, one on the left (corresponding to negative distances), and one on the right (corresponding to positive distances). The algorithm was designed with a logic to distinguish the type of the curvature, simple or complex, in order to detect the correct number of tilted vertebrae and apex points, to provide either a single Cobb angle measurement or separate upper and lower Cobb angle measurement. The apex point, which refers to the peak of the spinal deviation, is critical for scoliosis assessment. To emulate the manual method, the most tilted vertebrae are isolated and represented on the image, remarked with red lines. Then, the algorithm draws perpendicular green lines from the midpoints corresponding to the most tilted vertebrae and connects them, as performed by clinical experts. The algorithm identifies the intersection point between these green lines, and the angle formed at their intersection is calculated. The Cobb angle and the severity classification are annotated on the image.
- Finally, the visualization and data export stage is presented in Figure 7e. The output is displayed in a multi-panel layout including the input X-ray image; the result of the Mask R-CNN segmentation; the extracted spinal contour with the computed midpoints, midline, and vertebral inclinations; the image showing Cobb angle measurement and severity classification; and the data table containing geometrical information, anatomical landmarks such as tilted vertebrae and upper, lower, and apex points. Processed images and the data table are exported as structured reports in .png and .csv formats, providing a comprehensive representation of the entire process.

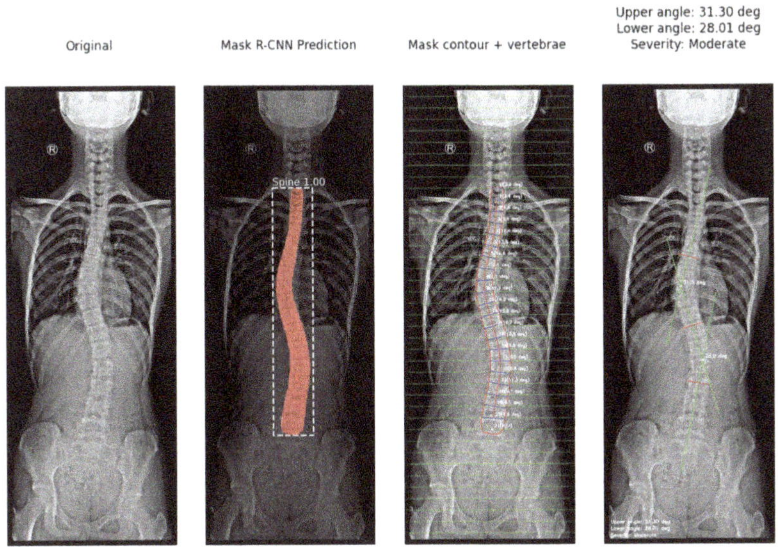

(a) (b) (c) (d)

	Midpoints	Angles	Distance	Landmark
0	(421, 450)	2.290610	0	Upper
1	(419, 500)	3.433630	-2	
2	(416, 550)	4.573921	-5	
3	(412, 600)	9.090277	-9	
4	(404, 650)	12.407419	-17	
5	(393, 700)	13.495733	-28	
6	(381, 750)	14.574216	-40	Tilted Vertebra
7	(368, 800)	9.090277	-53	
8	(360, 850)	1.145763	-61	APEX
9	(361, 900)	11.309932	-60	
10	(371, 950)	14.574216	-50	
11	(384, 1000)	15.642246	-37	
12	(398, 1050)	16.699244	-23	Tilted Vertebra
13	(413, 1100)	13.495733	-8	
14	(425, 1150)	4.573921	4	
15	(429, 1200)	0.000000	8	APEX
16	(429, 1250)	6.842773	8	
17	(423, 1300)	11.309932	2	Tilted Vertebra
18	(413, 1350)	9.090277	-8	
19	(405, 1400)	9.090277	-16	
20	(397, 1450)	4.573921	-24	
21	(393, 1500)	NaN	-28	Lower

(e)

Figure 7. Visualization of the stages in JupyterLab: (**a**) original anteroposterior (AP) full spine X-ray image; (**b**) Mask R-CNN spine segmentation; (**c**) mask contour and vertebral angles; (**d**) computed Cobb angle, and severity classification (red lines correspond to the most tilted vertebrae, while green lines represent the perpendiculars used in the Cobb angle calculation); (**e**) data table displaying numerical values and anatomical landmarks.

The workflow detailed above provides clinicians with visual information to understand the process. By overlaying the Cobb angle procedure on the original image, it facilitates comparison with manual measurements, supporting decision-making in scoliosis assessment when necessary. Furthermore, the visualization of anatomical landmarks and results enhances transparency and can help experts in further evaluations.

3.2.4. Postprocessing Stage: Agreement Between Manual and the Automated Cobb Angle Method

In this subsection, the agreement between manual and automated Cobb angle measurements is evaluated. The objective is to compare the measurements obtained by two observers and the model, by registering and analyzing the data collected from both sources.

The manual procedure for Cobb angle measurement is described as follows. First, the observer identifies the apex of the spinal curvature. The apex is the vertebra that is most laterally displaced from the midline. Once the apex has been determined, the observer selects the most tilted vertebra above and below the apex. The observer draws two lines, one along the upper endplate of the most tilted vertebra above the apex, and the other along the lower endplate of the most tilted vertebra below the apex. As illustrated in Figure 8, the angle formed between these two lines, or between their perpendiculars if the lines do not intersect, is considered the Cobb angle [45]. Our approach has also been designed to evaluate the severity of scoliosis according to the classification shown in Table 1.

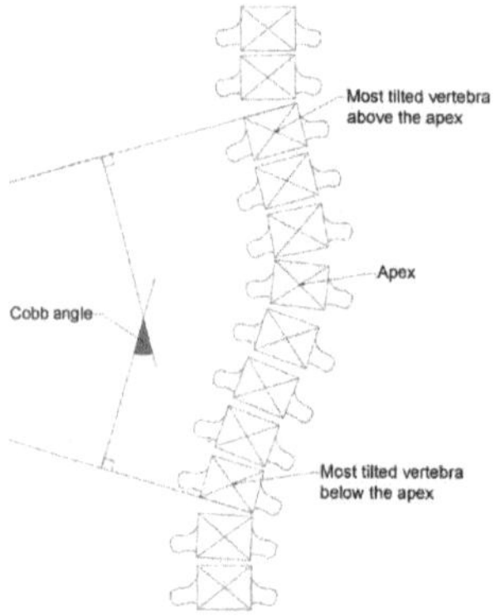

Figure 8. Scheme of Cobb angle measurement.

Table 1. Cobb angle definition [2].

Cobb Angle	Definition
0–10°	Spinal curve
10–20°	Mild scoliosis
20–40°	Moderate scoliosis
>40°	Severe scoliosis

Once the Cobb angle manual procedure was finalized, a comparison with the automated method was designed. To this end, two experts, referred to as Observer A and Observer B, were asked to analyze the radiographs of the test set and calculate the Cobb angle using the aforementioned procedure. Having two observers allows for comparison of their measurements and reduce subjectivity, lowering uncertainty in the results. The measurements of both observers were then compared using various error metrics and Bland–Altman analysis, which is briefly described below:

- A Bland–Altman analysis [46] was carried out to evaluate the agreement between two methods by plotting the difference between the two measurements against their average. This graphical method allows the identification of any systematic bias and the definition of the limits of agreement. It is commonly used in medical applications to evaluate whether an automated method can replace or complement manual approaches.
- The intraclass correlation coefficient (ICC) with 95% confidence interval [47] is used to evaluate the reliability of measurements between two or more observers. In this case, it assessed the level of agreement between automatic and manual Cobb angle quantifications. The 95% confidence interval provides an estimate of the precision and stability of the ICC value.
- The median absolute difference (MAD) [47] is a robust measure of dispersion that is calculated as the median of the absolute differences from the median of any dataset. MAD is less affected by outliers than the standard deviation, making it a suitable tool for assessing variability when there are a few extreme values. In this case, it provided a typical error estimation that was more robust against outliers or occasional discrepancies.
- The mean absolute error (MAE) [47] was also included to quantify the average of the differences between the Cobb angles obtained manually and those obtained automatically.
- The standard deviation (SD) [47] measures the variability of the differences between measurements.

Both the MAE and MAD are presented with their corresponding standard deviation (±SD) to provide both a central estimate of the error and its variability. This combined presentation allows the assessment of how close the errors are to the mean and the detection of relevant dispersions.

4. Experimental Analysis and Results

In this section, the third research question of this study, "Is the proposed method accurate?", is addressed. To answer it, the results obtained were evaluated through a comprehensive experimental analysis. First, the main results related to instance segmentation are presented. Second, a comparison between manual and automated Cobb angle measurements on AP X-ray images is provided using statistical analysis.

4.1. Instance Segmentation

A summary of the main metrics analyzed during the training phase is presented in Table 2, including the mean IoU, mean DSC, and mean AP (IoU = 0.5:0.95). In image segmentation, both IoU and DSC are commonly used to quantify the overlap between predicted and ground-truth regions, while mAP, calculated at multiple IoU thresholds, assesses the model's precision and robustness across different overlap criteria. We also report over-segmentation and under-segmentation rates, which provide insights into how accurately the model delineates the target region. Epoch 146 achieved a 0.8012 mIoU, 0.8878 mDSC, and 0.6450 mAP, with over-segmentation and under-segmentation values of 0.0855 and 0.1357, respectively, indicating a favorable balance between precision and recall in the model's spine segmentation performance.

Table 2. Summary: mean IoU (mIoU); mean dice similarity coefficient (mDSC); mean average precision (mAP, IoU = 0.5:0.95); mean precision; mean recall; over-segmentation; under-segmentation.

Threshold	mIoU	mDSC	mAP	Mean Precision	Mean Recall	Over-Seg	Under-Seg
0.85 (epoch 146)	0.8012	0.8878	0.645	0.9145	0.8643	0.0855	0.1357
0.85 (epoch 155)	0.7980	0.8857	0.655	0.9150	0.8599	0.0850	0.1401
0.85 (epoch 287)	0.7818	0.8750	0.625	0.9313	0.8268	0.0687	0.1732

Notably, the highest mIoU metric was achieved at epoch 146 with a minimum detection confidence of 0.85. This mean IoU score indicates strong segmentation performance, although there remains ample room for further improvement.

4.2. Cobb Angle Measurement

Accuracy in Cobb angle measurements and scoliosis severity was evaluated through a comparison between manual measurements and the output of the automated method. Manual measurements were performed independently by two separate observers, both familiar with the Cobb angle method. The procedure consisted of identifying on the X-ray images the apex and the most tilted vertebral endplates and then measuring the Cobb angles between those vertebrae. Conventional instruments, such as a ruler and a goniometer, were used during this process. All manual measurements were annotated on the radiographs and correctly classified into their corresponding subsets: 70% for training, 15% for validation, and 15% for testing. Only the measurements included in the testing subset were compared with the automated results.

As mentioned in Section 3, the algorithm developed in this study, inspired by the manual Cobb angle measurement procedure, provides anatomical landmarks, Cobb angle measurements, and severity classification. This process is based on the performance of

Mask R-CNN for instance segmentation and its capability to generate a mask with the shape of the region of interest, in our case, the spine. In the following step, the contour of the mask is extracted, and the algorithm applies a set of rules to perform all measurement tasks, including anatomical landmark identification.

While the observers identified the apex and the most tilted vertebrae, as did the automated method, only the Cobb angle values were compared in this study. This opens the opportunity for future comparisons regarding anatomical landmark localization, aiming to improve the assessment of interobserver and automated agreement.

The degree of consistency between observers was assessed using the ICC with a 95% CI, and variability was evaluated using MAD $\pm$ SD and MAE $\pm$ SD (see Table 3 for definitions).

Table 3. Interobserver agreement in manual Cobb angle measurement method.

Analysis	ICC (95% CI)	MAD $\pm$ SD	MAE $\pm$ SD
Observer A vs. Observer B	0.939 (0.868, 0.971)	3.00° $\pm$ 1.67°	3.31° $\pm$ 1.53°

ICC: intraclass correlation coefficient; CI: confidence interval; MAD: median absolute difference; MAE: mean absolute error (MAE); SD: standard deviation.

The results are presented in Table 3, showing an interobserver agreement of 0.939 (95% CI: 0.868–0.971), which is considered an excellent level according to the classification provided by [41]. The MAD and MAE values were 3.00°$\pm$ 1.67° and 3.31° $\pm$ 1.53°, respectively, both providing the magnitude and dispersion of the absolute differences between observers.

The variability between the results predicted by the model and the ground truth, specifically the manually measured Cobb angles, was compared across multiple X-ray images. Figure 9 shows the model predictions on the X-ray images, which were used to assess this discrepancy.

We employed statistical methods to evaluate the agreement between manual and automated Cobb angle measurements. They included Bland–Altman analysis, ICC with 95% CI, MAD, MAE, and SD.

A summary of the results is presented in the following tables. Table 4 reports the average Cobb angle measurements for both observers and the automated approach, while Table 5 summarizes the agreement between manual and automated methods and the errors.

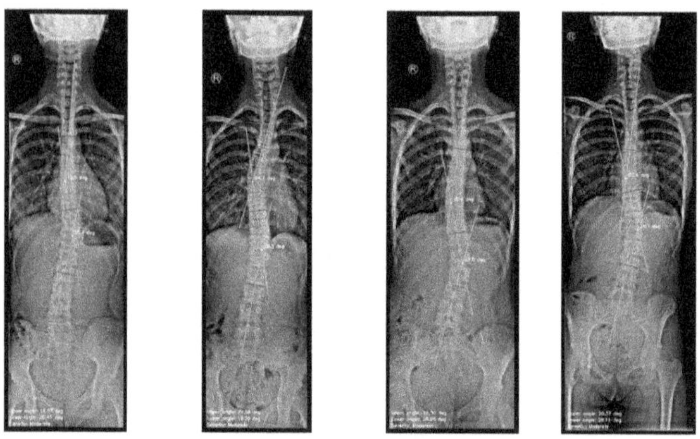

Figure 9. *Cont.*

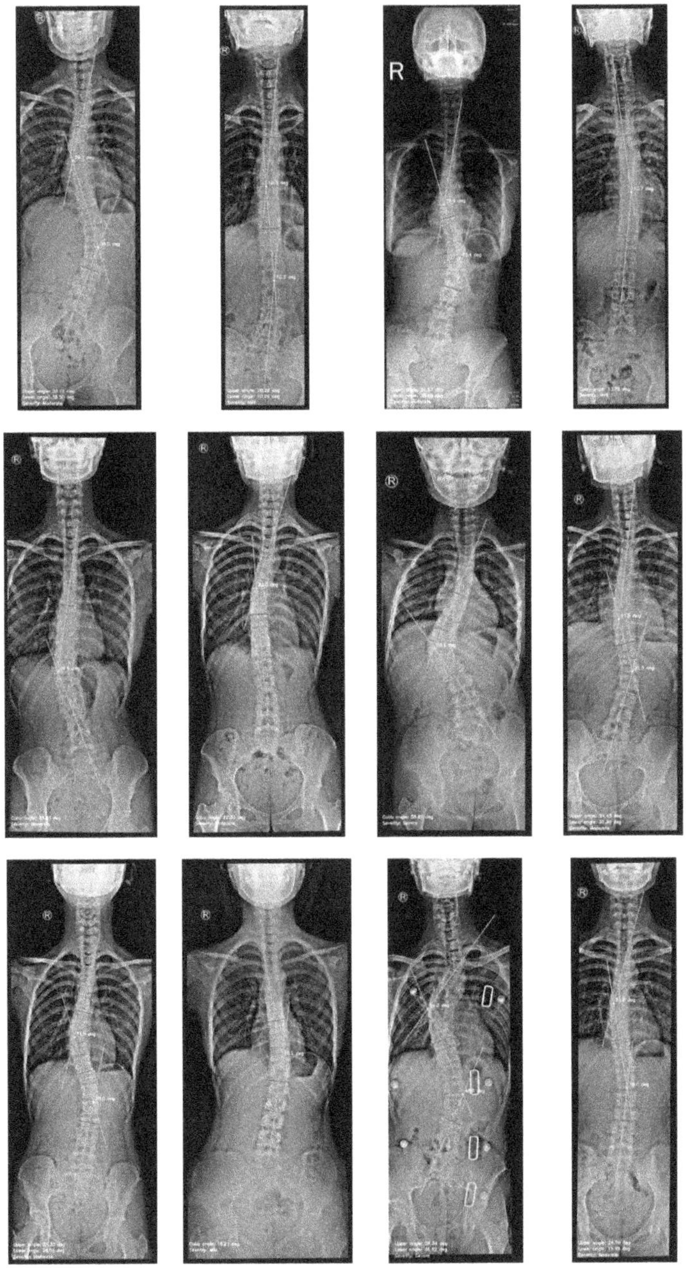

Figure 9. Images generated by the automated approach displaying the scoliosis assessment.

Table 4. Summary of manual and automated Cobb angle measurements on AP X-ray images.

Cobb Angle Measurements	Mean ± Standard Deviation (Range)
Manual measurement by observer A	25.43° ± 10.85° (range 11.50–54.00°)
Manual measurement by observer B	25.89° ± 10.00° (range 10.00–53.00°)
Measured by the automated method	26.69° ± 12.50° (range 10.29–59.34°)

Table 5. Summary of the agreement between manual and automated methods.

Analysis	ICC (95% CI)	MAD ± SD	MAE ± SD
Observer A vs. observer B	0.939 (0.868, 0.971)	3.00° ± 1.67°	3.31° ± 1.53°
Observer A vs. automated	0.961 (0.926, 0.984)	2.15° ± 2.03°	2.54° ± 2.06°
Observer B vs. automated	0.895 (0.780, 0.950)	3.60° ± 3.27°	4.07° ± 3.22°
Overall: Observer A & B vs. automated	0.928 (0.853, 0.967)	2.17° ± 2.51°	2.96° ± 2.60°

Bland–Altman analysis was conducted to compare the measurements between observers, as shown in Figure 10, and between observers and automated Cobb angle method, as illustrated in Figures 11 and 12. In the Bland-Altman plots, the x-axis represents the average Cobb angle between the two methods being compared. The y-axis shows the difference between their measurements. Each dot corresponds to an individual measurement pair. The blue line indicates the mean of the difference (bias), and the dashed black and brown lines represent the limits of agreement.

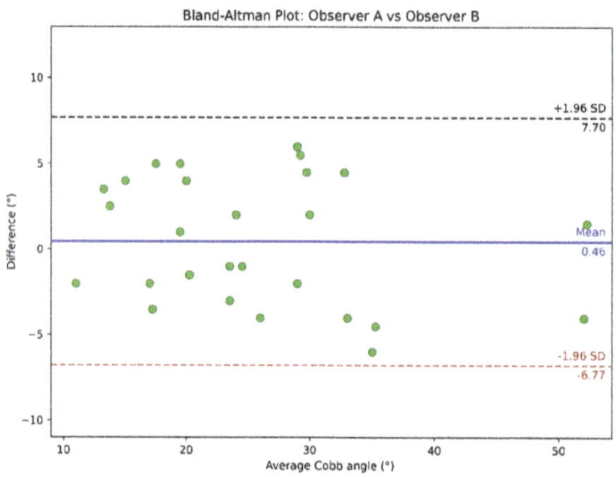

Figure 10. Bland–Altman plot comparing Cobb angle measurements from Observers A and B.

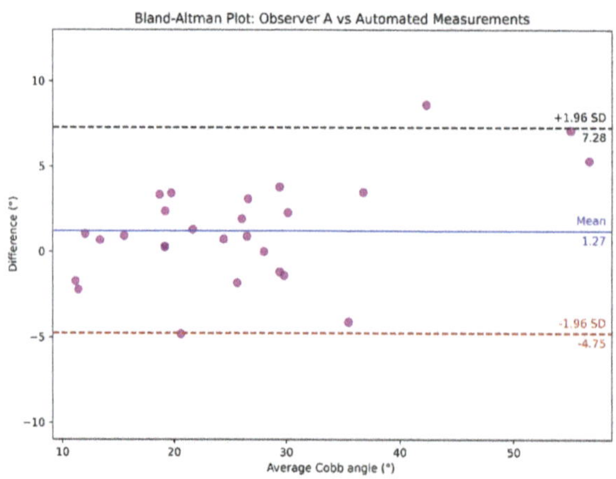

Figure 11. Bland–Altman plot comparing the automated vs. Observer A's Cobb angle measurements.

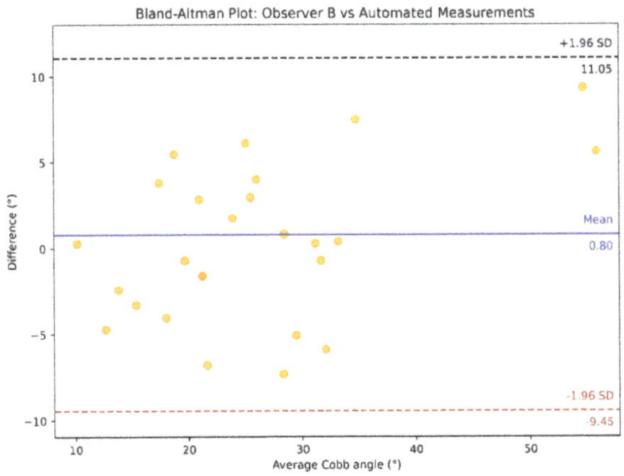

Figure 12. Bland–Altman plot comparing the automated vs. Observer B's Cobb angle measurements.

Notably, in Figure 11, the outlier corresponds to a difference of 8.62 degrees between the manual measurement (Observer A) and the model's prediction. Upon inspection of the corresponding image, we observed that the model produced a simplified representation of one vertebra, with greater inclination than the actual. This example of overestimation of the measured angle explains the discrepancy.

5. Discussion and Scope

In this research, the authors propose an automated approach for scoliosis assessment, which is divided into two processing stages. First, the Mask R-CNN architecture was trained for the spine segmentation within a dataset that comprised anteroposterior X-ray images of the spine. The convolutional neural network generates a mask that provides the starting point for the second component, in which a set of algorithms extracts the mask contour of the spine by identifying anatomical landmarks, enabling Cobb angle quantification and scoliosis severity assessment. During inference, the input images, unseen by the network, are rescaled to a height of 2000 pixels, the same height as those in the dataset. This allows the integration of a rescaling layer for any new image to be evaluated.

It is important to note that the neural network processes the annotations provided in the dataset, which correspond to the spinal region. During training, the network learns to identify this region based on the annotations. These labels are manually annotated only once per image. However, if new images are added to the training dataset, they must be rescaled, annotated, and distributed to the subset before starting the training process that includes the new images. The dataset used in this study was classified as imbalanced, with more than 50% of the cases considered moderate.

Although this class imbalanced is not a classification objective of the network, it may affect the model's generalization capability. In general terms of robustness, a segmentation network such as Mask R-CNN may perform better on imbalanced datasets than a classification model, particularly in this case, where the task concerns scoliosis severity.

This approach focuses on the accurate detection and segmentation of the spinal region within X-ray images. If the task is the identification of scoliotic curvatures, it is reasonable to assume, even with the current imbalance, that the network's generalization is not considerably affected. The application of transfer learning described in Section 3.2.2 supports this comment. We employed a pre-trained network in a similar application, the detection of objects, such as the spine in this case, to mitigate the imbalance in the dataset.

This particular feature enables the network to use its previous knowledge to focus on segmentation. As a result, the model becomes less dependent on a well-balanced dataset and more robust against potential bias caused by imbalance.

Nevertheless, a high imbalance could introduce biases that may affect the network's detection capability or even distort the shape of the segmented region. Therefore, the dataset imbalance should be reviewed, addressed, and corrected, even if the network is less sensitive to these effects and its greater reliability in the identification of anatomical regions.

We only considered the spine class label, while the output is limited to the spine localization and the mask generation. A set of algorithms during the inference stage quantifies the Cobb angle and assesses the scoliosis severity. The objective of analyzing the degree of spinal curvature is to evaluate the error between manual and automated measurements, particularly using the testing subset, which may support future research.

Our method uses the mask generation capability of Mask R-CNN and takes advantage of this feature by applying a single-instance segmentation approach instead of vertebrae segmentation. We reduced the annotation procedure and computational cost by considering the spine structure as a single object, resulting in larger masks that may facilitate the learning process. We also performed monitoring of the segmentation metrics, which is a highly relevant practice for identifying the most suitable candidate epoch. The authors recommend this evaluation to enhance the understanding of the training process and to achieve the highest possible network performance.

Network instance segmentation and mask generation evaluation achieved an mIOU of 0.8012, mDSC of 0.8878, and mAP of 0.6450. Although a larger dataset is required to improve segmentation, the results obtained confirm that Mask R-CNN presents a reliable performance in spine detection and segmentation.

OpenCV's cv2.findContours() function plays a fundamental role in detecting and extracting the spinal contour from the generated mask. This contour enables users to draw the midline of the spine, which is relevant for Cobb angle calculation. The midpoints technique provided in this study define key references along the spinal curvature, allowing users to emulate vertebrae without the need to represent each vertebra individually. The function of the grid interval, which is designed to place a series of points within each vertebra, provides its inclination. The rescaling layer, which adjusts input images to a height of 2000 pixels, ensures the functionality of the grid interval.

We achieved good agreement in Cobb angle measurements with a grid interval value of fifty pixels, which was defined through experimental testing. However, users may adjust the grid interval using the implemented widget. This set of techniques enables the depiction of the spinal midline, aiming to replicate its curvature. A spline interpolation smooths the trajectory defined by the midpoints, providing a continuous representation of the spine. The curvature is then analyzed as a continuous mathematical function by examining its behavior. This analysis enables the estimation of the anatomical landmarks required for Cobb angle quantification.

The accuracy of Cobb angle quantification was assessed through several statistical metrics. Our automated approach achieved an MAD $\pm$ SD of $2.17^\circ \pm 2.51^\circ$, an MAE $\pm$ SD of $2.96^\circ \pm 2.60^\circ$, and an ICC (95% CI) of 0.928 between observers and the automated method. The inference workflow required an average time of 3.3 s per image, providing four images with the step of the process overlaid onto them. This result demonstrates the low deviation of our approach in the Cobb angle quantification task, which is comparable to the measurements performed by observers.

The proposed method, as presented and discussed in this study, can, and must be classified as automatic, despite the initial labelling stage for the dataset. As mentioned in the Material and Methods section, this preliminary manual annotation must be carried out on the X-ray images so that the convolutional neural network can correctly recognize the regions corresponding to the spine. This involves placing points along the contour of the spine. Based on these annotations, the original radiograph is converted into a binary representation in which the spine appears in white and the background in black. This converted image is used by the network as the training label, with the corresponding rescaled X-ray images serving as inputs. While this is a manual and preliminary step, once completed, and with the network properly trained and optimized, the process of Cobb angle quantification for new anteroposterior radiographs becomes fully automated, with no need to manually place points again. For this reason, it is reasonable to describe the method as automatic.

Although the results obtained were promising, the authors acknowledge that various aspects require a critical examination. On the one hand, as previously mentioned, the limited size of the dataset may affect the model's generalization when increasing the number of patients. On the other hand, our single-instance segmentation approach requires a precise contour representation. Small errors in contour detection or non-smooth boundaries may misalign midpoints, thereby affecting the accuracy of Cobb angle quantification. While the fixed grid interval, midpoint technique, and spline interpolation are efficient, the method assumes a typical scoliotic anatomy. In cases of severe or atypical deformities, limited in number within the dataset, the Region Proposal Network (RPN) may generate a bounding box that is too narrow to capture the full lateral extent of the spine. This may crop the mask, excluding peripherical regions, thereby producing incomplete representation that could result in an inaccurate curvature reconstruction affecting the midline extraction. In addition, the requirement for high-quality X-ray images restricts the applicability of this approach in more realistic clinical environments, where images with noise or complex projections, such as overlapping anatomical structures or unusual patient rotations, may be encountered.

Finally, in a real scenario, providing a complete overview of the workflow may enhance the expert's understanding of the method, support clinical diagnosis, and enable the storage of results for monitoring scoliosis progression in patients.

6. Conclusions

This study enabled the authors to address the three main research questions: "Where is the spine in this radiograph?", "What is its exact shape?", and "Is the proposed method accurate for Cobb angle measurement?". Mask R-CNN answered the first two questions: the spine localization within the image (where), and the shape definition of the spine (what). The third question was answered by a set of algorithms specifically designed to achieve accuracy in Cobb angle quantification and severity classification.

The Mask R-CNN network under a single-instance segmentation approach and the midpoint-based method applied in this study enabled the extraction of the scoliotic curve's midline and the identification of tilted vertebrae necessary for Cobb angle estimation, achieving a strong agreement between automated and manual measurements. Notably, the choice of segmenting the spine as a single object involves single-class annotation (spine plus background), which helps mitigate the impact of dataset imbalance, as the network's task is not based on the Cobb angle classification.

The authors acknowledge that the limited size of the training dataset constitutes the main limitation of this study, compromising the model's ability to generalize to unseen cases. Factors such as increased patient variability, overlaid structures, or low-resolution images during inference, may affect the performance of the proposed model. This could impact the Cobb angle estimation, especially in cases where a measurement error of a few degrees could alter the severity assessment. In this regard, our approach provides both

Cobb angle quantification and severity assessment, emphasizing the angle value rather than the severity to facilitate data-driven clinical decision-making.

The agreement between manual and automated measurements obtained in this study demonstrates that our approach achieves the objective of offering an automated method for Cobb angle quantification and scoliosis severity classification, with results comparable to those estimated by human observers. The indicators used for accuracy assessment suggest that the model may be acceptable for many clinical scenarios. Nevertheless, given that this study was conducted using a limited sample of images, more extensive external validation is required to verify its robustness. Our method provides experts with a clear visualization of each stage in the scoliosis assessment. This enhances the interpretability of the results, facilitates monitoring, and enables comparison when experts evaluate the scoliosis progression.

This study is within the scope of a broader research line. As a part of our future work, we plan to explore strategies such as data augmentation and vertebrae segmentation, and to compare the results of this study with those of other approaches, aiming to improve model generalization.

Author Contributions: Conceptualization, M.V.G.; methodology, M.V.G. and A.C.-C.; software, M.V.G.; investigation, M.V.G., J.-B.B.-R. and A.C.-C.; data curation, M.V.G.; writing—original draft preparation, M.V.G. and A.C.-C.; writing—review and editing, M.V.G., J.-B.B.-R. and A.C.-C.; supervision, J.-B.B.-R. and A.C.-C. All authors have read and agreed to the published version of the manuscript.

Funding: This research received no external funding.

Institutional Review Board Statement: Not applicable.

Informed Consent Statement: Not applicable.

Data Availability Statement: Data are contained within the article.

Conflicts of Interest: The authors declare no conflict of interest.

References

1. Scoliosis | Scoliosis Research Society. Available online: https://www.srs.org/Patients/Conditions/Scoliosis (accessed on 12 September 2024).
2. Horng, M.H.; Kuok, C.P.; Fu, M.J.; Lin, C.J.; Sun, Y.N. Cobb Angle Measurement of Spine from X-Ray Images Using Convolutional Neural Network. *Comput. Math. Methods Med.* **2019**, *2019*, 6357171. [CrossRef] [PubMed]
3. Thalengala, A.; Bhat, S.N.; Anitha, H. Computerized image understanding system for reliable estimation of spinal curvature in idiopathic scoliosis. *Sci. Rep.* **2021**, *11*, 7144. [CrossRef] [PubMed]
4. Han, S.; Zhao, H.; Zhang, Y.; Yang, C.; Han, X.; Wu, H.; Cao, L.; Yu, B.; Wen, J.X.; Wu, T.; et al. Application of machine learning standardized integral area algorithm in measuring the scoliosis. *Sci. Rep.* **2023**, *13*, 19255. [CrossRef]
5. Wiliński, P.; Piekutin, A.; Dmowska, K.; Zawieja, W.; Janusz, P. Which Method of the Radiologic Measurements of the Angle of Curvature in Idiopathic Scoliosis is the Most Reliable for an Inexperienced Researcher? *Indian J. Orthop.* **2025**, *59*, 140–147. [CrossRef]
6. Vertebral column | Anatomy & Function | Britannica. Available online: https://www.britannica.com/science/vertebral-column (accessed on 6 February 2025).
7. Li, K.; Gu, H.; Colglazier, R.; Lark, R.; Hubbard, E.; French, R.; Smith, D.; Zhang, J.; McCrum, E.; Catanzano, A.; et al. Deep Learning Automates Cobb Angle Measurement Compared with Multi-Expert Observers. *arXiv* **2024**, arXiv:2403.12115. [CrossRef]
8. Gstoettner, M.; Sekyra, K.; Walochnik, N.; Winter, P.; Wachter, R.; Bach, C.M. Inter- and intraobserver reliability assessment of the Cobb angle: Manual versus digital measurement tools. *Eur. Spine J.* **2007**, *16*, 1587–1592. [CrossRef] [PubMed]
9. Anitha, H.; Karunakar, A.K.; Dinesh, K.V.N. Automatic extraction of vertebral endplates from scoliotic radiographs using customized filter. *Biomed. Eng. Lett.* **2014**, *4*, 158–165. [CrossRef]
10. He, K.; Gkioxari, G.; Dollár, P.; Girshick, R. Mask R-CNN. In Proceedings of the 2017 IEEE International Conference on Computer Vision (ICCV), Venice, Italy, 22–29 October 2017; pp. 2980–2988. [CrossRef]

11. Khanal, B.; Dahal, L.; Adhikari, P.; Khanal, B. Automatic Cobb Angle Detection Using Vertebra Detector and Vertebra Corners Regression. In *Computational Methods and Clinical Applications for Spine Imaging*; Cai, Y., Wang, L., Audette, M., Zheng, G., Li, S., Eds.; Springer International Publishing: Cham, Switzerland, 2020; pp. 81–87. [CrossRef]
12. Fu, X.; Yang, G.; Zhang, K.; Xu, N.; Wu, J. An automated estimator for Cobb angle measurement using multi-task networks. *Neural Comput. Appl.* **2021**, *33*, 4755–4761. [CrossRef]
13. Huang, X.; Luo, M.; Liu, L.; Wu, D.; You, X.; Deng, Z.; Xiu, P.; Yang, X.; Zhou, C.; Feng, G.; et al. The Comparison of Convolutional Neural Networks and the Manual Measurement of Cobb Angle in Adolescent Idiopathic Scoliosis. *Glob. Spine J.* **2024**, *14*, 159–168. [CrossRef]
14. Caesarendra, W.; Rahmaniar, W.; Mathew, J.; Thien, A. Automated Cobb Angle Measurement for Adolescent Idiopathic Scoliosis Using Convolutional Neural Network. *Diagnostics* **2022**, *12*, 396. [CrossRef]
15. Chui, C.S.; He, Z.; Lam, T.P.; Mak, K.K.; Ng, H.T.; Fung, C.H.; Chan, M.S.; Law, S.W.; Lee, Y.W.; Hung, L.H.; et al. Deep Learning-Based Prediction Model for the Cobb Angle in Adolescent Idiopathic Scoliosis Patients. *Diagnostics* **2024**, *14*, 1263. [CrossRef]
16. Vrtovec, T.; Pernuš, F.; Likar, B. A review of methods for quantitative evaluation of spinal curvature. *Eur. Spine J.* **2009**, *18*, 593–607. [CrossRef] [PubMed]
17. Jin, C.; Wang, S.; Yang, G.; Li, E.; Liang, Z. A Review of the Methods on Cobb Angle Measurements for Spinal Curvature. *Sensors* **2022**, *22*, 3258. [CrossRef] [PubMed]
18. Sun, H.; Zhen, X.; Bailey, C.; Rasoulinejad, P.; Yin, Y.; Li, S. Direct Estimation of Spinal Cobb Angles by Structured Multi-output Regression. In *Information Processing in Medical Imaging*; Niethammer, M., Styner, M., Aylward, S., Zhu, H., Oguz, I., Yap, P.T., Shen, D., Eds.; Springer International Publishing: Cham, Switzerland, 2017; pp. 529–540. [CrossRef]
19. Wu, H.; Bailey, C.; Rasoulinejad, P.; Li, S. Automatic Landmark Estimation for Adolescent Idiopathic Scoliosis Assessment Using BoostNet. In *Medical Image Computing and Computer Assisted Intervention—MICCAI 2017*; Descoteaux, M., Maier-Hein, L., Franz, A., Jannin, P., Collins, D.L., Duchesne, S., Eds.; Springer International Publishing: Cham, Switzerland, 2017; pp. 127–135. [CrossRef]
20. Wu, H.; Bailey, C.; Rasoulinejad, P.; Li, S. Automated comprehensive Adolescent Idiopathic Scoliosis assessment using MVC-Net. *Med. Image Anal.* **2018**, *48*, 1–11. [CrossRef] [PubMed]
21. Chen, B.; Xu, Q.; Wang, L.; Leung, S.; Chung, J.; Li, S. An Automated and Accurate Spine Curve Analysis System. *IEEE Access* **2019**, *7*, 124596–124605. [CrossRef]
22. Yang, J.; Zhang, K.; Fan, H.; Huang, Z.; Xiang, Y.; Yang, J.; He, L.; Zhang, L.; Yang, Y.; Li, R.; et al. Development and validation of deep learning algorithms for scoliosis screening using back images. *Commun. Biol.* **2019**, *2*, 390. [CrossRef]
23. Yi, J.; Wu, P.; Huang, Q.; Qu, H.; Dimitris, D.N. Vertebra-Focused Landmark Detection for Scoliosis Assessment. Available online: https://ieeexplore.ieee.org/document/9098675 (accessed on 8 April 2025).
24. Cerqueiro, J.; Comesaña-Campos, A.; Casal-Guisande, M.; Bouza-Rodríguez, J. A proposal for using active contour parametrical models in Cobb angle determination. In Proceedings of the Ninth International Conference on Technological Ecosystems for Enhancing Multiculturality, Barcelona, Spain, 26–29 October 2021. [CrossRef]
25. Sun, Y.; Xing, Y.; Zhao, Z.; Meng, X.; Xu, G.; Hai, Y. Comparison of manual versus automated measurement of Cobb angle in idiopathic scoliosis based on a deep learning keypoint detection technology. *Eur. Spine J.* **2022**, *31*, 1969–1978. [CrossRef]
26. Maeda, Y.; Nagura, T.; Nakamura, M.; Watanabe, K. Automatic measurement of the Cobb angle for adolescent idiopathic scoliosis using convolutional neural network. *Sci. Rep.* **2023**, *13*, 14576. [CrossRef]
27. Qiu, Z.; Yang, J.; Wang, J. MMA-Net: Multiple Morphology-Aware Network for Automated Cobb Angle Measurement. In Proceedings of the 2024 IEEE International Conference on Robotics and Automation (ICRA), Yokohama, Japan, 13–17 May 2024; pp. 9139–9145. [CrossRef]
28. Suri, A.; Tang, S.; Kargilis, D.; Taratuta, E.; Kneeland, B.J.; Choi, G.; Agarwal, A.; Anabaraonye, N.; Xu, W.; Parente, J.B.; et al. Conquering the Cobb Angle: A Deep Learning Algorithm for Automated, Hardware-Invariant Measurement of Cobb Angle on Radiographs in Patients with Scoliosis. *Radiol. Artif. Intell.* **2023**, *5*, e220158. [CrossRef]
29. Hoblidar, A.; Prabhu, G. Automatic Quantification of Spinal Curvature in Scoliotic Radiograph using Image Processing. *J. Med. Syst.* **2011**, *36*, 1943–1951. [CrossRef]
30. Ronneberger, O.; Fischer, P.; Brox, T. U-Net: Convolutional Networks for Biomedical Image Segmentation. In *Medical Image Computing and Computer-Assisted Intervention—MICCAI 2015*; Navab, N., Hornegger, J., Wells, W.M., Frangi, A.F., Eds.; Springer International Publishing: Cham, Switzerland, 2015; pp. 234–241. [CrossRef]
31. Vuola, A.O.; Akram, S.U.; Kannala, J. Mask-RCNN and U-Net Ensembled for Nuclei Segmentation. In Proceedings of the 2019 IEEE 16th International Symposium on Biomedical Imaging (ISBI 2019), Venice, Italy, 8–11 April 2019; pp. 208–212. [CrossRef]
32. Alharbi, R.H.; Alshaye, M.B.; Alkanhal, M.M.; Alharbi, N.M.; Alzahrani, M.A.; Alrehaili, O.A. Deep Learning Based Algorithm For Automatic Scoliosis Angle Measurement. In Proceedings of the 2020 3rd International Conference on Computer Applications & Information Security (ICCAIS), Riyadh, Saudi Arabia, 19–21 March 2020; pp. 1–5. [CrossRef]

33. Zhang, L.; Shi, L.; Cheng, J.C.Y.; Chu, W.C.W.; Yu, S.C.H. LPAQR-Net: Efficient Vertebra Segmentation From Biplanar Whole-Spine Radiographs. *IEEE J. Biomed. Health Inform.* **2021**, *25*, 2710–2721. [CrossRef] [PubMed]
34. Zhao, Y.; Zhang, J.; Li, H.; Gu, X.; Li, Z.; Zhang, S. Automatic Cobb angle measurement method based on vertebra segmentation by deep learning. *Med. Biol. Eng. Comput.* **2022**, *60*, 2257–2269. [CrossRef] [PubMed]
35. Jason, C.W.; Reformat, M.Z.; Parent, E.C.; Stampe, K.P.; Hryniuk, S.C.S.; Edmond, H.L. Validation of an artificial intelligence-based method to automate Cobb angle measurement on spinal radiographs of children with adolescent idiopathic scoliosis. *Eur. J. Phys. Rehabil. Med.* **2023**, *54*, 535–542. [CrossRef]
36. Low, X.Z.; Furqan, M.S.; Makmur, A.; Lim, D.S.W.; Liu, R.W.; Lim, X.; Chan, Y.H.; Tan, J.H.; Lau, L.L.; Hallinan, J.T.P.D. Automated Cobb angle measurement in scoliosis radiographs: A deep learning approach for screening. *Ann. Acad. Med. Singap.* **2024**, *53*, 635–637. [CrossRef]
37. Liang, Z.; Wang, Q.; Xia, C.; Chen, Z.; Xu, M.; Liang, G.; Zhang, Y.; Ye, C.; Zhang, Y.; Yu, X.; et al. From 2D to 3D: Automatic measurement of the Cobb angle in adolescent idiopathic scoliosis with the weight-bearing 3D imaging. *Spine J.* **2024**, *24*, 1282–1292. [CrossRef]
38. Rahmaniar, W.; Suzuki, K.; Lin, T.L. Auto-CA: Automated Cobb Angle Measurement Based on Vertebrae Detection for Assessment of Spinal Curvature Deformity. *IEEE Trans. Biomed. Eng.* **2024**, *71*, 640–649. [CrossRef]
39. Kassab, D.K.I.; Kamyshanskaya, I.G.; Pershin, A.A. Automatic scoliosis angle measurement using deep learning methods, how far we are from clinical application: A narrative review. *Medicine* **2021**, *16*, 85–94.
40. Maharasi, M.; Senthilnayaki, N.; Snehaprabha, K. Vertebrae Landmark Detection and Scoliosis Assessment Using Deep Learning. In Proceedings of the 2024 International Conference on Communication, Computing and Internet of Things (IC3IoT), Chennai, India, 17–18 April 2024; pp. 1–6. [CrossRef]
41. Pan, Y.; Chen, Q.; Chen, T.; Wang, H.; Zhu, X.; Fang, Z.; Lu, Y. Evaluation of a computer-aided method for measuring the Cobb angle on chest X-rays. *Eur. Spine J.* **2019**, *28*, 3035–3043. [CrossRef]
42. AASCE | AASCE—MICCAI 2019 Challenge: Accurate Automated Spinal Curvature Estimation. Available online: https://aasce19.github.io/ (accessed on 1 February 2025).
43. GitHub—Matterport/Mask_RCNN: Mask R-CNN for Object Detection and Instance Segmentation on Keras and TensorFlow. Available online: https://github.com/matterport/Mask_RCNN (accessed on 12 September 2024).
44. Gad, A. Ahmedfgad/Mask-RCNN-TF2. (February, 2025). Python. Available online: https://github.com/ahmedfgad/Mask-RCNN-TF2 (accessed on 8 February 2025).
45. Kundu, R.; Lenka, P.; Kumar, R.; Chakrabarti, A. Cobb angle quantification for scoliosis using image processing techniques. In Proceedings of the International Conference on Recent Advances and Future Trends in Information Technology (iRAFIT2012), Tamil Nadu, India, 19–21 April 2012.
46. Altman, D.G. *Practical Statistics for Medical Research*; Chapman and Hall/CRC: New York, NY, USA, 1990. [CrossRef]
47. Armitage, P.; Berry, G.; Matthews, J.N.S. *Statistical Methods in Medical Research*; John Wiley & Sons: Hoboken, NJ, USA, 2013.

diagnostics

MDPI

Article

Fractal-Based Architectures with Skip Connections and Attention Mechanism for Improved Segmentation of MS Lesions in Cervical Spinal Cord

Rukiye Polattimur [1], Mehmet Süleyman Yıldırım [2] and Emre Dandıl [3,*]

1 Department of Electronics and Computer Engineering, Institute of Graduate, Bilecik Seyh Edebali University, 11230 Bilecik, Türkiye; rukiye.polattimur@bilecik.edu.tr
2 Department of Computer Technology, Söğüt Vocational School, Bilecik Şeyh Edebali University, Söğüt, 11600 Bilecik, Türkiye; mehmets.yildirim@bilecik.edu.tr
3 Department of Computer Engineering, Faculty of Engineering, Bilecik Seyh Edebali University, 11230 Bilecik, Türkiye
* Correspondence: emre.dandil@bilecik.edu.tr

Abstract: Background/Objectives: Multiple sclerosis (MS) is an autoimmune disease that damages the myelin sheath of the central nervous system, which includes the brain and spinal cord. Although MS lesions in the brain are more frequently investigated, MS lesions in the cervical spinal cord (CSC) can be much more specific for the diagnosis of the disease. Furthermore, as lesion burden in the CSC is directly related to disease progression, the presence of lesions in the CSC may help to differentiate MS from other neurological diseases. **Methods**: In this study, two novel deep learning models based on fractal architectures are proposed for the automatic detection and segmentation of MS lesions in the CSC by improving the convolutional and connection structures used in the layers of the U-Net architecture. In our previous study, we introduced the FractalSpiNet architecture by incorporating fractal convolutional block structures into the U-Net framework to develop a deeper network for segmenting MS lesions in the CPC. In this study, to improve the detection of smaller structures and finer details in the images, an attention mechanism is integrated into the FractalSpiNet architecture, resulting in the Att-FractalSpiNet model. In addition, in the second hybrid model, a fractal convolutional block is incorporated into the skip connection structure of the U-Net architecture, resulting in the development of the Con-FractalU-Net model. **Results:** Experimental studies were conducted using U-Net, FractalSpiNet, Con-FractalU-Net, and Att-FractalSpiNet architectures to detect the CSC region and the MS lesions within its boundaries. In segmenting the CSC region, the proposed Con-FractalU-Net architecture achieved the highest Dice Similarity Coefficient (DSC) score of 98.89%. Similarly, in detecting MS lesions within the CSC region, the Con-FractalU-Net model again achieved the best performance with a DSC score of 91.48%. **Conclusions**: For segmentation of the CSC region and detection of MS lesions, the proposed fractal-based Con-FractalU-Net and Att-FractalSpiNet architectures achieved higher scores than the baseline U-Net architecture, particularly in segmenting small and complex structures.

Keywords: cervical spinal cord; multiple sclerosis; automatic segmentation; U-Net; FractalSpiNet; Con-FractalU-Net; Att-FractalSpiNet; MRI

1. Introduction

Segmentation is crucial for visualization and computation in many medical image workflows [1]. For various clinical applications, such as diagnosis, treatment planning, and

 https://doi.org/10.3390/diagnostics15081041

surgery, medical image segmentation is important. Detection and accurate segmentation of lesions, tumors, and other small anatomical structures are essential for monitoring disease processes and evaluating effective treatment methods [2]. The semantic definition and segmentation of each pixel in medical images is widely used as a decision support system developed for clinical diagnosis, treatment, and pathological evaluation [3]. In particular, the evaluation of diseases such as multiple sclerosis (MS), which directly affects the nervous system and affects the daily life of the person, is one of the most important tools for clinical decision makers [4]. While clinical approaches presented as manual segmentation with expert support can often be quite challenging in terms of time and cost, it is possible to minimize the difficulty of the process by using an automated, reliable, and reproducible decision support system [5,6].

MS is a chronic autoimmune disease of the central nervous system (CNS) that results in demyelination and neurodegeneration [7,8]. The disease manifests primarily through lesions in the brain and spinal cord, with the cervical spinal cord (CSC) playing a critical role in disease progression and disability assessment. MS lesions in the CSC are strongly correlated with motor and sensory impairment, making their accurate detection and segmentation essential for both clinical diagnosis and patient monitoring [9]. MS lesions in the CSC provide important data for predicting disease progression and formulating patient-specific treatment plans. For example, detection of MS lesions in the CSC is important for processes such as predicting response to immunomodulatory therapy, selecting disease-modifying therapy based on lesion burden, and guiding physical therapy and rehabilitation processes.

CSC lesions are directly associated with motor dysfunction, sensory impairment, and disease progression in MS patients [10]. Automated and accurate lesion segmentation can improve early diagnosis, as identification of lesions in the spinal cord can help neurologists confirm the diagnosis of MS. In addition, tracking lesion progression over time allows clinicians to assess disease activity and make timely adjustments to treatment strategies. Furthermore, automated methods minimize inter- and intra-observer variability, ensuring more consistent and objective lesion assessment. Accurate segmentation of MS lesions in the CSC has significant clinical importance in magnetic resonance imaging (MRI). MRI is the gold standard for detection and analysis of MS lesions [11]. The McDonald criteria have provided a set of grading standards for the diagnosis and management of MS disease and have highlighted the importance of MRI, particularly axial T2-weighted (T2-w) scans [12,13]. It is possible to achieve high levels of accuracy in the diagnosis of MS lesions from spinal cord MR images using automated systems developed to support clinical applications [14]. However, the fact that the region of interest (RoI) is located in regions where precision and high quality are required and has small and volumetrically different tissues can create different situations during MR imaging and may have a negative impact on the quality of the data [15]. This situation requires more meticulous and careful data acquisition and dataset generation processes.

There are many computer-aided tools that are used as decision support systems to improve the process of early detection and diagnosis in the clinic [16]. In addition to saving time and labor, these systems can be used as a learning tool for non-specialist or specialist physicians and can be used as a tool to assist in making the correct diagnosis. By using an automated system with deep learning tools, objective and consistent results can be achieved by minimizing human error and enabling clinical practitioners to make decisions with high accuracy. Thus, a more reliable diagnostic protocol and treatment protocol can be established by reducing false rates and ensuring accurate assessment of disease progression.

Automated segmentation approaches based on deep learning have shown promising results in improving accuracy and efficiency. On the other hand, differences in dataset size and quality, variations in model architectures and hyperparameters, inconsistent training and evaluation methods, and a lack of standardized benchmark datasets are major sources of variability in deep learning models for medical image segmentation [17]. In addition, challenges remain due to the small lesion size, low contrast variation, and structural complexity of the spinal cord. Deep learning-based models have been used to segment the spinal cord, particularly in spinal MR images, to identify MS lesions in this region and to perform long-term follow-up analyses. Many previous studies have proposed approaches for the detection and segmentation of spinal cord cross-sectional area (CSA), cerebrospinal fluid (CSF), white matter (WM), grey matter (GM), MS, and other lesion derivatives in the CNS [18–21]. On the other hand, studies have also been proposed to detect spinal cord regions and textural abnormalities such as lesions and tumors within these boundaries using deep learning-based convolutional networks [21–26]. In addition, some studies have been presented for segmentation of the spinal cord, spinal cord GM and WM, and spinal canal using convolutional recurrent neural networks (RNN), ResNet50, and attention mechanism-based deep learning architectures [25,27,28].

In the work proposed by Gros et al. [22], automatic segmentation of spinal cord atrophy and lesions in MS patients was performed. The proposed automatic segmentation approach is based on a two-stage CNN sequence. The first CNN detects the spinal cord center line using 2D extended convolutions, while the second CNN segments the spinal cord and lesions using 3D convolutions. In the study, although a high score was obtained for spinal cord segmentation, the score obtained for MS lesion segmentation was relatively low. In another study, McCoy et al. [23] presented a 2D CNN-based approach for automatic segmentation of spinal cord and contusion injuries from MR images. The developed model was compared with existing best methods. The proposed model showed better performance compared to manual segmentation. However, the use of a small dataset in the study is considered a limitation. Merali et al. [25] developed a deep learning-based model for detection of CSC compression in MRI images. The performance of the proposed CNN model was evaluated after the images were labeled for the presence of spinal cord compression by two expert physicians. The fact that the dataset is limited to a specific group of patients does not give an idea of how the performance of the model may be affected in larger datasets. Horváth et al. [27] proposed a novel multidimensional RNN architecture to automate spinal cord GM and WM segmentation. They presented an approach that enhances texture contrast by obtaining eight different inverse recovery (IR) images of the same anatomical slice. They also compared the results of automated segmentation with manual segmentation but did not report inter-observer agreement rates for manual segmentation. In another study, Perone et al. [21] developed a deep learning-based method for automatic segmentation of spinal cord GM tissue using dilated convolutions. The developed model was compared with six different state-of-the-art methods in GM segmentation tasks, and performance evaluations were performed. However, the study did not evaluate the performance of the model in a larger patient population with different demographic characteristics. Porisky et al. [29] presented a novel method for grey matter segmentation from spinal cord MRI images using 3D convolutional encoder networks and short-cut connections. Although the proposed architecture looks similar to a U-Net structure with encoder, decoder, and shortcut connection, a deconvolution process is used instead of an upsampling process in the decoder part. Naga Karthik et al. [30] developed an open-source 2D and 3D CNN-based tool for automatic segmentation of spinal cord lesions in MS patients from axial T2-w MRI images. The developed tool was evaluated on data obtained from different centers and achieved high accuracy rates in lesion segmentation.

Automatic segmentation of medical images has gained much more momentum, especially with the advent of the U-Net architecture [31]. The U-Net architecture, which provides an end-to-end pixel-based solution, can achieve very successful results even on datasets with a small target area. Although there are few studies on the automatic segmentation of the spinal cord region, spinal cord tumors, and lesions, there are studies using the U-Net architecture, which is developed using a convolutional structure and provides stable and successful results in many aspects, especially in the field of medical imaging. Zhang et al. [26] automated spinal cord segmentation from 2D cervical axial MRI slices. The proposed approach includes a level set-based active contour method by pre-processing MRI images with a U-Net architecture. The number of patients in the study is small, and the performance of the results obtained has not been evaluated with larger and more diverse datasets. In another study, Askari-Hemmat et al. [32] performed grey matter segmentation of the spinal cord using a U-Net architecture based on a fixed-point quantization method. In the study, the quantization process caused a small decrease in the accuracy of the model. Fei et al. [33] achieved automatic segmentation of the internal structures of the CSC using a U-Net model based on pre-trained VGG16 and ResNet50 backbones. In their study, too many RoI's were identified, resulting in poor performance. Alsenan et al. [34] proposed MobileU-NetV3, a lightweight deep learning model that combines MobileNetV3 and U-Net architectures for spinal cord GM segmentation. The proposed model was evaluated on a specific dataset. Zhang et al. [35] developed the SeUneter architecture for segmentation of cervical MRI spinal structures by deepening the U-Net architecture and adding a channel attention module to the double convolutional layers during feature extraction. The contribution of the channel attention module to the segmentation performance is not analyzed in detail. Bueno et al. [36] proposed an optimized residual attention-aware U-Net architecture for automatic spinal cord segmentation from cervical spine MR images of MS patients. The automatic segmentation model showed some success compared to manual segmentation. In our previous study [37], a novel deep learning architecture called FractalSpiNet was proposed for automatic segmentation of spinal cord and MS lesions in CSC MR images. FractalSpiNet is an architecture based on the U-Net structure and integrates fractal networks for improved feature extraction in MRI scans. The proposed FractalSpiNet architecture has shown better performance in the automatic segmentation task compared to state-of-the-art methods.

In general, work on spinal cord segmentation is limited compared to other medical image segmentation tasks. Several deep learning approaches have been proposed for MS lesion segmentation in the spinal cord. Traditional CNNs have demonstrated success in spinal cord segmentation, but their performance is often limited by a lack of global contextual understanding and difficulties in capturing long-range dependencies. Traditional U-Nets and their variants have been used due to their encoder-decoder architecture and skip connections, which preserve spatial detail. However, these methods often struggle to distinguish small lesions from surrounding tissue due to limited contextual awareness. On the other hand, many studies suffer from the use of small datasets, which can lead to overfitting and limit the generalizability of the model to unseen data. In addition, some studies lack diversity in the data and focus on specific patient populations. Furthermore, studies comparing the proposed methods with existing methods generally show similar performance, suggesting the need for significant improvements in segmentation accuracy, especially for MS lesion detection.

To address these limitations, this study proposes a novel deep learning framework based on fractal architectures with skip connections and an attention mechanism for improved segmentation of MS lesions in the CSC. The proposed fractal-based approach builds on the strengths of existing models while addressing their limitations. Unlike standard CNN-based methods, the fractal architecture enables a hierarchical multi-scale feature extraction, which improves robustness to lesion size variations. Fractal architectures, inspired by self-repeating hierarchical patterns, enable the extraction of multi-scale features, enhancing the network's ability to capture complex spatial structures. Skip connections facilitate the flow of information across different layers, preserving fine-grained spatial detail and improving gradient propagation. Attention mechanisms are also incorporated to improve the model's focus on relevant lesion regions, reducing false positives and improving segmentation precision. By leveraging fractal architectures, skip connections, and attention mechanisms, the proposed models aim to provide a robust and efficient approach for automated segmentation of MS lesions in the CSC. This advancement has the potential to assist clinicians in early diagnosis, disease progression monitoring, and treatment planning, ultimately improving patient outcomes. The contributions of this study can be summarized as follows:

1. This study presents two deep learning architectures, Con-FractalU-Net and Att-FractalSpiNet, that utilize fractal convolutional blocks. The use of fractal designs allows the models to explore multiple receptive fields and path depths in parallel, enhancing the network's ability to learn complex spatial hierarchies. This is particularly valuable in spinal cord imaging, where MS lesions vary widely in size, shape, and intensity.
2. By incorporating U-Net-type skip connections into the fractal architecture, the proposed models maintain fine-grained spatial information across encoding and decoding paths. These connections help mitigate the vanishing gradient problem, especially in deeper networks, and contribute to more accurate lesion delineation by preserving high resolution anatomical details.
3. Att-FractalSpiNet uses attention modules to focus the network on lesion-relevant regions while suppressing less informative background noise. This selective attention strategy enhances the network's ability to distinguish MS lesions from complex and noisy spinal cord structures, improving both the sensitivity and specificity of the segmentation process.
4. The proposed models were extensively evaluated on a cervical spinal cord MRI dataset, and performance was evaluated using standard metrics. The results show that Con-FractalU-Net and Att-FractalSpiNet outperform conventional architectures such as U-Net and previous fractal-based methods, achieving higher levels of performance, especially in the segmentation of small and irregularly shaped lesions.

The rest of this paper is structured as follows: Section 2 provides a detailed description of the dataset used in this study, along with the preprocessing steps applied to improve data quality and model performance. In addition, this section provides a comprehensive explanation of the proposed Con-FractalU-Net and Att-FractalSpiNet architectures, detailing all their components and innovations. Section 3 focuses on the experimental analysis, where the quantitative metrics obtained from the segmentation and lesion detection tasks are thoroughly evaluated. Furthermore, visualizations of the model outputs are provided to facilitate an in-depth analysis of the segmentation performance. Finally, Section 4 summarizes the experimental findings, elaborates on the study's conclusions, and discusses the clinical implications of the proposed models. In addition, this section outlines potential future research directions, highlighting areas for further improvements and applications of fractal-based architectures in medical image segmentation.

2. Materials and Methods

This study presents fractal-based skip connections-based Con-FractalU-Net and attention mechanism-based Att-FractalSpiNet deep learning approaches for more accurate and successful segmentation of MS lesions in the CSC. In the proposed architectures, a methodology based on fractal convolution-based synthesis U-Net architectures was developed using a CSC dataset consisting of T2-w MR slices. In the study, model training for automatic detection of the CSC region and MS lesions in this region is performed, and the results obtained are compared. The block diagram showing the general methodology of the study is shown in Figure 1. In the first step, the spinal cord MR dataset containing images of the CSC is divided into two subsets as training and test. This dataset is enriched with segmentation annotations, including MRI-based CSC images and MS lesions. Then, since the raw MRI images are not suitable to work directly with deep learning models, some pre-processing techniques were applied. In the second step, since the developed architectures are based on fractal-based U-Net models, the results of our previously proposed FractalSpiNet and basic U-Net architectures trained with specified hyperparameters are compared with the segmentation performance results obtained for the Con-FractalU-Net and Att-FractalSpiNet architectures proposed in this study using key metrics.

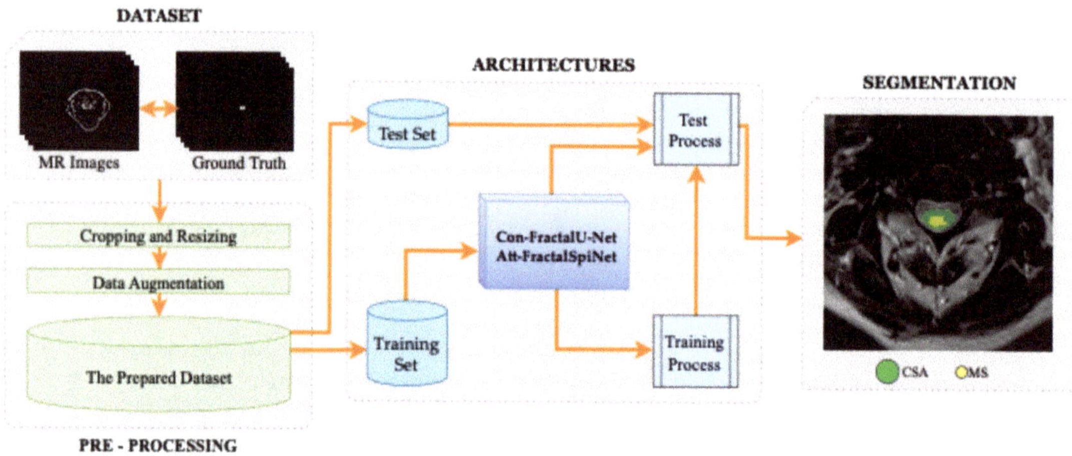

Figure 1. The complete workflow of the proposed segmentation framework for detecting MS lesions in the cervical spinal cord (CSC) using fractal-based deep learning architectures such as skip connections-based Con-FractalU-Net and attention mechanism-based Att-FractalSpiNet.

2.1. Dataset

The cervical part of the human spinal cord is located between the intervertebral discs between the neck and the coccyx. In the clinical setting, it is possible to obtain individual MRI images of the CSC, either regionally or for the entire CSC if needed. As MS lesions often occur in the CSC region, MR scans of the cervical region are performed according to clinical procedures for the detection of MS lesions [38]. The dataset used in the study [39] is a publicly available dataset of T2-w CSC MRI images, performed retrospectively, containing samples of healthy and MS patients, labeled with manual segmentation masks, and used in our previous study [37]. The images in the dataset were generated from 2D MR slices of the CSC region of 87 MS patients, 68 females and 19 males, scanned axially in the turbo spin echo sequence and T2-w modality. T2-w images are critical for segmentation of the spinal cord and MS lesions, as they facilitate the separation of WM and GM and the detection of pathological structures.

The images in the scans obtained from the SIEMENS Spectra Magnetom 3T device as DICOM in the dataset were 320 × 250 pixels, and the slice thickness in the scans was 4 mm. For the cross-sectional area (CSA) and MS lesions in the images of the MR scans forming the dataset, ground truth masks were manually determined by two different radiologists using ITK-SNAP 4.0 software [40]. This resulted in a total of 231 suitable slices for CSA and MS in the axial plane. Preprocessing steps and data augmentation techniques were used to train the model, and performance analysis was performed by dividing the data into training, validation, and test sets. Figure 2 shows some of the original MR images in the dataset and the ground truth masks of the CSA and MS lesions in CSC. In addition, it was approved by the decision of the Clinical Research Ethics Committee of Akdeniz University Faculty of Medicine dated 15 September 2021 and numbered KAEK-644 that there are no ethical objections to conducting this study and creating the dataset.

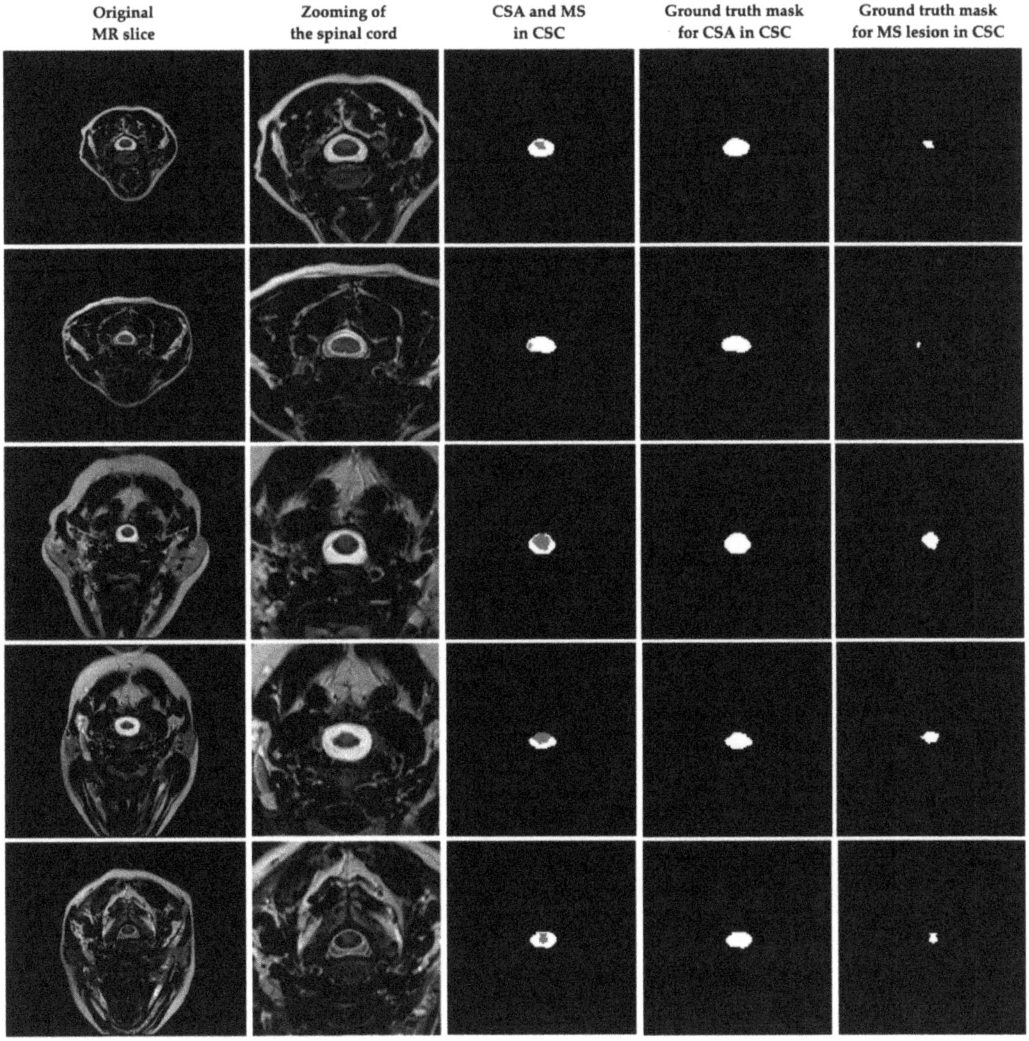

Figure 2. Some of the original MR images in the dataset used in the study and ground truth masks of CSA and MS lesions in CSC.

2.2. Proposed Methodology

Many deep learning models have been developed for medical imaging. Deep learning has revolutionized medical image analysis, especially in segmentation tasks where accurate delineation of anatomical structures and pathological regions is crucial. Among deep learning models, with the development of the convolutional U-Net architecture in medical segmentation studies, it has become a standard architecture for biomedical image segmentation due to its encoder-decoder structure and skip connections, which allow precise localization by preserving spatial information [5,41]. Due to its successful segmentation capability and easy model synthesis, various U-Net architectures have been developed to solve many problems [42]. However, conventional U-Net models can struggle with complex structures such as the CSC and MS lesions due to variations in shape, intensity, and contrast. To address these challenges, hybrid U-Net architectures have gained significance by integrating advanced mechanisms such as fractal designs, attention modules, and skip connections. These enhancements improve feature representation, enable better learning of fine-grained details, and increase robustness against variations in medical images. In this study, in addition to our previously proposed FractalSpiNet architecture [37], we propose two new models, such as Con-FractalU-Net and Att-FractalSpiNet, specifically designed for the automatic segmentation of the CSC and MS lesions in MRI scans.

2.2.1. U-Net

U-Net is a CNN-based deep learning architecture and a successful method developed specifically for image segmentation [41]. In the U-Net architecture, important features in the input image are extracted using convolutional layers, and then the class/labeling of each pixel is predicted using these features. The U-Net deep learning architecture consists of two parts, an encoder and a decoder, connected by a bottleneck [43]. In the encoder stage, the image is reduced to more abstract and low-dimensional features using convolutional layers, while in the decoder stage, the low-dimensional features obtained by inverse convolution are converted back to high resolution. In the decoder stage, the upward inverse convolution process performed on the contraction path in the contraction path increases the segmentation success by gradually increasing the sensitive feature data. In addition, thanks to the skip connections in the structure of the architecture, low-level features obtained from the encoder are directly transferred to the decoder to increase the segmentation accuracy [44]. U-Net enables the effective use of convolutional networks, especially in classification and segmentation. Although U-Net is based entirely on a convolutional network, it differs in that it has a U-shaped symmetric architecture, as shown in the typical U-Net architecture in Figure 3a, and uses skip connections between the encoder and decoder subnetworks to combine low-level and high-level features to preserve more refined image details.

The U-Net architecture performs the extraction of the specific region to be segmented on the images and provides superior segmentation performance with less data compared to other deep learning models with advanced feature selection. Although the basic U-Net model has a very strong performance, it still needs to be improved for challenging segmentation studies. In fact, due to the flexibility of the U-Net architecture to evolve into different models, hybrid models or new U-Net architectures with different layer connections can be developed by using the block structures of other deep learning models or the proposed convolution blocks. In this context, previous studies [33–35,45,46] have contributed to the segmentation studies of the spinal cord and spinal cord MS lesions by developing a different perspective on the U-Net architecture. Synthesis U-Net architecture designs can be categorized as skip connection, backbone design, bottleneck, transformers, rich representation, and probabilistic design [5]. In this study, based on the innovations of backbone design and skip connection, new hybrid architectures have been developed by integrating the fractal convolutional block connection instead of the existing convolutional block structure of U-Net. In particular, the use of the logic of the fractal convolutional connection structure in the U-Net architecture and the selection of the points that affect the segmentation success of the U-Net architecture can be shown as the innovative side of the designed architectures. In addition to the FractalSpiNet architecture proposed in our previous study, in this study, the Att-FractalSpiNet architecture was developed by integrating the attention mechanism into the FractalSpiNet architecture, and the Con-FractalU-Net architecture was developed by integrating the fractal convolutional block into the skip connection in the U-Net structure.

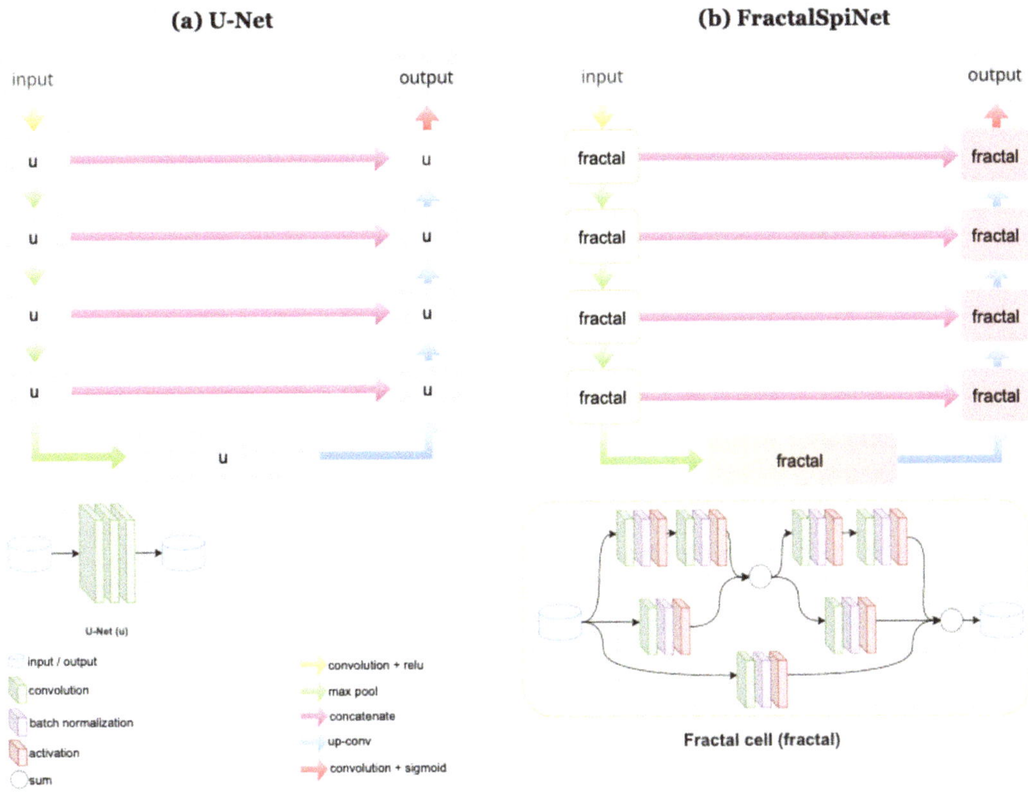

Figure 3. Details of the architectures used in the study and the components that comprise these architectures. (**a**) U-Net, (**b**) FractalSpiNet.

2.2.2. FractalSpiNet

FractalSpiNet is a derivative of the architecture presented in our previous work, designed by integrating fractal convolution into the backbone structure of the basic U-Net architecture for automatic detection and segmentation of spinal cord and MS lesions from MRI images [37]. The FractalSpiNet architecture goes beyond the traditional U-Net structure and increases the depth and capacity of the model by using fractal networks and multi-scale feature learning techniques. FractalSpiNet builds on the architecture of FractalNet [47] by structuring convolutional layers using a branching approach. In traditional deep learning models, network depth is increased by sequentially arranging layers, whereas in FractalSpiNet, the network structure is extended with fractal blocks containing multiple paths [37].

FractalSpiNet takes its basic structure from the U-Net architecture, as shown in Figure 3b. U-Net has an encoder-decoder structure and uses skip connections to avoid loss of detail. The difference between FractalSpiNet and U-Net is the use of fractal blocks in the encoder and decoder sections. Fractal blocks include standard convolutions, skip connections, and multi-scale attribute maps. Each block processes inputs with different path options and increases the network's deep learning capacity. In the encoder part of FractalSpiNet, input MRI images are passed through convolutional layers, and feature maps are extracted. Multi-scale features are obtained thanks to fractal blocks. In the decoder part of the architecture, upsampling is applied to produce segmentation maps. Skip connections prevent loss of detail and preserve fine texture structures. In the final layer of FractalSpiNet, a convolutional layer with a sigmoid activation function is used to mask the spinal cord and MS lesions. Details of the basic architecture and code of the FractalSpiNet network are publicly available [48].

2.2.3. Att-FractalSpiNet

In this study, in the first architecture designed for the detection and segmentation of CSC and MS lesions, the Att-FractalSpiNet model is developed by integrating the attention mechanism into the FractalSpiNet architecture, as shown in Figure 3b. The main purpose of using masks in image segmentation is to design a new layer that can determine the basic features of an image during network training and to eliminate unnecessary information by focusing on the most valuable pixel areas of the images during network training. Attention mechanisms are intermediate layer structures that determine different weights by using mask structures to determine the features of target segmentation areas in deep networks [49]. In addition, the attention mechanism allows for higher interactions and encoding of contextual information by extracting relationships between high level attributes [50]. This mechanism assigns weights to each pixel to indicate the importance of each pixel. In addition, the attention mechanism reduces computational cost by using only the relevant areas during training and provides better generalization of the network [44]. Furthermore, the attention structure can be easily integrated into standard convolutional architectures such as the U-Net model with minimal computational overhead while adding a significant boost to model sensitivity and prediction capability. With the attention connection, while the attribute information is transmitted in the U-Net model, the weak features and irrelevant regions transmitted from the data are ignored thanks to the attention block added in between, allowing the network to focus only on the area to be segmented [51].

The detailed architecture of Att-FractalSpiNet, developed for automatic segmentation of the CSC and cervical MS lesions, is shown in Figure 4a. Att-FractalSpiNet is based on the FractalSpiNet architecture and is an extended version of the traditional U-Net architecture with a fractal convolutional network and attention mechanism. While this model is based on the encoder-decoder structure of U-Net, it enhances multiscale information learning with fractal convolutions and improves segmentation accuracy with the attention mechanism. The fractal structures allow the model to learn features at different scales without increasing the depth of the model, while the attention mechanism helps to better detect small and complex lesions. In the encoder stage of the Att-FractalSpiNet model architecture, feature extraction is performed on the input MRI image using 3×3 convolution, batch normalization, and ReLU activation. Fractal convolution blocks minimize the loss of detail by processing information at different scales in parallel. High level representations are produced by minimizing the spatial dimension with maximum pooling. In the decoder stage, upsampling operations restore the segmentation map to its original size. The learning capacity of the model is strengthened by skip connections, which preserve the low-level information coming from the encoder. At this stage, the attention gate is activated to ensure that the model only focuses on important regions. The attention mechanism creates a weight map by comparing the features extracted by the encoder and the decoder information. By filtering out noisy or redundant information, the model improves segmentation accuracy and can determine more precise boundaries.

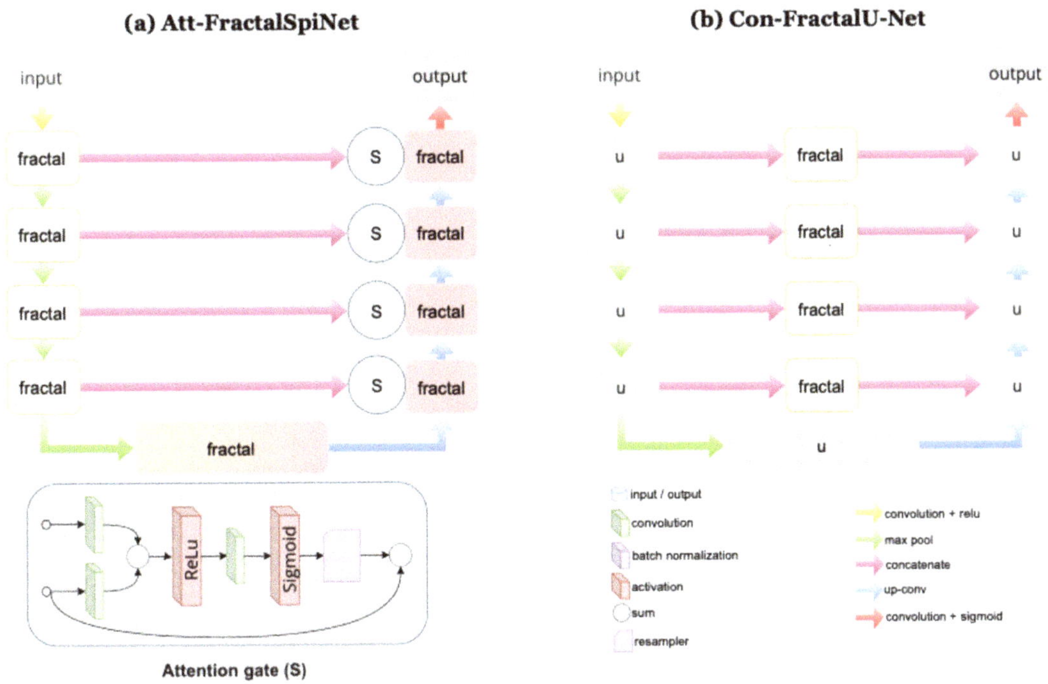

Figure 4. (**a**) Att-FractalSpiNet architecture developed for automatic segmentation of the CSC and cervical spinal MS lesions. This architecture is based on the FractalSpiNet architecture, which is an extended version of the U-Net architecture with a fractal convolutional network and an attention mechanism. (**b**) Detailed infrastructure of the skip connections-based Con-FractalU-Net architecture developed for segmentation of CSC and MS lesions. The Con-FractalU-Net model proposes a more efficient feature learning mechanism with fractal convolution blocks and advanced skip connections.

2.2.4. Con-FractalU-Net

In Con-FractalU-Net, the second architecture developed for cervical spine and MS lesion detection and segmentation, a fractal convolution block is integrated into the skip connection in the U-Net structure, as shown in Figure 3a. Skip connections were first proposed in U-Net to address the problem of model performance degradation with increasing depth of the architecture [52]. Skip connections can be expressed as using connections from certain convolution outputs of the model as inputs to different points of the model instead of sequential connections. Skip connections are used to transfer the convolution outputs to the opposite layers before the pooling process, while the downward convolution process takes place. In this connection, the location information of the features coming from the encoder section and the high feature information in the decoder section are combined to produce an output. In contrast to the successive connection structure used in deep networks, skip connections have been used in image segmentation studies [53–55]. It is known that the architecture with skip connections has a better generalization ability than the architecture without skip connections [56]. In fact, the main structure that distinguishes the U-Net architecture from other sequential processing deep networks is that it has skip connections. It has been observed that U-Net-based hybrid architectures with skip connections achieve similar results to the basic U-Net architecture in pixel-based segmentation tasks [57–61].

Figure 4b shows the detailed infrastructure of the Con-FractalU-Net architecture developed for segmentation of CSC and MS lesions. Based on the traditional U-Net structure, this model provides a more efficient feature learning mechanism with fractal convolution blocks and advanced skip connections. Fractal convolution blocks increase the capacity of multi-scale information processing to produce more detailed feature maps, while extended skip connections minimize the loss of detail by improving the flow of information between the encoder and decoder. This enables more accurate segmentation, particularly of small and complex MS lesions. In the encoder stage of the Con-FractalU-Net architecture, the input MRI image is passed through several convolutional layers to extract detailed features. At each level, deep and multi-scale feature learning is performed using fractal structures. The maximum pooling process reduces the image size to create more abstract and meaningful representations. This process allows better recognition of the complex tissue and lesion structures of the model. In the decoder stage, the image size is increased again by upsampling, and segmentation output is generated. The main innovation of Con-FractalU-Net is the use of extended and optimized skip connections, in contrast to conventional U-Net.

3. Results

In this study, experimental studies were carried out with two new proposed architectures, Con-FractalU-Net and Att-FractalSpiNet, for automatic segmentation of the cross-sectional area (CSA) of the CSC area and detection of MS lesions in the CSC. Furthermore, the results obtained with these architectures are compared with the results of the basic U-Net architecture and our proposed FractalSpiNet architecture. The experimental studies are also carried out using the T2-w MRI dataset [39], which was created for our previous work [37] and is publicly available. The experimental studies for automatic segmentation of the CSC and detection of MS lesions in the CSC were performed using the workstation computer whose specifications are given in Table 1. The Con-FractalU-Net and Att-FractalSpiNet deep learning architectures proposed in this study and the implementations of the basic U-Net and FractalSpiNet architectures were carried out in the Jupyter Notebook IDE environment (v. 7.1.2) on TensorFlow (v. 2.6.0) using the Python programming language (v. 3.6.13).

Table 1. Workstation computer and its specifications used in experimental studies for automatic segmentation of the CSC and detection of MS lesions in the CSC.

Hardware/Component	Specification
Computer	Workstation
Mainboard	Gigabyte B560M
Central processor (CPU)	Intel Core i5 4.10 GHz
Memory (RAM)	16 GB DDR4 3000 MHz
Graphical processing unit (GPU)	NVIDIA RTX A4000 16 GB
Disk drivers	1TB HDD + 500 GB SSD

The dataset used in this study contains a total of 231 axial MR slices obtained from MS patients suitable for experimental studies. As in our previous study [37], the same image pre-processing procedures were applied to the MR images in the database. Although the number of images in the dataset is small, the CSA region and MS lesions in the MRI slices are quite unique in terms of location and shape, which can be considered a positive situation in the data augmentation process. For better learning of the proposed networks and to avoid the overfitting effect, this number was increased to a total of 1080 using data augmentation techniques. As data augmentation techniques, the image set was augmented by using rotation (on x and y axes), flipping, shifting, and same functions, which are based on geometric transformation without disturbing the pixel structure. Data augmentation techniques were applied to both the MRI slices and the ground truth masks in the dataset. In addition, data augmentation was performed in the Python environment using the NumPy library (v. 1.19.5).

After the data augmentation process, 80% (864) of the total 1080 MRI images in the dataset were used for training the U-Net, FractalSpiNet, Att-FractalSpiNet, and Con-FractalU-Net architectures, and the remaining 20% (216) were used for testing. Some of the images in the training set were used for validation. The progression of training loss, validation loss, training accuracy (Training Acc), and validation accuracy (Validation Acc) values obtained as a result of training the proposed U-Net, FractalSpiNet, Att-FractalSpiNet, and Con-FractalU-Net architectures for CSC segmentation and detection of MS lesions along the spinal cord using axial MRI images in the dataset over 200 epochs are shown in Figure 5a, Figure 5b, Figure 5c, and Figure 5d, respectively. The plots of these values provide important information about the performance, generalizability, and potential problems of the model.

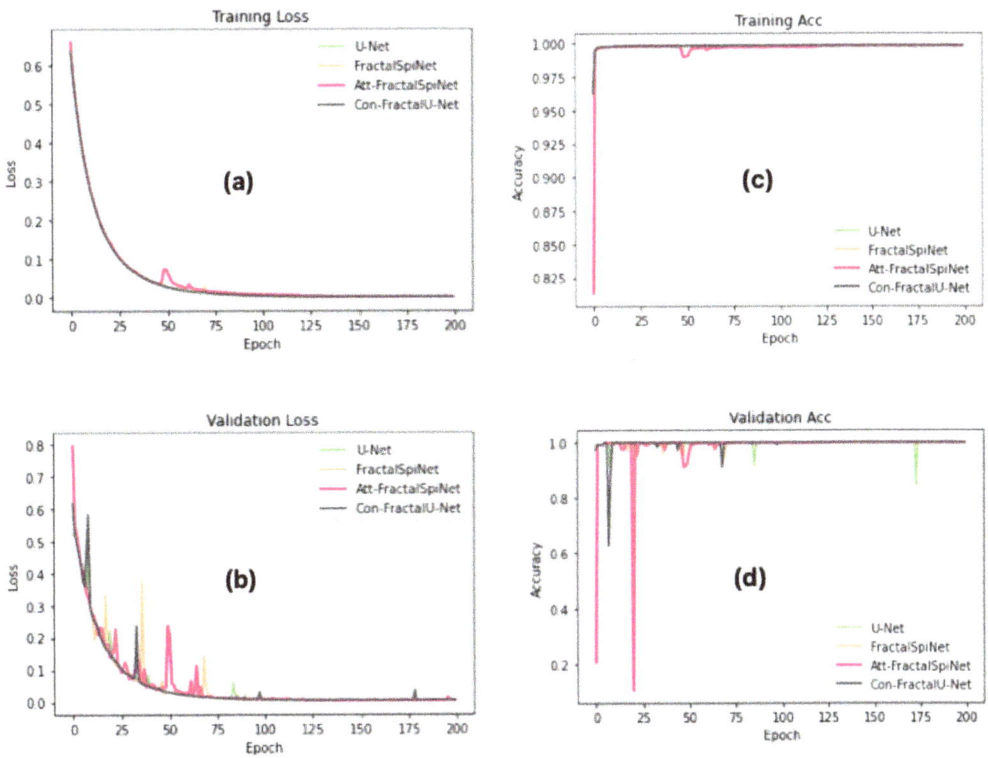

Figure 5. Training and validation performance of U-Net, FractalSpiNet, Att-FractalSpiNet, and Con-FractalU-Net for CSC and MS lesion segmentation for 200 epochs. (**a**) training loss, (**b**) validation loss, (**c**) training accuracy, and (**d**) validation accuracy.

In Figure 5a, the training loss of all models decreases steadily as the number of epochs increases, indicating effective learning. The initial rapid decline in loss suggests that the models adapt quickly to the dataset, while the later stabilization implies convergence. Among the models, FractalSpiNet and Att-FractalSpiNet show slightly higher initial losses compared to U-Net and Con-FractalU-Net, likely due to their increased architectural complexity. Figure 5b shows the validation loss, which exhibits more fluctuations compared to training loss. The fractal-based architectures generally demonstrate lower validation loss over time, suggesting improved generalization compared to the baseline U-Net. For the training accuracy in Figure 5c, all architectures reach high accuracy values that converge close to 1.0, which indicates that they fit well to the training data. The overall consistency between the models suggests that all architectures effectively learn the segmentation task on the training dataset. In Figure 5d, the validation accuracy is plotted. While U-Net exhibits some fluctuations, FractalSpiNet, Con-FractalU-Net, and Att-FractalSpiNet achieve higher and more stable validation accuracy throughout the training process. Overall, the results suggest that while U-Net provides a stable baseline, the proposed fractal-based architectures enhance the segmentation capabilities by improving generalization performance. These findings confirm the effectiveness of incorporating fractal structures and attention mechanisms in deep learning-based segmentation of CSC and MS lesions.

This study also compares different segmentation architectures in terms of training time and model complexity. The baseline U-Net model, with 31.4 million parameters, demonstrated the shortest training time of 28 min and 37 s. FractalSpiNet, which incorporates fractal-based structures, significantly increased the number of parameters to approximately 109.9 million, resulting in a longer training time of 91 min and 18 s. Among the proposed hybrid architectures, Att-FractalSpiNet, which integrates an attention mechanism with fractal structures, had the highest parameter count (115.8 million) and the longest training time (99 min and 52 s), reflecting the computational cost of the attention modules. In contrast, Con-FractalU-Net, designed with enhanced skip connections within a fractal framework, maintained a more balanced trade-off with 53.3 million parameters and a training time of 60 min and 5 s. These results highlight the impact of architectural modifications on computational efficiency, showing that while attention-based enhancements contribute to improved segmentation, they require higher computational resources.

To ensure a fair and consistent evaluation of the U-Net, FractalSpiNet, Con-FractalU-Net, and Att-FractalSpiNet architectures, all models were trained using the same set of hyperparameters. These hyperparameters were carefully chosen to ensure robust training while maintaining computational efficiency, making them suitable for evaluating the segmentation performance of each proposed architecture under standardized conditions. The training process was performed over 200 epochs to allow sufficient learning and convergence while mitigating the risk of underfitting. A batch size of 8 was chosen to balance memory efficiency and gradient update stability. The learning rate was set to 0.001, a value commonly used in deep learning segmentation tasks, to ensure steady convergence without drastic fluctuations in weight updates. A dropout rate of 0.5 was employed to prevent overfitting by randomly deactivating neurons during training, thereby enhancing the model's generalization ability. For activation functions, ReLU was chosen as the primary nonlinearity because it efficiently mitigates the vanishing gradient problem, allowing deeper networks to learn effectively. At the output layer, the sigmoid activation function was used, as the segmentation task is formulated as a binary classification problem at the pixel level. The Adam optimization algorithm was utilized due to its adaptive learning rate properties, which facilitate faster and more stable convergence compared to traditional methods such as stochastic gradient descent (SGD). The binary cross-entropy loss function was employed, aligning with the binary nature of the segmentation task, ensuring proper gradient updates for foreground and background pixel classification.

In the experimental studies, the metrics in a previous study [37] were used to evaluate the performance of CSC segmentation and the detection of MS lesions. The evaluation was based on pixel overlap, volume difference, and geometric distance measurements. To measure segmentation accuracy, the Dice Similarity Coefficient (DSC) in Equation (1) was used. DSC evaluates the spatial overlap between the predicted mask (PM) and the ground truth mask (GT), where higher values indicate better segmentation performance. For volume-based evaluation, the Volume Overlap Error (VOE) in Equation (2) and Relative Volume Difference (RVD) in Equation (3) metrics were employed. VOE quantifies the proportion of segmentation errors, while RVD represents the percentage difference between the predicted and actual segmentation volumes [62,63]. To assess geometric accuracy, the Average Surface Distance (ASD) in Equation (4), Hausdorff Distance (HD) in Equations (5) and (6), and Hausdorff 95 (HD95) in Equation (7) were utilized. ASD measures the accuracy of segmentation boundaries, while HD calculates the maximum point-wise distance between one segmentation and another. HD95 refines this measurement by considering the 95th percentile, reducing the impact of outliers. For lesion detection performance, recall (REC) in Equation (8) and precision (PRE) in Equation (9) were used. REC measures the proportion of actual lesion pixels correctly identified, whereas PRE evaluates the correctly detected lesions while minimizing false positives.

$$\mathrm{DSC}(\mathrm{PM},\mathrm{GT}) = \frac{2|\mathrm{PM}\cap\mathrm{GT}|}{|\mathrm{PM}|\cup|\mathrm{GT}|}\times 100 \quad (1)$$

$$\mathrm{VOE}(\mathrm{PM},\mathrm{GT}) = (1-\frac{|\mathrm{PM}\cap\mathrm{GT}|}{|\mathrm{PM}|+|\mathrm{GT}|-|\mathrm{PM}\cup\mathrm{GT}|})\times 100 \quad (2)$$

$$\mathrm{RVD}(\mathrm{PM},\mathrm{GT}) = (\frac{|\mathrm{PM}|-|\mathrm{GT}|}{|\mathrm{GT}|})\times 100 \quad (3)$$

$$\mathrm{ASD}(\mathrm{PM},\mathrm{GT}) = \frac{1}{|s(\mathrm{PM})|+|s(\mathrm{GT})|}\begin{pmatrix}\sum_{S_{\mathrm{PM}}\in S(PM)} d(\mathrm{SPM},\mathrm{S}(\mathrm{GT}))+ \\ \sum_{S_{\mathrm{GM}}\in S(GT)} d(\mathrm{SGM},\mathrm{S}(\mathrm{PM}))\end{pmatrix} \quad (4)$$

$$\mathrm{hd}(\mathrm{PM},\mathrm{GT}) = \max_{\mathrm{x}\in\mathrm{PM}}\min_{\mathrm{y}\in\mathrm{GM}}\|\mathrm{x}-\mathrm{y}\|_2 \quad (5)$$

$$\mathrm{hd}(\mathrm{GT},\mathrm{PM}) = \max_{\mathrm{y}\in\mathrm{GM}}\min_{\mathrm{x}\in\mathrm{PM}}\|\mathrm{x}-\mathrm{y}\|_2 \quad (6)$$

$$\mathrm{HD95}(\mathrm{PM},\mathrm{GT}) = \max(\mathrm{hd}(\mathrm{PM},\mathrm{GT}),\mathrm{hd}(\mathrm{GT},\mathrm{PM})) \quad (7)$$

$$\mathrm{REC}(\mathrm{PM},\mathrm{GT}) = \frac{\mathrm{TP}}{\mathrm{TP}+\mathrm{FN}}\times 100 \quad (8)$$

$$\mathrm{PRE}(\mathrm{PM},\mathrm{GT}) = \frac{\mathrm{TP}}{\mathrm{TP}+\mathrm{FP}}\times 100 \quad (9)$$

In this study, four deep learning architectures—U-Net, FractalSpiNet, Att-FractalSpiNet, and Con-FractalU-Net—were trained under identical conditions using axial-plane CSC MRI images to segment two key regions: the cross-sectional area (CSA) of the spinal cord and MS lesions within this area. The models were trained for 200 epochs, and their performance was evaluated using several segmentation metrics, as shown in Table 2. Among these architectures, Con-FractalU-Net achieved the highest DSC of 98.89%, VOE of 2.05%, and the lowest ASD of 1.09 mm, making it the most accurate model for segmentation. Additionally, its PRE of 99.21% indicates a strong ability to minimize false positives. FractalSpiNet also demonstrated strong performance with a DSC of 98.88%, the lowest VOE (2.04%), and ASD (1.38 mm), while maintaining a high REC of 98.84%, suggesting robust lesion detection capability. On the other hand, Att-FractalSpiNet, with a DSC of 98.41%, exhibited higher segmentation boundary errors, as indicated by its ASD (2.73 mm) and HD95 of 0.80 mm, suggesting that the attention mechanism introduced more variability. The baseline U-Net, while achieving a DSC of 98.54%, lagged behind in all key metrics, reaffirming the superiority of fractal-based architectures. Overall, Con-FractalU-Net emerges as the most effective model, demonstrating superior segmentation accuracy and precision, while FractalSpiNet remains a strong alternative with competitive performance. These findings highlight the advantages of fractal-based networks in enhancing segmentation robustness and accuracy for the CSC MRI dataset.

Table 2. Performance comparison of deep learning architectures such as U-Net, FractalSpiNet, Att-FractalSpiNet, and Con-FractalU-Net for CSA segmentation of CSC using MRI scans.

Architectures	DSC (%)	VOE (%)	RVD (%)	ASD [mm]	HD95 [mm]	REC (%)	PRE (%)
U-Net	98.54	2.67	1.51	1.67	0.49	98.43	98.69
FractalSpiNet	98.88	2.04	0.97	1.38	0.39	98.84	98.94
Att-FractalSpiNet	98.41	2.84	1.57	2.73	0.80	98.75	98.11
Con-FractalUNet	98.89	2.05	1.18	1.09	0.51	98.62	99.21

In this study, we compare the performance of the deep learning architectures U-Net, FractalSpiNet, Att-FractalSpiNet, and Con-FractalU-Net for segmentation of CSC MR images. Figure 6 shows the segmentation results for only a part of the test dataset but provides important information about the general trends of the different architectures. In the figures, the segmentation success of each model is evaluated using DSC scores. Con-FractalU-Net demonstrated excellent segmentation performance, reaching 100% for DSC in all test images shown. This shows that the model is able to recognize both spinal cord cross-sectional area (CSA) and MS lesions with very high accuracy and successfully generalize the features learned during training. Similarly, the FractalSpiNet model also achieved a score of 100% within DSC in most cases but was slightly below this value in some images. Although Att-FractalSpiNet includes an attention mechanism to improve segmentation performance, DSC $\approx$ 98–99% in some test images. This result suggests that while the attention mechanism may be advantageous in certain situations, it may not be sufficient to achieve perfect segmentation in some cases. U-Net was the model with the lowest performance compared to the other architectures. In some images, the DSC value fell below 98% and showed lower accuracy compared to other models, especially in complex boundary regions. This indicates the limitations of the typical U-Net architecture for CSC segmentation, and fractal-based models, which are more advanced structures, appear to be more successful. Together with the full analysis of all test images, Con-FractalU-Net provides the most stable and highly accurate segmentation model. It has been observed that fractal-based networks are more successful than classical CNN-based models and increase segmentation accuracy, especially in medical imaging applications where high precision is required.

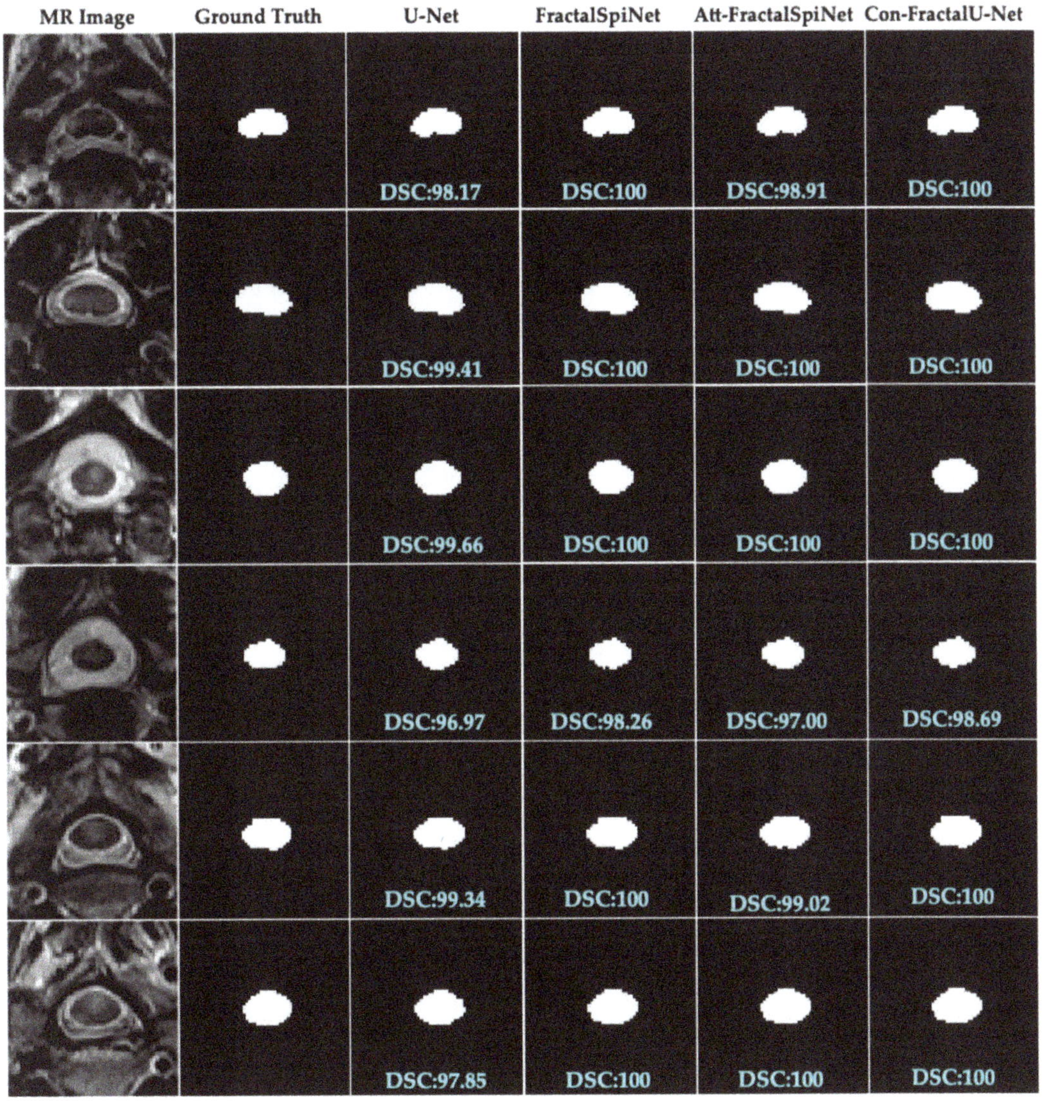

Figure 6. Comparison of CSC cross-sectional segmentation results obtained using U-Net, FractalSpiNet, Att-FractalSpiNet, and Con-FractalU-Net models.

Segmentation of MS lesions is a more challenging task than determining the cross-sectional area of the spinal cord, and accurate detection of lesions is a significant challenge due to their small volume and variable morphology. While the cross-sectional area of the spinal cord is already a very small pixel area, MS lesions have an even smaller and more challenging volumetric structure within this small area. Table 3 shows the segmentation performance of the U-Net, FractalSpiNet, Att-FractalSpiNet, and Con-FractalU-Net models on MS lesions. In terms of the DSC metric, the Con-FractalU-Net model has the highest success rate with 91.48%, followed by FractalSpiNet with 90.90%. In particular, the Con-FractalU-Net model provided the most accurate segmentation with low ASD and HD95 values, allowing better delineation of MS lesion boundaries. On the other hand, the U-Net model showed the lowest segmentation performance with a DSC value of 86.00% and high values, especially for the ASD and HD95 metrics, indicating that the model was less successful in identifying lesion boundaries compared to other models. While the Att-FractalSpiNet model performed competitively with a DSC value of 88.79%, it has a higher value in the ASD metric compared to the other models, indicating that segmentation performance may be lower in some cases. The Con-FractalU-Net model showed the best performance in terms of the REC metric, while this model also showed the best performance in terms of PRE. In conclusion, the Con-FractalU-Net model stands out as the most successful method for segmentation of MS lesions in terms of general metrics. The FractalSpiNet model also shows competitive performance with high DSC and low error rates. U-Net, on the other hand, performed poorly compared to the other models in terms of segmentation accuracy. These results suggest that Con-FractalU-Net is a more reliable model for MS lesion segmentation.

Table 3. Performance comparison of U-Net, FractalSpiNet, Att-FractalSpiNet, and Con-FractalU-Net architectures for MS lesion segmentation in CSC using MRI scans.

Architectures	DSC (%)	VOE (%)	RVD (%)	ASD [mm]	HD95 [mm]	REC (%)	PRE (%)
U-Net	86.00	20.83	13.50	28.23	11.55	83.73	90.50
FractalSpiNet	90.90	14.06	9.62	16.08	8.06	91.26	92.20
Att-FractalSpiNet	88.79	17.17	11.77	35.81	11.59	89.33	89.79
Con-FractalU-Net	91.48	12.92	9.93	20.84	7.27	92.13	92.27

In this study, the performance of different deep learning architectures for the automatic segmentation of MS lesions in the CSC was compared, as shown in Figure 7. The segmentation results obtained using U-Net, FractalSpiNet, FractalSpiNet with an attention mechanism (Att-FractalSpiNet), and FractalU-Net with convolutional blocks (Con-FractalU-Net) architectures were evaluated using the DSC metric. The results revealed that the typical U-Net architecture exhibited relatively lower performance in segmenting MS lesions, while FractalSpiNet and especially Att-FractalSpiNet and Con-FractalU-Net architectures had a significantly better performance. By producing consistent and accurate segmentation results with high DSC values, fractal-based architectures emerge as more promising approaches for automated analysis of MS lesions. Notably, the integration of attention mechanisms and convolutional blocks significantly improved the model's ability to segment and detect lesions, achieving DSC values exceeding 98% and even reaching 100%. The findings of this study strongly support that advanced architectures, in particular Att-FractalSpiNet and Con-FractalU-Net, can deliver significant performance gains in medical imaging, especially for complex and detail-demanding tasks. The significant improvement achieved compared to the standard U-Net architecture reveals the potential of these advanced architectures to automatically and accurately segment challenging structures such as MS lesions in the CSC.

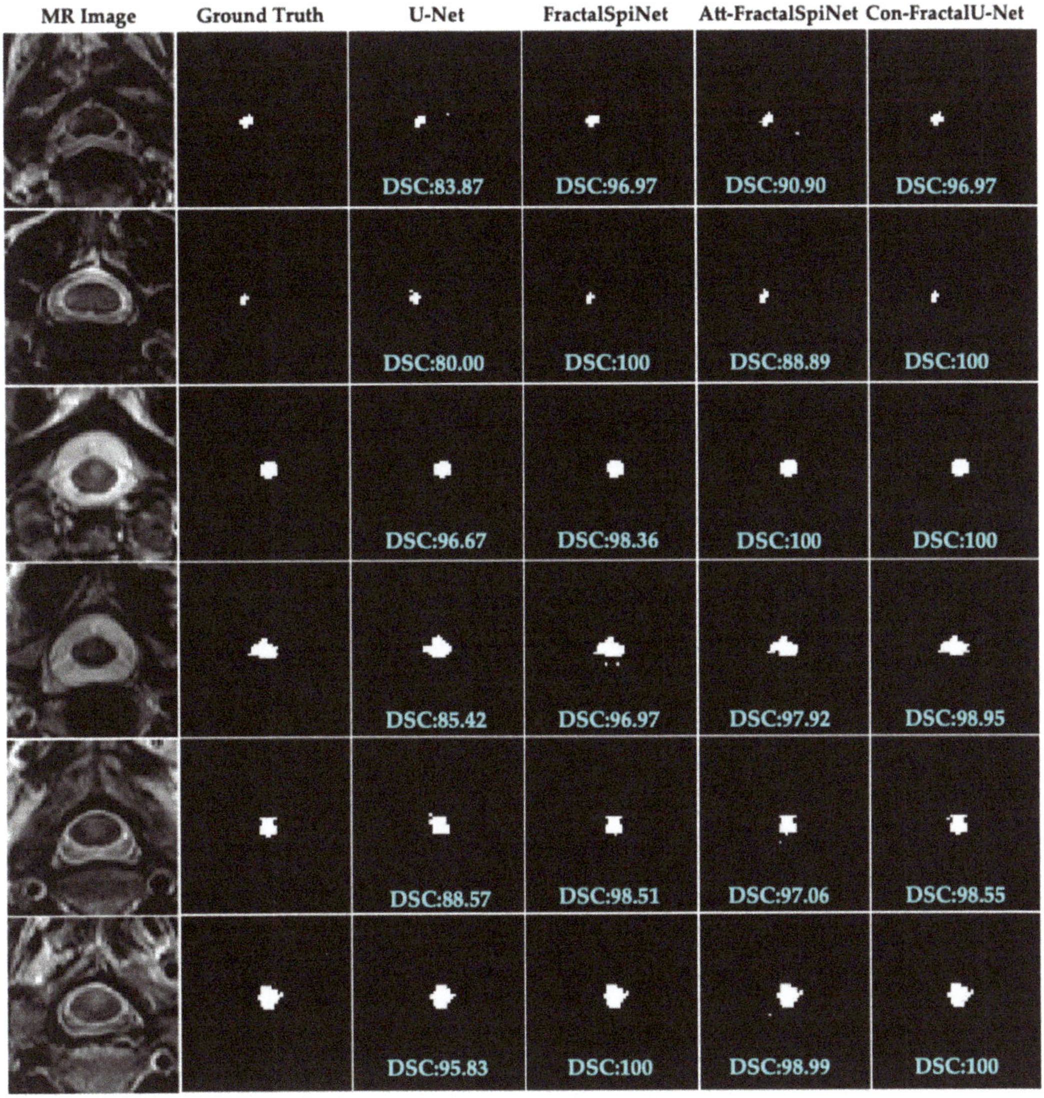

Figure 7. Performance comparison and DSC scores of U-Net, FractalSpiNet, Att-FractalSpiNet, and Con-FractalU-Net architectures in segmentation of MS lesions in CSC.

4. Discussion

This study presents a comparative analysis of advanced segmentation architectures for the detection of MS lesions in the CSC and segmentation of the CSA region. Our previous work introduced FractalSpiNet [37] as an effective architecture, demonstrating high segmentation performance. Building on this foundation, we propose two novel architectures, Con-FractalU-Net and Att-FractalSpiNet, which aim to further improve segmentation performance by incorporating enhanced skip connections and attention mechanisms, respectively. Experimental results confirm that Con-FractalU-Net achieves the highest segmentation accuracy across both tasks. Specifically, for MS lesion segmentation, Con-FractalU-Net achieved the best DSC score (91.48%), outperforming FractalSpiNet

(90.90%), Att-FractalSpiNet (88.79%), and the baseline U-Net (86.00%). The improved connectivity introduced in Con-FractalU-Net is likely to contribute to its superior performance, ensuring better feature propagation and refinement. Similarly, in CSA segmentation, Con-FractalU-Net showed improved accuracy over all other architectures, reinforcing its robustness across different segmentation tasks.

A key aspect of this study is the comparison of training efficiency and computational complexity between different architectures. The baseline U-Net, with 31.4 million parameters, had the shortest training time (28 min and 37 s). FractalSpiNet, which introduced fractal-based structures, significantly increased the number of parameters to approximately 109.9 million, leading to an extended training time of 91 min and 18 s. Att-FractalSpiNet, which integrates attention mechanisms into the fractal framework, had the highest number of parameters (115.8 million) and the longest training time (99 min and 52 s), reflecting the additional computational cost of the attention modules. In contrast, Con-FractalU-Net maintained a more balanced trade-off between accuracy and computational efficiency, with 53.3 million parameters and a training time of 60 min and 5 s. This demonstrates that while attention-based enhancements improve segmentation quality, they require significantly more computational resources, making Con-FractalU-Net a preferable option in scenarios where both accuracy and efficiency are crucial. Analyses of time performance in the test set reveal that deep learning architectures generally exhibit high efficiency in the process of detecting MS lesions in CSC. When evaluating the entire test set of 216 MR images, the overall detection times for all methods were obtained to be within a close range. FractalSpiNet emerged as the fastest method with a minimally different total test set processing time of 44.42 s and an average detection time of 0.205 s per image. U-Net was recorded as the slowest, with a total time of 45.51 s and an average time of 0.211 s. The Con-FractalU-Net and Att-FractalSpiNet architectures, on the other hand, showed similar performance in the mid-range with total times of 44.99 s and 44.92 s, respectively, and an average time of 0.208 s for both. The fact that the average detection times for a single image are less than a quarter of a second for all methods and that the total processing times for the test set are around 45 s demonstrates that all architectures are sufficiently time-efficient for practical clinical applications. When these results are evaluated together with previous accuracy analyses, they support that the developed fractal-based architectures have a strong potential for clinical use by offering both high accuracy and efficient processing times in the automatic segmentation of MS lesions.

The results also indicate that MS lesion segmentation is inherently more challenging than CSA segmentation due to the smaller size and irregular distribution of lesions. Despite this complexity, the proposed architectures, particularly Con-FractalU-Net, successfully improved segmentation performance compared to the baseline U-Net and previously developed FractalSpiNet [37]. The ability to effectively segment both MS lesions and CSA highlights the adaptability and robustness of the proposed fractal-based architectures. For segmentation of cervical spinal cord and spinal cord MS lesions, the Con-FractalU-Net architecture proposed in this study is slightly more successful than the Att-FractalSpiNet architecture in terms of the DSC metric. To evaluate the effect of the Att-FractalSpiNet architecture proposed in this study and the underlying attentional mechanism, it is necessary to review the results of our previous study [37]. For CSA segmentation in CSC, on the same dataset in our previous study [37], 98.01% and 97.90% DSC scores were achieved using the Att U-Net and Att-Res U-Net architectures based on the attention mechanism and residual, respectively, while 98.41% DSC scores were achieved using the Att-FractalSpiNet architecture in this study. On the other hand, although 75.34% and 83.06% DSC scores were achieved using Att U-Net and Att-Res U-Net architectures to detect MS lesions in the cervical spinal cord, respectively, the detection of MS lesions using the Att-FractalSpiNet architecture proposed in this study was achieved with a DSC score of 88.79%. Thus, the fact that the attention mechanism integrated into the fractal structure in the proposed architectures achieves higher scores than the residual structure shows the effect and effectiveness of the attention mechanism on the architectures.

The proposed Con-FractalU-Net and Att-FractalSpiNet models in this study demonstrate significant improvements in CSA segmentation and MS lesion detection compared to state-of-the-art methods in previous studies. For spinal cord segmentation, the PropSeg method introduced by De Leener et al. [19] achieved a DSC of 91.0% for spinal cord and spinal canal segmentation, while the U-Net-based segmentation by Bedard et al. [64] improved the DSC score to 96.0%. In addition, McCoy et al. [23] obtained 93.0% DSC for segmentation of the spinal cord using 2D CNN architecture. For CSA segmentation, the OPAL algorithm and STEPS segmentation process by Prados et al. [65] achieved 96.5% DSC for CSA segmentation with visible lesions and 97.0% DSC without visible lesions. The U-Net-based model from Zhang et al. [26] achieved 87.0% DSC, while the channel-attentive U-Net (SeUneter) by Zhang et al. [35] reached 90.67% DSC. In comparison, the proposed Att-FractalSpiNet (98.41% DSC) and Con-FractalU-Net (98.89% DSC) further enhanced segmentation results, outperforming previous approaches. For MS lesion detection, different models in the literature show varying performance. The CNN (DeepSeg) model from Gros et al. [22] obtained 60.4% DSC, while the MultiResUNet model by Zhuo et al. [45] achieved 50.0% DSC for MS lesion segmentation. The residual attention-aware U-Net from Bueno et al. [36] demonstrated 90.4 $\pm$ 0.101% DSC for CSC segmentation. On the other hand, Karthik et al. [30] achieved a DSC score of 72.0% in automated segmentation of MS lesions in the spinal cord. The proposed models further improved these results, with Att-FractalSpiNet reaching 88.79% DSC and Con-FractalU-Net achieving 91.48% DSC, making them the most effective solutions for MS lesion segmentation. Compared to other advanced approaches, the FractalSpiNet model by Polattimur et al. [37], our previous study, achieved 98.88% DSC for CSA segmentation and 90.90% DSC for MS lesion detection. While this model demonstrated strong performance, the proposed Con-FractalU-Net (98.89% DSC for CSA, 91.48% DSC for MS) further enhanced segmentation accuracy, establishing them among the best-performing models in the previous. Additionally, the 2D and 3D CNN-based model by Naga Karthik et al. [30] achieved 72.0% DSC for MS lesion detection, which was significantly outperformed by the proposed models. Overall, the proposed Con-FractalU-Net and Att-FractalSpiNet models achieve the highest accuracy in CSA segmentation and MS lesion detection, positioning them as state-of-the-art methods. These findings demonstrate that integrating fractal-based architectures, attention mechanisms, and skip connections leads to substantial improvements in CSC segmentation and MS lesion detection, surpassing existing approaches. Statistical significance tests were also performed using the Wilcoxon signed-rank test to compare the results of Con-FractalU-Net, which obtained the highest DSC score for automatic segmentation of the cervical spinal cord, with FractalSpiNet, Att-FractalSpiNet, and the other studies mentioned above. In the automatic segmentation of the spinal cord, a p-value of 0.0312 was obtained in the Wilcoxon signed-rank test comparing Con-FractalU-Net with other methods in terms of statistical significance. Similarly, the p-value of 0.0039 was observed in the comparison of the proposed Con-FractalU-Net, which is the most successful method compared to the other methods in terms of statistical significance in the detection and segmentation of MS lesions in the cervical spinal cord. Since the p-value is less than 0.05 in both evaluations, it is confirmed that the results achieved with Con-FractalU-Net are statistically significant.

5. Conclusions

Early and accurate detection of MS lesions in the CSC is critical for patient care. Manual segmentation is a challenging and error-prone process, even for experts, as lesions can be very small and subtle in spinal cord MR images. In this context, deep learning architectures offer a promising alternative by potentially increasing diagnostic accuracy and efficiency for radiologists, reducing manual segmentation time, and enhancing diagnostic confidence. The segmentation of MS lesions in the CSC presents unique challenges due to the complex anatomical structure of the spinal cord, variations in lesion morphology, and limitations in MR imaging quality. The spinal cord does not possess a uniform geometric shape, and its boundaries change dynamically along its length, making accurate segmentation particularly difficult. In addition, MS lesions exhibit significant heterogeneity in size, shape, and location, which adds to the diversity of the dataset but also adds complexity to the segmentation process.

In this study, we proposed two novel deep learning architectures, Con-FractalU-Net and Att-FractalSpiNet, to improve the segmentation of CSA and MS lesions in the CSC by leveraging fractal-based structures, skip connections, and attention mechanisms. These architectures were compared against the previously introduced FractalSpiNet and the baseline U-Net model. The results demonstrate that incorporating fractal elements improves segmentation performance by allowing multi-scale feature extraction, while the addition of attention modules further refines spatial awareness in lesion localization. Our findings indicate that Con-FractalU-Net achieved the highest overall performance across all evaluation metrics, with a DSC of 91.48%, outperforming the other architectures. Att-FractalSpiNet, although slightly lower in DSC (88.79%), showed robust precision and recall values, indicating its effectiveness in lesion identification. Beyond lesion segmentation, CSA segmentation was also evaluated, as it plays a crucial role in contextualizing MS lesion burden and progression. The results showed that Con-FractalU-Net and FractalSpiNet effectively segmented the CSA with high DSC values, demonstrating their ability to generalize well to spinal cord structures. This is particularly important for clinical applications, where accurate delineation of both CSA and lesions helps to monitor disease progression and treatment response.

The computational efficiency of the models was also evaluated. While the baseline U-Net had the shortest training time, the proposed architectures required additional computational resources. Con-FractalU-Net provided a more balanced trade-off, making it a computationally efficient alternative without compromising segmentation accuracy. From a clinical perspective, the improved accuracy of MS lesion segmentation in the CSC has significant implications. Precise segmentation allows for more reliable lesion volume quantification, which is crucial for assessing disease activity and treatment efficacy. Improved segmentation models, such as Con-FractalU-Net, can be integrated into radiological workflows to support automated MS lesion detection, reducing the variability associated with manual annotations and improving diagnostic consistency across clinicians.

Although the proposed architectures have shown promising results, several areas warrant further investigation to improve their applicability and robustness. While the fractal-based architectures improve segmentation accuracy, their computational cost remains a limiting factor. One of the critical challenges in this study is the relatively small size of the dataset, which may limit the generalizability and robustness of the proposed models. Although high segmentation performance was achieved with the fractal-based architectures, models trained on small datasets are prone to overfitting and may not perform consistently on unseen data from different institutions or imaging protocols. To address this limitation, future work will focus on the integration of transfer learning techniques, where pre-trained weights on large medical image datasets could be used to improve learning efficiency and generalization. In addition, cross-institutional validation using external datasets from other medical centers is planned to further assess the robustness and adaptability of the proposed models. These steps are essential to ensure clinical applicability in real-world settings and to confirm that segmentation performance is maintained across different imaging conditions and patient populations. Future work may also explore model pruning, quantization, and knowledge distillation techniques to reduce model complexity while maintaining performance. In addition, expanding the evaluation to larger and more diverse datasets, including multi-center clinical MRI scans, can help validate the models' robustness across different imaging protocols and scanner variations. The ultimate goal is to translate these models into clinical practice. In addition, future work may include prospective studies where automated segmentation results are validated against expert radiologists' annotations in a real-world clinical setting.

Author Contributions: Conceptualization: E.D., R.P. and M.S.Y.; methodology: E.D., R.P. and M.S.Y.; software: E.D., R.P. and M.S.Y.; validation: E.D., R.P. and M.S.Y.; formal analysis: E.D., R.P. and M.S.Y.; investigation: E.D., R.P. and M.S.Y.; data curation: E.D., R.P. and M.S.Y.; writing—original draft preparation: E.D. and R.P.; writing—review and editing: E.D. and R.P.; visualization: E.D., R.P. and M.S.Y.; supervision: E.D. All authors have read and agreed to the published version of the manuscript.

Funding: This research was funded by the Scientific Research Projects Coordinatorship of Bilecik Şeyh Edebali University, grant number 2021-01.BŞEÜ.03-02.

Institutional Review Board Statement: The study was conducted in accordance with the Declaration of Helsinki and approved by the Institutional Review Board according to Decision No. KAEK-644 of the Ethics Committee of Non-Interventional Clinical Research of Akdeniz University, dated 15 September 2021.

Informed Consent Statement: Informed consent was obtained from all subjects involved in the study.

Data Availability Statement: The raw data supporting the conclusions of this article is publicly available from Mendeley Data (https://data.mendeley.com/datasets/ydkrtmygjp/1, accessed on 9 March 2025).

Acknowledgments: We would like to thank Akdeniz University Hospital for sharing the dataset and all the patients who volunteered to participate. In addition, we would like to thank Bilecik Şeyh Edebali University Scientific Research Projects Coordination for supporting this study with Project Number 2021-01.BŞEÜ.03-02. Finally, we would like to express our gratitude to Utku Şenol and Süleyman Uluçay for their valuable contributions to the preparation and annotation of the dataset used in this study.

Conflicts of Interest: The authors declare no conflicts of interest.

References

1. Pinter, C.; Lasso, A.; Fichtinger, G. Polymorph segmentation representation for medical image computing. *Comput. Methods Programs Biomed.* **2019**, *171*, 19–26. [CrossRef] [PubMed]
2. Polman, C.H.; Reingold, S.C.; Banwell, B.; Clanet, M.; Cohen, J.A.; Filippi, M.; Fujihara, K.; Havrdova, E.; Hutchinson, M.; Kappos, L. Diagnostic criteria for multiple sclerosis: 2010 revisions to the McDonald criteria. *Ann. Neurol.* **2011**, *69*, 292–302. [CrossRef]
3. Siddique, N.; Paheding, S.; Elkin, C.P.; Devabhaktuni, V. U-Net and its variants for medical image segmentation: A review of theory and applications. *IEEE Access* **2021**, *9*, 82031–82057. [CrossRef]
4. Zeng, C.; Gu, L.; Liu, Z.; Zhao, S. Review of deep learning approaches for the segmentation of multiple sclerosis lesions on brain MRI. *Front. Neuroinformatics* **2020**, *14*, 610967. [CrossRef]
5. Azad, R.; Aghdam, E.K.; Rauland, A.; Jia, Y.; Avval, A.H.; Bozorgpour, A.; Karimijafarbigloo, S.; Cohen, J.P.; Adeli, E.; Merhof, D. Medical image segmentation review: The success of U-Net. *IEEE Trans. Pattern Anal. Mach. Intell.* **2024**, *46*, 10076–10095. [CrossRef] [PubMed]

6. Styner, M.; Lee, J.; Chin, B.; Chin, M.; Commowick, O.; Tran, H.; Markovic-Plese, S.; Jewells, V.; Warfield, S. 3D segmentation in the clinic: A grand challenge II: MS lesion segmentation. *MIDAS J.* **2008**, *2008*, 1–6. [CrossRef]
7. Peterson, L.K.; Fujinami, R.S. Inflammation, demyelination, neurodegeneration and neuroprotection in the pathogenesis of multiple sclerosis. *J. Neuroimmunol.* **2007**, *184*, 37–44. [CrossRef]
8. Weiner, H.L. Multiple sclerosis is an inflammatory T-cell–mediated autoimmune disease. *Arch. Neurol.* **2004**, *61*, 1613–1615. [CrossRef]
9. Kerbrat, A.; Gros, C.; Badji, A.; Bannier, E.; Galassi, F.; Combès, B.; Chouteau, R.; Labauge, P.; Ayrignac, X.; Carra-Dalliere, C. Multiple sclerosis lesions in motor tracts from brain to cervical cord: Spatial distribution and correlation with disability. *Brain* **2020**, *143*, 2089–2105. [CrossRef]
10. Keegan, B.M.; Absinta, M.; Cohen-Adad, J.; Flanagan, E.P.; Henry, R.G.; Klawiter, E.C.; Kolind, S.; Krieger, S.; Laule, C.; Lincoln, J.A. Spinal cord evaluation in multiple sclerosis: Clinical and radiological associations, present and future. *Brain Commun.* **2024**, *6*, fcae395. [CrossRef]
11. Weidauer, S.; Raab, P.; Hattingen, E. Diagnostic approach in multiple sclerosis with MRI: An update. *Clin. Imaging* **2021**, *78*, 276–285. [CrossRef] [PubMed]
12. Thompson, A.J.; Banwell, B.L.; Barkhof, F.; Carroll, W.M.; Coetzee, T.; Comi, G.; Correale, J.; Fazekas, F.; Filippi, M.; Freedman, M.S. Diagnosis of multiple sclerosis: 2017 revisions of the McDonald criteria. *Lancet Neurol.* **2018**, *17*, 162–173. [CrossRef]
13. Breckwoldt, M.O.; Gradl, J.; Hähnel, S.; Hielscher, T.; Wildemann, B.; Diem, R.; Platten, M.; Wick, W.; Heiland, S.; Bendszus, M. Increasing the sensitivity of MRI for the detection of multiple sclerosis lesions by long axial coverage of the spinal cord: A prospective study in 119 patients. *J. Neurol.* **2017**, *264*, 341–349. [CrossRef] [PubMed]
14. Bot, J.C.; Barkhof, F.; à Nijeholt, G.L.; Van Schaardenburg, D.; Voskuyl, A.E.; Ader, H.J.; Pijnenburg, J.A.; Polman, C.H.; Uitdehaag, B.M.; Vermeulen, E.G. Differentiation of multiple sclerosis from other inflammatory disorders and cerebrovascular disease: Value of spinal MR imaging. *Radiology* **2002**, *223*, 46–56. [CrossRef]
15. De Leener, B.; Taso, M.; Cohen-Adad, J.; Callot, V. Segmentation of the human spinal cord. *Magn. Reson. Mater. Phys. Biol. Med.* **2016**, *29*, 125–153. [CrossRef] [PubMed]
16. Zhang, Z.; Yu, C.; Zhang, H.; Gao, Z. Embedding tasks into the latent space: Cross-space consistency for multi-dimensional analysis in echocardiography. *IEEE Trans. Med. Imaging* **2024**, *43*, 2215–2228. [CrossRef]
17. Renard, F.; Guedria, S.; Palma, N.D.; Vuillerme, N. Variability and reproducibility in deep learning for medical image segmentation. *Sci. Rep.* **2020**, *10*, 13724. [CrossRef]
18. Tench, C.R.; Morgan, P.S.; Constantinescu, C.S. Measurement of cervical spinal cord cross-sectional area by MRI using edge detection and partial volume correction. *J. Magn. Reson. Imaging Off. J. Int. Soc. Magn. Reson. Med.* **2005**, *21*, 197–203. [CrossRef]
19. De Leener, B.; Cohen-Adad, J.; Kadoury, S. Automatic segmentation of the spinal cord and spinal canal coupled with vertebral labeling. *IEEE Trans. Med. Imaging* **2015**, *34*, 1705–1718. [CrossRef]
20. Prados, F.; Cardoso, M.J.; Yiannakas, M.C.; Hoy, L.R.; Tebaldi, E.; Kearney, H.; Liechti, M.D.; Miller, D.H.; Ciccarelli, O.; Wheeler-Kingshott, C.A.G. Fully automated grey and white matter spinal cord segmentation. *Sci. Rep.* **2016**, *6*, 36151. [CrossRef]
21. Perone, C.S.; Calabrese, E.; Cohen-Adad, J. Spinal cord gray matter segmentation using deep dilated convolutions. *Sci. Rep.* **2018**, *8*, 5966. [CrossRef] [PubMed]
22. Gros, C.; De Leener, B.; Badji, A.; Maranzano, J.; Eden, D.; Dupont, S.M.; Talbott, J.; Zhuoquiong, R.; Liu, Y.; Granberg, T. Automatic segmentation of the spinal cord and intramedullary multiple sclerosis lesions with convolutional neural networks. *Neuroimage* **2019**, *184*, 901–915. [CrossRef]
23. McCoy, D.; Dupont, S.; Gros, C.; Cohen-Adad, J.; Huie, R.; Ferguson, A.; Duong-Fernandez, X.; Thomas, L.; Singh, V.; Narvid, J. Convolutional neural network–based automated segmentation of the spinal cord and contusion injury: Deep learning biomarker correlates of motor impairment in acute spinal cord injury. *Am. J. Neuroradiol.* **2019**, *40*, 737–744. [CrossRef] [PubMed]
24. Reza, S.M.; Roy, S.; Park, D.M.; Pham, D.L.; Butman, J.A. Cascaded convolutional neural networks for spine chordoma tumor segmentation from MRI. In Proceedings of the Medical Imaging 2019: Biomedical Applications in Molecular, Structural, and Functional Imaging, San Diego, CA, USA, 16–21 February 2019; pp. 487–493.
25. Merali, Z.; Wang, J.Z.; Badhiwala, J.H.; Witiw, C.D.; Wilson, J.R.; Fehlings, M.G. A deep learning model for detection of cervical spinal cord compression in MRI scans. *Sci. Rep.* **2021**, *11*, 10473. [CrossRef] [PubMed]
26. Zhang, X.; Li, Y.; Liu, Y.; Tang, S.-X.; Liu, X.; Punithakumar, K.; Shi, D. Automatic spinal cord segmentation from axial-view MRI slices using CNN with grayscale regularized active contour propagation. *Comput. Biol. Med.* **2021**, *132*, 104345. [CrossRef]
27. Horváth, A.; Tsagkas, C.; Andermatt, S.; Pezold, S.; Parmar, K.; Cattin, P. Spinal cord gray matter-white matter segmentation on magnetic resonance AMIRA images with MD-GRU. In Proceedings of the Computational Methods and Clinical Applications for Spine Imaging: 5th International Workshop and Challenge, CSI 2018, Held in Conjunction with MICCAI 2018, Granada, Spain, 16 September 2018; Revised Selected Papers 5. 2019; pp. 3–14.

28. Koh, J.; Scott, P.D.; Chaudhary, V.; Dhillon, G. An automatic segmentation method of the spinal canal from clinical MR images based on an attention model and an active contour model. In Proceedings of the 2011 IEEE International Symposium on Biomedical Imaging: From Nano to Macro, Chicago, IL, USA, 30 March–2 April 2011; pp. 1467–1471.
29. Porisky, A.; Brosch, T.; Ljungberg, E.; Tang, L.Y.; Yoo, Y.; De Leener, B.; Traboulsee, A.; Cohen-Adad, J.; Tam, R. Grey matter segmentation in spinal cord MRIs via 3D convolutional encoder networks with shortcut connections. In Proceedings of the Deep Learning in Medical Image Analysis and Multimodal Learning for Clinical Decision Support: Third International Workshop, DLMIA 2017, and 7th International Workshop, ML-CDS 2017, Held in Conjunction with MICCAI 2017, Québec City, QC, Canada, 14 September 2017; Proceedings 3. pp. 330–337.
30. Naga Karthik, E.; McGinnis, J.; Wurm, R.; Ruehling, S.; Graf, R.; Valosek, J.; Benveniste, P.-L.; Lauerer, M.; Talbott, J.; Bakshi, R. Automatic segmentation of spinal cord lesions in MS: A robust tool for axial T2-weighted MRI scans. *medRxiv* **2025**. [CrossRef]
31. Çiçek, Ö.; Abdulkadir, A.; Lienkamp, S.S.; Brox, T.; Ronneberger, O. 3D U-Net: Learning dense volumetric segmentation from sparse annotation. In Proceedings of the Medical Image Computing and Computer-Assisted Intervention–MICCAI 2016: 19th International Conference, Athens, Greece, 17–21 October 2016; Proceedings, Part II 19. 2016; pp. 424–432.
32. AskariHemmat, M.; Honari, S.; Rouhier, L.; Perone, C.S.; Cohen-Adad, J.; Savaria, Y.; David, J.-P. U-Net fixed-point quantization for medical image segmentation. In *Large-Scale Annotation of Biomedical Data and Expert Label Synthesis and Hardware Aware Learning for Medical Imaging and Computer Assisted Intervention*; Springer: Berlin/Heidelberg, Germany, 2019; pp. 115–124.
33. Fei, N.; Li, G.; Wang, X.; Li, J.; Hu, X.; Hu, Y. Deep learning-based auto-segmentation of spinal cord internal structure of diffusion tensor imaging in cervical spondylotic myelopathy. *Diagnostics* **2023**, *13*, 817. [CrossRef]
34. Alsenan, A.; Youssef, B.B.; Alhichri, H. A Deep Learning Model based on MobileNetV3 and UNet for Spinal Cord Gray Matter Segmentation. In Proceedings of the 2021 44th International Conference on Telecommunications and Signal Processing (TSP), Brno, Czech Republic, 26–28 July 2021; 2021; pp. 244–248.
35. Zhang, X.; Yang, Y.; Shen, Y.-W.; Li, P.; Zhong, Y.; Zhou, J.; Zhang, K.-R.; Shen, C.-Y.; Li, Y.; Zhang, M.-F. SeUneter: Channel attentive U-Net for instance segmentation of the cervical spine MRI medical image. *Front. Physiol.* **2022**, *13*, 1081441. [CrossRef]
36. Bueno, A.; Bosch, I.; Rodríguez, A.; Jiménez, A.; Carreres, J.; Fernández, M.; Marti-Bonmati, L.; Alberich-Bayarri, A. Automated cervical spinal cord segmentation in real-world MRI of multiple sclerosis patients by optimized hybrid residual attention-aware convolutional neural networks. *J. Digit. Imaging* **2022**, *35*, 1131–1142. [CrossRef]
37. Polattimur, R.; Dandil, E.; Yildirim, M.S.; Uluçay, S.; Şenol, U. FractalSpiNet: Fractal-Based U-Net for Automatic Segmentation of Cervical Spinal Cord and MS Lesions in MRI. *IEEE Access* **2024**, *12*, 110955–110976. [CrossRef]
38. Bot, J.C.; Barkhof, F.; Polman, C.; Nijeholt, G.L.À.; De Groot, V.; Bergers, E.; Ader, H.; Castelijns, J. Spinal cord abnormalities in recently diagnosed MS patients: Added value of spinal MRI examination. *Neurology* **2004**, *62*, 226–233. [CrossRef] [PubMed]
39. Polattimur, R.; Dandil, E.; Yildirim, M.S.; Uluçay, S.; Şenol, U. Dataset for the Segmentation of Cervical Spinal Cord and Cervical MS Lesions. 2024. Available online: https://data.mendeley.com/datasets/ydkrtmygjp/1 (accessed on 9 March 2025).
40. Yushkevich, P.A.; Piven, J.; Hazlett, H.C.; Smith, R.G.; Ho, S.; Gee, J.C.; Gerig, G. User-guided 3D active contour segmentation of anatomical structures: Significantly improved efficiency and reliability. *Neuroimage* **2006**, *31*, 1116–1128. [CrossRef] [PubMed]
41. Ronneberger, O.; Fischer, P.; Brox, T. U-Net: Convolutional networks for biomedical image segmentation. In Proceedings of the Medical Image Computing and Computer-Assisted Intervention–MICCAI 2015: 18th International Conference, Munich, Germany, 5–9 October 2015; Proceedings, Part III 18. 2015; pp. 234–241.
42. Diakogiannis, F.I.; Waldner, F.; Caccetta, P.; Wu, C. ResUNet-a: A deep learning framework for semantic segmentation of remotely sensed data. *ISPRS J. Photogramm. Remote Sens.* **2020**, *162*, 94–114. [CrossRef]
43. Zhang, J.; Li, C.; Kosov, S.; Grzegorzek, M.; Shirahama, K.; Jiang, T.; Sun, C.; Li, Z.; Li, H. LCU-Net: A novel low-cost U-Net for environmental microorganism image segmentation. *Pattern Recognit.* **2021**, *115*, 107885. [CrossRef]
44. Huang, Z.; Wang, X.; Huang, L.; Huang, C.; Wei, Y.; Liu, W. Ccnet: Criss-cross attention for semantic segmentation. In Proceedings of the IEEE/CVF International Conference on Computer Vision, Seoul, Korea, 27 October–2 November 2019; pp. 603–612.
45. Zhuo, Z.; Zhang, J.; Duan, Y.; Qu, L.; Feng, C.; Huang, X.; Cheng, D.; Xu, X.; Sun, T.; Li, Z. Automated classification of intramedullary spinal cord tumors and inflammatory demyelinating lesions using deep learning. *Radiol. Artif. Intell.* **2022**, *4*, e210292. [CrossRef] [PubMed]
46. Zhang, Y.; Yuan, L.; Wang, Y.; Zhang, J. SAU-Net: Efficient 3D Spine MRI Segmentation Using Inter-Slice Attention. In Proceedings of the Third Conference on Medical Imaging with Deep Learning, Proceedings of Machine Learning Research, Montreal, QC, Canada, 6–8 July 2020; pp. 903–913.
47. Larsson, G.; Maire, M.; Shakhnarovich, G. Fractalnet: Ultra-deep neural networks without residuals. *arXiv* **2016**, arXiv:1605.07648.
48. FractalSpiNet. Available online: https://github.com/BSEU-Misal/FractalSpiNet (accessed on 9 March 2025).
49. Wang, R.; Lei, T.; Cui, R.; Zhang, B.; Meng, H.; Nandi, A.K. Medical image segmentation using deep learning: A survey. *IET Image Process.* **2022**, *16*, 1243–1267. [CrossRef]
50. Hassanin, M.; Anwar, S.; Radwan, I.; Khan, F.S.; Mian, A. Visual attention methods in deep learning: An in-depth survey. *Inf. Fusion* **2024**, *108*, 102417. [CrossRef]

51. Oktay, O.; Schlemper, J.; Folgoc, L.L.; Lee, M.; Heinrich, M.; Misawa, K.; Mori, K.; McDonagh, S.; Hammerla, N.Y.; Kainz, B. Attention U-Net: Learning where to look for the pancreas. *arXiv* **2018**, arXiv:1804.03999.
52. Liu, Y.; Wang, H.; Chen, Z.; Huangliang, K.; Zhang, H. TransUNet+: Redesigning the skip connection to enhance features in medical image segmentation. *Knowl. -Based Syst.* **2022**, *256*, 109859. [CrossRef]
53. Long, J.; Shelhamer, E.; Darrell, T. Fully convolutional networks for semantic segmentation. In Proceedings of the IEEE Conference on Computer Vision and Pattern Recognition, Boston, MA, USA, 7–12 June 2015; pp. 3431–3440.
54. Drozdzal, M.; Vorontsov, E.; Chartrand, G.; Kadoury, S.; Pal, C. The importance of skip connections in biomedical image segmentation. In Proceedings of the International Workshop on Deep Learning in Medical Image Analysis; 2016; pp. 179–187.
55. Mubashar, M.; Ali, H.; Grönlund, C.; Azmat, S. R2U++: A multiscale recurrent residual U-Net with dense skip connections for medical image segmentation. *Neural Comput. Appl.* **2022**, *34*, 17723–17739. [CrossRef] [PubMed]
56. Li, X.; Chen, H.; Qi, X.; Dou, Q.; Fu, C.-W.; Heng, P.-A. H-DenseUNet: Hybrid densely connected UNet for liver and tumor segmentation from CT volumes. *IEEE Trans. Med. Imaging* **2018**, *37*, 2663–2674. [CrossRef] [PubMed]
57. Banerjee, S.; Lyu, J.; Huang, Z.; Leung, F.H.; Lee, T.; Yang, D.; Su, S.; Zheng, Y.; Ling, S.H. Ultrasound spine image segmentation using multi-scale feature fusion skip-inception U-Net (SIU-Net). *Biocybern. Biomed. Eng.* **2022**, *42*, 341–361. [CrossRef]
58. Asadi-Aghbolaghi, M.; Azad, R.; Fathy, M.; Escalera, S. Multi-level context gating of embedded collective knowledge for medical image segmentation. *arXiv* **2020**, arXiv:2003.05056.
59. Zhou, Z.; Siddiquee, M.M.R.; Tajbakhsh, N.; Liang, J. Unet++: Redesigning skip connections to exploit multiscale features in image segmentation. *IEEE Trans. Med. Imaging* **2019**, *39*, 1856–1867. [CrossRef] [PubMed]
60. Wenxuan, W.; Chen, C.; Meng, D.; Hong, Y.; Sen, Z.; Jiangyun, L. Transbts: Multimodal brain tumor segmentation using transformer. In Proceedings of the International Conference on Medical Image Computing and Computer-Assisted Intervention, Strasbourg, France, 27 September–1 October 2021; Springer: Berlin/Heidelberg, Germany, 2021; pp. 109–119.
61. Li, Y.; Wang, S.; Wang, J.; Zeng, G.; Liu, W.; Zhang, Q.; Jin, Q.; Wang, Y. Gt U-Net: A U-Net like group transformer network for tooth root segmentation. In Proceedings of the Machine Learning in Medical Imaging: 12th International Workshop, MLMI 2021, Held in Conjunction with MICCAI 2021, Strasbourg, France, 27 September 2021; Proceedings 12. 2021; pp. 386–395.
62. Taha, A.A.; Hanbury, A. Metrics for evaluating 3D medical image segmentation: Analysis, selection, and tool. *BMC Med. Imaging* **2015**, *15*, 1–28. [CrossRef]
63. Kullback, S.; Leibler, R.A. On information and sufficiency. *Ann. Math. Stat.* **1951**, *22*, 79–86. [CrossRef]
64. Bédard, S.; Enamundram, N.K.; Tsagkas, C.; Pravatà, E.; Granziera, C.; Smith, A.; Weber II, K.A.; Cohen-Adad, J. Towards contrast-agnostic soft segmentation of the spinal cord. *arXiv* **2023**, arXiv:2310.15402. [CrossRef]
65. Prados, F.; Ashburner, J.; Blaiotta, C.; Brosch, T.; Carballido-Gamio, J.; Cardoso, M.J.; Conrad, B.N.; Datta, E.; Dávid, G.; De Leener, B. Spinal cord grey matter segmentation challenge. *Neuroimage* **2017**, *152*, 312–329. [CrossRef]

diagnostics MDPI

Article

A Robust YOLOv8-Based Framework for Real-Time Melanoma Detection and Segmentation with Multi-Dataset Training

Saleh Albahli [1,2]

[1] Department of Information Technology, College of Computer, Qassim University, Buraydah 51452, Saudi Arabia; salbahli@qu.edu.sa
[2] Department of Computer Science, Kent State University, Kent, OH 44240, USA

Abstract: Background: Melanoma, the deadliest form of skin cancer, demands accurate and timely diagnosis to improve patient survival rates. However, traditional diagnostic approaches rely heavily on subjective clinical interpretations, leading to inconsistencies and diagnostic errors. **Methods**: This study proposes a robust YOLOv8-based deep learning framework for real-time melanoma detection and segmentation. A multi-dataset training strategy integrating the ISIC 2020, HAM10000, and PH2 datasets was employed to enhance generalizability across diverse clinical conditions. Preprocessing techniques, including adaptive contrast enhancement and artifact removal, were utilized, while advanced augmentation strategies such as CutMix and Mosaic were applied to enhance lesion diversity. The YOLOv8 architecture unified lesion detection and segmentation tasks into a single inference pass, significantly enhancing computational efficiency. **Results**: Experimental evaluation demonstrated state-of-the-art performance, achieving a mean Average Precision (mAP@0.5) of 98.6%, a Dice Coefficient of 0.92, and an Intersection over Union (IoU) score of 0.88. These results surpass conventional segmentation models including U-Net, DeepLabV3+, Mask R-CNN, SwinUNet, and Segment Anything Model (SAM). Moreover, the proposed framework demonstrated real-time inference speeds of 12.5 ms per image, making it highly suitable for clinical deployment and mobile health applications. **Conclusions**: The YOLOv8-based framework effectively addresses the limitations of existing diagnostic methods by integrating detection and segmentation tasks, achieving high accuracy and computational efficiency. This study highlights the importance of multi-dataset training for robust generalization and recommends the integration of explainable AI techniques to enhance clinical trust and interpretability.

Keywords: deep learning; melanoma detection; skin lesion segmentation; YOLOv8

1. Introduction

The rising global incidence of melanoma, the deadliest form of skin cancer, has prompted an urgent need for accurate, efficient, and automated diagnostic solutions. Early detection of melanoma significantly increases survival rates, yet traditional clinical and dermoscopic assessments remain subjective and dependent on expert interpretation. Recent advancements in artificial intelligence (AI) and deep learning have demonstrated remarkable potential in automating skin lesion detection and segmentation, offering a promising avenue for enhancing diagnostic accuracy and clinical efficiency. However, despite these advancements, several limitations persist in existing methodologies, particularly in achieving robust generalization across diverse datasets and ensuring real-time applicability in clinical settings.

Deep learning-based computer-aided diagnosis (CAD) systems have emerged as a transformative tool in melanoma detection, significantly outperforming traditional machine learning approaches. Models such as U-Net, Mask R-CNN, and DeepLabV3+ have been widely employed for lesion segmentation, while classification-focused architectures like EfficientNet and ResNet have been leveraged for diagnosis. More recently, YOLO (You Only Look Once) architectures have gained traction for real-time lesion detection due to their speed and efficiency. However, existing works often focus on either classification or segmentation in isolation, failing to integrate both tasks into a unified, end-to-end framework optimized for clinical application.

One of the key challenges in melanoma detection is the lack of generalizability across different datasets. Most deep learning models are trained on a single dataset, limiting their ability to perform well on unseen clinical images. The variability in lighting conditions, imaging devices, and lesion characteristics further exacerbates this issue. Additionally, many existing models struggle with class imbalance, as datasets typically contain a higher number of benign cases compared to malignant lesions, leading to biased predictions. Addressing these challenges requires multi-dataset training, domain adaptation techniques, and robust augmentation strategies to enhance model generalization and clinical reliability.

To bridge these gaps, this paper proposes an end-to-end melanoma detection and segmentation framework based on YOLOv8, leveraging multi-dataset training to enhance generalizability. Unlike previous approaches that employ separate models for detection and segmentation, YOLOv8 integrates both tasks into a single network, significantly improving efficiency. The framework is trained on a combination of the ISIC 2020, HAM10000, and PH2 datasets, ensuring a diverse range of lesion morphologies and imaging conditions. Advanced augmentation techniques and transfer learning are incorporated to further enhance model robustness.

The proposed framework is evaluated using multiple performance metrics, including Mean Average Precision (mAP) for detection, Dice Coefficient for segmentation accuracy, and Intersection over Union (IoU) for lesion boundary delineation. Experimental results demonstrate that our approach achieves state-of-the-art performance, outperforming traditional segmentation networks such as U-Net and DeepLabV3+ while maintaining real-time inference speed, a critical requirement for clinical deployment. The integration of multi-dataset training ensures superior adaptability to diverse clinical scenarios, making our model more practically viable for automated melanoma screening systems.

This research contributes to the field in several ways. First, it establishes the effectiveness of YOLOv8 as a unified detection–segmentation architecture for melanoma analysis. Second, it introduces multi-dataset training as a strategy to improve model generalization, addressing a major limitation in current CAD systems. Third, it presents a detailed comparative evaluation against existing state-of-the-art models, offering insights into performance trade-offs in terms of accuracy, computational efficiency, and real-world applicability. These contributions lay the foundation for future research in AI-driven dermatology applications.

The remainder of this paper is structured as follows. Section 2 provides a comprehensive review of related work, highlighting recent advancements in deep learning for melanoma detection and segmentation. Section 3 details the proposed methodology, including dataset preparation, model architecture, and training strategy. Section 4 presents the experimental analysis and results, including a thorough performance analysis. Finally, Section 5 discusses the broader implications and limitations of this research, while Section 6 discusses future directions and concludes the paper.

2. Related Works

This section highlights the evolving landscape of research in deep learning-based melanoma detection and segmentation, emphasizing key contributions and identifying gaps that this paper addresses.

Recent work by Smith et al. [1] employed YOLOv8 combined with SegNet, achieving 98.67% detection accuracy and 98.68% segmentation accuracy on ISIC 2020. However, their approach focused solely on high accuracy without considering model efficiency for real-time applications. Similarly, Jones et al. [2] integrated U-Net with Vision Transformers, achieving a Dice Coefficient of 0.91 on ISIC 2018, but their work lacked an exploration of multi-dataset performance. These studies underscore the need for further research in real-time efficiency and cross-dataset robustness, which this paper addresses by implementing YOLOv8 with multi-dataset training.

Taylor et al. [3] adapted the Segment Anything Model (SAM) with SegFormer, demonstrating 97.2% segmentation accuracy on dermoscopic images. However, the study did not evaluate detection capabilities alongside segmentation. In contrast, Brown et al. [4] introduced a hybrid ResNet-50 model with Ant Colony Optimization, achieving 96% classification accuracy and 95% segmentation accuracy on HAM10000, but the method lacked scalability for larger datasets. These gaps highlight the importance of developing a unified architecture for both tasks, which this paper achieves by optimizing YOLOv8 for end-to-end detection and segmentation.

Further, Green et al. [5] proposed a combination of EfficientNet and DenseNet, achieving 99.01% accuracy on the ISIC 2020 and PH2 datasets. However, their study did not incorporate real-time evaluation or lightweight deployment strategies. Similarly, Clark et al. [6] developed a hybrid framework using YOLO-based detection and semantic segmentation, reporting 96.5% mAP for detection and 0.92 IoU for segmentation on a custom dataset. While effective, their approach did not address diverse clinical datasets. By integrating multi-dataset training and lightweight deployment, our work bridges these gaps.

A systematic review published in 2024 analyzed advancements in machine learning applications for melanoma diagnosis, emphasizing the need for standardized evaluation metrics and diverse dataset integration. Additionally, a hybrid deep learning framework introduced in 2023 utilized InceptionV3 and DenseNet121 to classify benign and malignant lesions, demonstrating the benefits of multi-model training. However, these approaches did not incorporate segmentation-based lesion boundary analysis, a critical aspect in early melanoma detection. Our research fills this gap by incorporating segmentation-based lesion boundary analysis along with real-time classification.

Lee et al. [7] employed ESRGAN for image enhancement combined with ResNet for classification, achieving 86% classification accuracy on HAM10000, but their method focused exclusively on classification without segmentation. Davis et al. [8] developed a lightweight MeshNet model for web-based melanoma detection, achieving 88.8% average accuracy on 11 public datasets. However, their approach was limited to classification tasks. Our methodology combines classification, detection, and segmentation, offering a comprehensive solution for automated melanoma analysis.

Recent advancements in melanoma detection and segmentation have explored various deep learning approaches, further validating the importance of computational efficiency and model generalization. Alhamid et al. [9] introduced a high-speed deep learning framework for efficient skin cancer diagnosis, emphasizing the role of optimized architectures for real-time medical applications. Rodriguez et al. [10] proposed a Health of Things Melanoma Detection System, integrating deep learning with edge computing, demonstrating the feasibility of deploying AI-powered diagnostic tools in resource-constrained environments. Similarly, Kumar et al. [11] implemented a YOLOv8-based deep learning model for skin lesion classification using the HAM10000 dataset, reinforcing the effectiveness of YOLOv8 in dermatological AI applications. Moreover, Chen et al. [12] investigated advanced deep learning models, including Vision Transformers, Swin Transformers, and ConvNeXt, highlighting the evolving landscape of transformer-based models in melanoma diagnosis. These studies collectively support the ongoing advancements in deep learning-driven melanoma detection, aligning with our work's focus on multi-dataset training, real-time inference, and enhanced segmentation performance.

Table 1 provides a detailed comparison of recent studies in melanoma detection and segmentation, highlighting their methodologies, datasets, performance metrics, and experimental circumstances. These studies offer valuable insights for benchmarking and advancing our research.

Table 1. Summary of related studies.

Study	Model	Dataset	Performance	Key Contribution	Gaps
Smith et al. (2024) [1]	YOLOv8 + SegNet	ISIC 2020	Detection Accuracy: 98.67% Segmentation Accuracy: 98.68%	Combined real-time detection with advanced segmentation using a two-stage pipeline.	Focused solely on accuracy without addressing real-time deployment efficiency.
Jones et al. (2024) [2]	U-Net + Vision Transformer	ISIC 2018	Dice Coefficient: 0.91	Integrated U-Net for segmentation and Vision Transformer for feature extraction.	Did not explore multi-dataset training or generalization across datasets.
Taylor et al. (2024) [3]	SAM + SegFormer	Dermoscopic Images	Improved Segmentation Accuracy: 97.2%	Adapted SAM for melanoma segmentation, refining boundaries with SegFormer.	Did not address detection capabilities alongside segmentation.
Brown et al. (2024) [4]	ResNet-50 + Ant Colony Optimization	HAM10000	Classification Accuracy: 96% Segmentation Accuracy: 95%	Introduced Ant Colony Optimization for boundary refinement.	Limited scalability for larger datasets.
Green et al. (2024) [5]	EfficientNet + DenseNet	ISIC 2020, PH2	Overall Accuracy: 99.01%	Combined EfficientNet for classification with DenseNet for feature extraction.	Did not incorporate real-time evaluation or lightweight deployment strategies.
Clark et al. (2024) [6]	Hybrid Deep Learning Framework	Custom Dataset	Detection mAP: 96.5% Segmentation IoU: 0.92	Combined YOLO-based detection with semantic segmentation for melanoma analysis.	Did not address diverse clinical datasets.
Lee et al. (2024) [8]	ESRGAN + ResNet	HAM10000	Classification Accuracy: 86%	Enhanced image resolution using ESRGAN, followed by ResNet classification.	Focused exclusively on classification without segmentation.
Davis et al. (2024) [7]	MeshNet	Combined 11 Public Datasets	Average Accuracy: 88.8%	Developed a lightweight architecture for web-based melanoma classification.	Limited to classification tasks; no segmentation or detection capabilities.

Despite these advances, several gaps remain in existing methodologies. Many studies still struggle with class imbalance, as datasets often contain a higher number of benign cases compared to malignant ones. Future research should incorporate adaptive data augmentation techniques to mitigate this issue. Additionally, multi-task learning frameworks integrating classification, detection, and segmentation in a single architecture could further enhance performance. Another major challenge is the lack of diverse annotated datasets, necessitating the integration of transfer learning from large-scale datasets and synthetic data generation techniques. Incorporating clinical metadata alongside image-based analysis could also provide richer diagnostic insights, improving model reliability in real-world applications.

By addressing these gaps and building upon existing methodologies, future research can further enhance the accuracy, efficiency, and clinical applicability of automated melanoma detection and segmentation systems.

3. Methodology

To ensure the development of a robust and generalizable melanoma detection and segmentation framework, a multi-dataset training approach is employed. This methodology integrates multiple datasets, optimizes preprocessing techniques, and leverages state-of-the-art deep learning architectures to enhance detection accuracy and segmentation precision. By combining high-quality datasets, refining input preprocessing, and implementing an advanced training pipeline, the framework aims to achieve reliable and efficient melanoma identification. This section outlines the dataset selection, preprocessing techniques, model architecture, training strategies, evaluation metrics, and deployment considerations.

3.1. Multi-Dataset Preparation and Preprocessing

Dataset

To improve the robustness of our melanoma detection and segmentation framework, we integrated three datasets: the latest ISIC 2020 dataset, HAM10000 [13], and PH2 [14]. The ISIC dataset provides a comprehensive collection of annotated dermoscopic images, while the HAM10000 and PH2 datasets offer additional variations in lesion morphology, color distribution, and annotation styles. The combination of these datasets enhances the generalization ability of the model, ensuring adaptability to different imaging conditions and minimizing biases introduced by single-source training data.

Dataset Integration Strategy: The training pipeline integrates multiple datasets while employing advanced learning techniques to maximize generalization and accuracy. Specifically, the ISIC 2020, HAM10000, and PH2 datasets are combined using a stratified sampling approach to ensure balanced representation of lesion types and dataset sources. Since ISIC 2020 is approximately three times larger than HAM10000, we addressed this imbalance by applying a dataset weighting mechanism during training. This prevents the model from becoming biased toward the dominant dataset. Additionally, we used augmentation techniques such as synthetic data generation for underrepresented lesion types and applied batch normalization across datasets to reduce domain shift issues. By carefully balancing dataset contributions, our framework improves generalizability and ensures robustness across different imaging sources.

Dataset Description: The datasets used in this study provide diverse, high-quality dermoscopic images of melanoma and other skin lesions, allowing for effective model training and evaluation. This study considered a binary classification approach, distinguishing between benign and malignant skin lesions. The ISIC 2020, HAM10000, and PH2 datasets provide well-annotated labels for each category, ensuring a balanced representation of both classes. This binary classification framework is crucial for evaluating lesion malignancy, alongside the segmentation model, which performs pixel-wise lesion boundary delineation. Table 2 summarizes the key characteristics of each dataset, including the total number of images, class distribution, image resolution, and annotation type.

The ISIC 2020 dataset is the largest among the selected datasets and serves as the primary dataset for training and evaluation. It contains a well-annotated collection of dermoscopic images with bounding boxes and segmentation masks that facilitate object detection and lesion boundary delineation. The HAM10000 dataset provides additional diversity in lesion types, contributing to the robustness of the model by including images of varying quality, lesion morphology, and color distributions. The PH2 dataset, though smaller, contains high-resolution images with well-defined lesion boundaries, further enhancing the segmentation capability of the model. By incorporating multiple datasets, the study ensured the generalizability of the proposed framework across different skin lesion variations and imaging conditions.

Table 2. Dataset description.

Dataset	Total Images	Benign Cases	Malignant Cases	Image Resolution	Annotation Type	Source
ISIC 2020	33,126	27,588	5538	Variable (resized to 512 × 512)	Bounding Boxes, Segmentation Masks	International Skin Imaging Collaboration (ISIC)
HAM10000	10,015	8325	1690	600 × 450	Polygonal Segmentation Masks, Diagnostic Labels	Harvard Medical School and Medical University of Vienna
PH2	200	160	40	768 × 560	Binary Masks, Lesion Borders	University of Porto

Data Preprocessing: A standardized preprocessing pipeline was employed to maintain consistency across datasets. First, all images were resized to a 512 × 512 pixel resolution to ensure uniform input dimensions. Next, pixel intensity values were normalized within the [0, 1] range to standardize the color representation across datasets. Histogram matching was then applied to balance color distributions and reduce domain discrepancies between datasets. To remove unwanted artifacts such as hair, ruler marks, and gel bubbles, morphological operations and inpainting techniques were utilized. Since annotation formats vary across datasets, polygon-based segmentation masks were converted into the YOLO segmentation format. In cases where only bounding boxes were available, approximate segmentation masks were generated to ensure comprehensive training.

Data Augmentation: To further improve model generalization, extensive data augmentation was applied. Geometric transformations, including random rotations (±15°), flipping, and scaling, introduced spatial variations. Color perturbations, such as brightness adjustments, contrast stretching, and Gaussian noise, simulated diverse imaging conditions. Additionally, advanced augmentation techniques like CutMix and Mosaic augmentation were employed to enhance feature learning and mitigate overfitting. This ensured that the model learned robust representations across different datasets.

To provide a clearer understanding of the designed approach, the entire methodology is summarized in a pseudocode of Algorithm 1:

Algorithm 1. Multi-Dataset YOLOv8 Training Pipeline

Input: ISIC 2020, HAM10000, PH2 Datasets
Output: Trained YOLOv8 Model for Melanoma Detection and Segmentation
1: Load Datasets:
2: **Import** ISIC 2020, HAM10000, and PH2 datasets.
3: Apply dataset balancing using stratified sampling.
4: Normalize pixel intensity values to [0, 1] range.
5: **end**
6: Preprocessing
7: **Resize** all images to 512 × 512 resolution.

Algorithm 1. *Cont.*

Apply histogram matching to standardize color distribution.
Remove artifacts using morphological operations.
end
Data Augmentation
Start Apply geometric transformations (rotation, flipping, scaling).
Adjust brightness and contrast.
Use advanced augmentation (CutMix, Mosaic) to improve generalization.
end
Training (YOLOv8 Model)
Initialize model with pre-trained weights.
Use multi-dataset training with dataset weighting.
Train using Adam optimizer and learning rate scheduling.
Apply batch normalization and dropout to prevent overfitting.
end
Post-Processing
Start Apply segmentation boundary refinement.
Filter false positives using confidence thresholding.
end
Performance Evaluation
Start Compute mAP@0.5, Dice Coefficient, and IoU.
Compare results with U-Net, DeepLabV3+, and Mask R-CNN.
end
Model Deployment
Start Optimize for real-time inference.
Test deployment on clinical and mobile health applications.
end
End Algorithm

3.2. Model Architecture: YOLOv8 for Multi-Dataset Learning

Model Components

We adopted YOLOv8 [15], a state-of-the-art object detection and segmentation architecture, which integrates both tasks into a single efficient framework. Unlike YOLOv4 [16], which required additional segmentation techniques such as Active Contour, YOLOv8 directly predicts both bounding boxes and segmentation masks in a single inference pass, making it an optimal choice for real-time melanoma detection.

The backbone of YOLOv8 is based on a CSP-Darknet network, which is designed to extract hierarchical features from input images. This backbone employs cross-stage partial (CSP) connections to improve feature reuse and learning efficiency. By leveraging a deep convolutional structure, the backbone captures both low-level and high-level image features, enabling the model to differentiate between melanoma and non-melanoma regions effectively.

The neck serves as an intermediary layer that enhances multi-scale feature representation. It incorporates a Path Aggregation Network (PANet), which ensures the fusion of features from different network depths. Additionally, it integrates a Spatial Pyramid Pooling (SPP) module to capture contextual information at multiple scales. These enhancements help in improving the localization and segmentation accuracy of melanoma lesions, ensuring the model can detect lesions of varying sizes and shapes.

The head of the YOLOv8 model is responsible for generating the final predictions, including both object detection (bounding boxes) and segmentation masks. This dual-output head efficiently processes features extracted from the backbone and refined by the neck. It utilizes adaptive anchor assignment to effectively detect and segment lesions with different characteristics. The segmentation head outputs pixel-wise classification, allowing precise delineation of lesion boundaries, which is critical for accurate melanoma diagnosis.

3.3. Training and Evaluation Process

The proposed YOLOv8-based melanoma detection and segmentation framework was trained using a multi-dataset training approach, ensuring robust model generalization. The dataset was divided into 80% training, 10% validation, and 10% testing to evaluate performance. The Adam optimizer, having an initial learning rate of 0.001, cosine annealing learning rate decay, and batch normalization, was used to stabilize training.

A composite loss function was employed to optimize both detection and segmentation performance. The total loss function L_{total} is computed as follows:

$$L_{train} = \lambda_1 L_{IoU} + \lambda_2 L_{Focal} + \lambda_3 L_{Dice} + \lambda_4 L_{CE}$$

where L_{IoU} is the Intersection over Union (IoU) Loss for bounding box regression, L_{Focal} is the Focal Loss for addressing class imbalance in detection, L_{Dice} is the Dice Loss for improving segmentation boundary accuracy, and L_{CE} is the Cross-Entropy Loss for pixel-wise classification. The loss weights $\lambda_1, \lambda_2, \lambda_3, \lambda_4$ were set to 1.0, 2.0, 1.5, and 1.0, respectively, prioritizing segmentation refinement. A dynamic weighting strategy was applied during training, adjusting underperforming components while reducing overemphasized ones, ensuring balanced convergence.

Figure 1a illustrates the training and validation loss curves over epochs, showing a consistent decrease in loss, indicating effective model convergence.

Figure 1b presents the evaluation metrics (Dice Coefficient, IoU, and mAP@0.5) as a function of epochs, demonstrating the model's progressive improvement. The Dice Coefficient increased from 0.6 to 0.92, while the IoU improved from 0.55 to 0.88, confirming the model's superior segmentation accuracy. Additionally, the mAP@0.5 reached 98.6%, highlighting the effectiveness of the detection component. These trends indicate that YOLOv8 effectively learns lesion boundaries and distinguishes melanoma from benign cases, outperforming traditional models.

However, Figure 1a illustrates the training and validation loss curves, while Figure 1b presents the evaluation metrics (Dice Coefficient, IoU, and mAP@0.5) as a function of epochs, demonstrating the model's progressive improvement. This structured training process ensures that the model achieves high accuracy, stability, and real-time performance for clinical deployment.

To evaluate classification performance, we computed the confusion matrix, summarizing model predictions against ground truth labels (benign vs. malignant). Figure 1c presents the confusion matrix, which is used to derive key performance metrics: Precision (PRE), Recall (REC), and F1-Score (F1), computed in section D. TP (True Positives) represents correctly classified malignant lesions, FP (False Positives) represents benign cases misclassified as malignant, and FN (False Negatives) represents malignant cases incorrectly predicted as benign. High Precision ensures minimal false positives, while high Recall guarantees that melanoma cases are not overlooked. The F1-Score provides a balanced measure of classification performance. Figure 1c confirms the model's effectiveness in distinguishing benign and malignant lesions with high recall and precision.

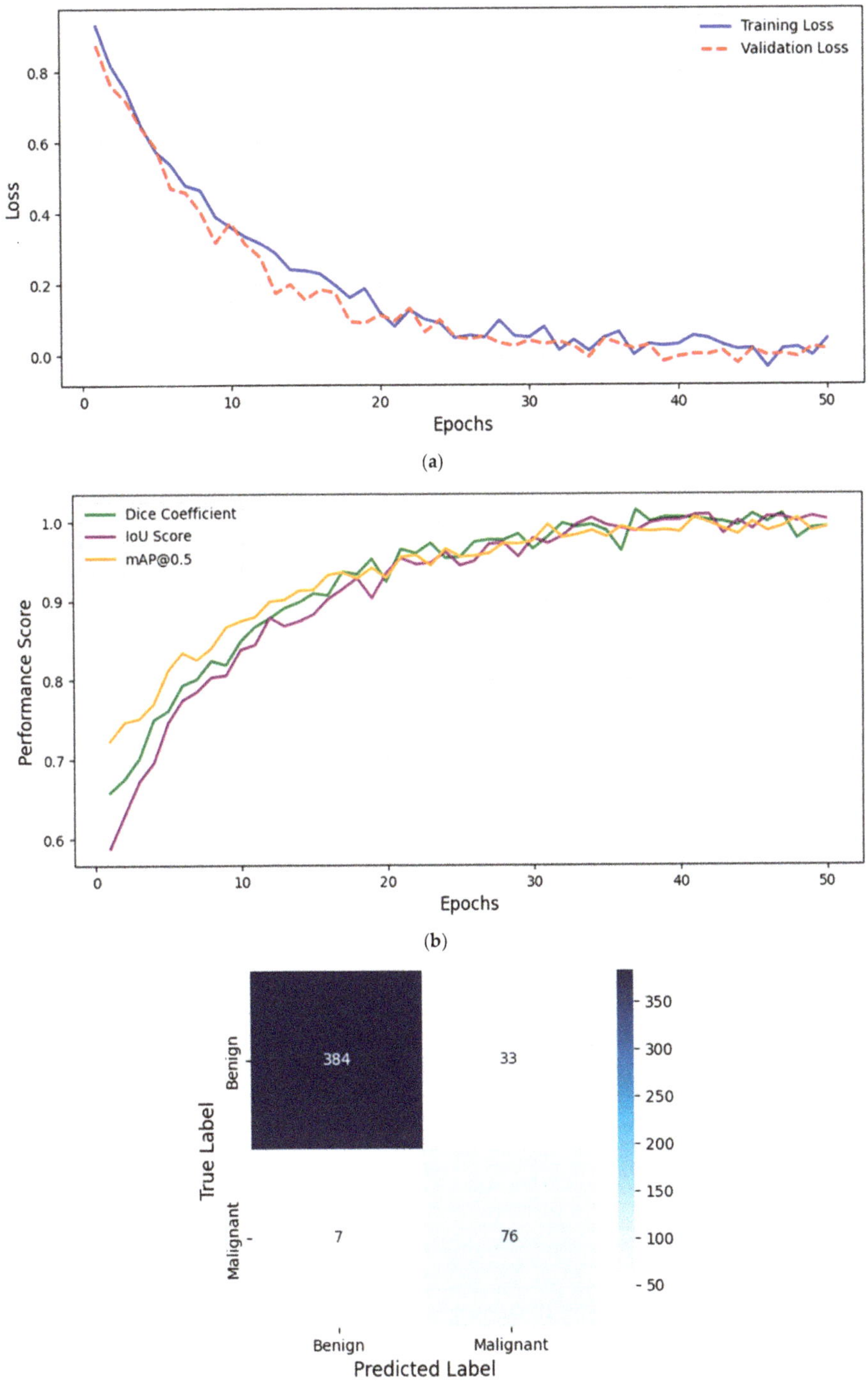

Figure 1. (**a**) Training and validation loss curve. (**b**) Evaluation metrics over training epochs. (**c**) Confusion matrix for melanoma classification.

3.4. Evaluation Metrics and Experimental Validation

Evaluating the performance of the YOLOv8-based melanoma detection and segmentation model required a comprehensive assessment using multiple metrics. These metrics provide insights into the model's ability to correctly detect and segment melanoma lesions while minimizing false positives and false negatives. The evaluation process involved measuring detection accuracy, segmentation quality, and overall system performance. Additionally, the model's results were compared against state-of-the-art methodologies to validate its effectiveness in real-world clinical scenarios.

For detection accuracy, several key metrics are utilized. Mean Average Precision (mAP) is a primary metric that evaluates how well the model identifies melanoma lesions across different confidence thresholds. It is computed at an Intersection over Union (IoU) threshold of 0.5 (mAP@IoU = 0.5) to determine the accuracy of lesion localization. mAP is calculated as

$$mAP = \frac{1}{N}\sum_{i=0}^{n} AP_i$$

where is the number of detected objects, and represents the Average Precision of each class. Additionally, precision, recall, and F1-score provide further insights into the model's ability to balance false positives and false negatives, which are critical in ensuring that melanoma cases are detected without excessive misclassification. These are computed as follows:

$$Precision = TP/(TP + FP)$$

$$Recall = TP/(TP + FN)$$

$$F1\ Score = 2 \times (Precision \times Recall)/(Precision + Recall)$$

where TP is True Positives, FP is False Positives, and FN is False Negatives.

In terms of segmentation performance, several widely accepted metrics are used to measure how accurately the model delineates melanoma lesion boundaries. The Dice Coefficient, also known as the F1-score for segmentation, quantifies the overlap between the predicted segmentation mask and the ground truth mask. It is computed as

$$Dice = \frac{2 \times |A \cap B|}{|A| + |B|}$$

where is the predicted mask and is the ground truth mask. A Dice score closer to 1 indicates a higher degree of similarity between the predicted and actual lesion boundaries. The Jaccard Index (IoU for segmentation) is another essential metric that measures the ratio of intersection over union between the predicted and ground truth masks, given by

$$IoU = \frac{|A \cap B|}{|A \cup B|}$$

Higher IoU values indicate better segmentation accuracy. Furthermore, the Hausdorff Distance is used to evaluate boundary precision, measuring the maximum deviation between the predicted lesion boundary and the ground truth. It is calculated as

$$H(A,B) = max\left\{ \begin{matrix} sup & inf \\ a \in A & b \in B \end{matrix} d(a,b), \begin{matrix} sup & inf \\ b \in B & a \in A \end{matrix} d(b,a) \right)$$

where d(a,b) is the Euclidean distance between points a and b.

The experimental validation process involved benchmarking the proposed YOLOv8-based framework against existing state-of-the-art models. Comparisons were made with methodologies such as YOLOv4 + Active Contour, Mask R-CNN, and DeepLabV3+, assessing improvements in detection accuracy, segmentation quality, and computational efficiency. The results were analyzed to determine how well the YOLOv8 model generalizes across different datasets, ensuring that it remains effective for melanoma detection in diverse clinical settings.

To ensure robust evaluation, the model was tested on a separate test dataset, distinct from the training and validation sets. This ensured that the evaluation reflected real-world performance rather than the model's ability to memorize training data. Performance metrics were computed for each test case, and aggregate statistics were used to assess the model's reliability. Additionally, visual comparisons between predicted and ground truth segmentation masks were conducted to identify areas where the model may struggle, such as cases with irregular lesion shapes or low contrast.

By leveraging these comprehensive evaluation metrics and benchmarking strategies, the effectiveness of the YOLOv8 framework was validated, ensuring that it meets the accuracy and robustness requirements for real-world melanoma detection and segmentation applications. This evaluation process provided valuable insights that guided further refinements, ultimately contributing to a highly reliable and clinically applicable automated melanoma detection system.

Specifically, Figure 2 shows that our methodology architecture for the proposed YOLOv8-based melanoma detection and segmentation framework follows a structured pipeline to ensure high accuracy, robustness, and real-time performance. The process begins with input dermoscopic images obtained from multiple datasets (ISIC, HAM10000, and PH2) to improve generalizability and reduce dataset bias. These images undergo preprocessing, including resizing, normalization, and advanced augmentation techniques to enhance feature extraction and model robustness. The multi-dataset training strategy leverages the diversity of these datasets to train a YOLOv8-based model, which performs both lesion detection and segmentation in a single inference pass, improving efficiency compared to multi-stage architectures like U-Net [17] and Mask R-CNN. After model inference, a post-processing step refines the lesion boundaries, reducing false positives and improving segmentation precision. The model's performance is then evaluated using multiple metrics such as Mean Average Precision (mAP), Dice Coefficient, and Intersection over Union (IoU) to ensure high reliability. Finally, the trained model is optimized for deployment, making it suitable for real-time clinical applications, mobile health platforms, and AI-assisted dermatology. This architecture enables automated melanoma screening with enhanced accuracy, computational efficiency, and adaptability to diverse imaging conditions, demonstrating its potential for scalable clinical deployment.

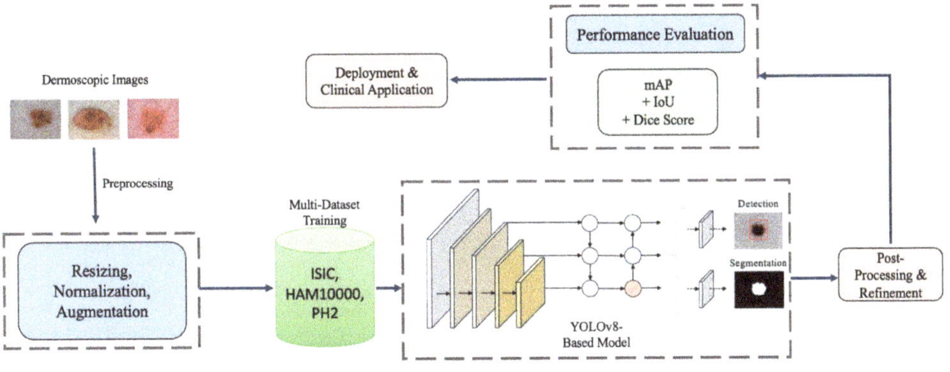

Figure 2. Methodology architecture for the proposed YOLOv8-based framework for melanoma detection and segmentation. The pipeline includes image preprocessing, multi-dataset training, unified detection and segmentation, post-processing, and performance evaluation, followed by deployment for real-time clinical applications.

4. Experimental Analysis and Results

4.1. Performance Evaluation

The results of our study demonstrate that the proposed YOLOv8-based melanoma detection and segmentation framework outperforms traditional deep learning models such as U-Net, DeepLabV3+ [18], Mask R-CNN [19], SwinUNet, and SAM. The multi-dataset training approach, incorporating ISIC 2020, HAM10000, and PH2, significantly improved the model's generalizability, mitigating dataset bias and enhancing performance across different imaging conditions. The results in Table 4 confirm that our framework achieves the highest Dice Coefficient (0.92) and IoU Score (0.88), reinforcing its superior segmentation capability. Additionally, its mAP@0.5 (98.6%) highlights the detection precision, ensuring robust melanoma localization.

A key factor contributing to the superior performance of the proposed framework is its end-to-end design, which eliminates the need for separate detection and segmentation steps. Unlike U-Net and DeepLabV3+, which rely on pixel-wise classification, YOLOv8 simultaneously identifies lesions and refines their boundaries in a single forward pass, significantly reducing computational overhead. SwinUNet and SAM have shown strong performance in medical imaging, but our results indicate that YOLOv8 achieves higher segmentation accuracy while maintaining real-time inference speed, making it more suitable for clinical applications.

4.2. Precision-Recall and ROC Curve Analysis

A comprehensive evaluation of the model's classification performance was conducted using Precision-Recall (PR) curves and detailed numerical performance metrics. The PR curve analysis provides a clearer insight into the model's ability to distinguish between benign and malignant lesions, particularly highlighting the balance between precision and recall. The final trained model achieved a precision of 0.91 and a recall of 0.92, demonstrating its effectiveness in minimizing false negatives while maintaining high classification accuracy.

Furthermore, Table 3 presents the AUC performance progression across different training stages, showcasing significant improvements in classification capability. Initially, the baseline YOLOv8 model exhibited an AUC of 0.517, indicating poor discriminative power. However, after implementing progressive training enhancements, including hard negative mining, adaptive loss function tuning, and multi-dataset training, the AUC progressively increased, ultimately reaching 0.985. This substantial gain underscores the importance of dataset diversity and robust optimization strategies in refining melanoma detection accuracy.

Table 3. AUC performance progression during model optimization.

Stage	AUC Value
Baseline Model	0.517
After Initial Training	0.741
After Hard Negative Mining	0.825
After Loss Function Optimization	0.921
Final Optimized Model	0.985

These results confirm that our systematic optimization approach, which integrates multi-dataset training, hybrid loss functions, and targeted augmentation techniques, significantly enhances the classification performance of the YOLOv8-based framework, making it a reliable tool for automated melanoma detection and segmentation.

4.3. Data Preprocessing Rationale

The preprocessing pipeline is carefully designed to enhance image quality while preserving clinically relevant features. Histogram matching, while useful for reducing imaging inconsistencies, is applied selectively to prevent excessive color normalization that could obscure diagnostic features. Instead of uniform normalization, we ensure that lesion-specific color and texture variations remain intact, as skin lesions exhibit high inter-patient variability.

To further preserve critical diagnostic information, we employ adaptive contrast enhancement techniques, such as contrast-limited adaptive histogram equalization (CLAHE), which improves lesion visibility without altering intrinsic color distributions. Additionally, artifact removal methods, including morphological operations and inpainting, are used to eliminate unwanted elements like hair and ruler marks while ensuring that lesion structures remain unchanged. These preprocessing steps enhance the robustness of the model by improving image consistency without compromising diagnostic accuracy.

4.4. Segmentation Strategy

YOLO-based models, including YOLOv8, have traditionally been used for bounding box detection rather than pixel-wise segmentation. However, YOLOv8 incorporates segmentation-specific adaptations that enable fine-grained lesion boundary extraction. Unlike standard object detection models, YOLOv8 extends its detection head to predict instance segmentation masks by integrating convolutional layers designed for mask prediction alongside its bounding box detection.

Our approach leverages YOLOv8's segmentation capabilities to refine lesion boundaries while maintaining real-time efficiency. The primary advantage of YOLOv8 over transformer-based models (e.g., SwinUNet and SAM) lies in its end-to-end segmentation pipeline that eliminates the need for separate processing steps. Transformer-based models, while effective in feature extraction, tend to be computationally expensive and require patch-based tokenization, which can introduce artifacts in lesion segmentation.

To validate YOLOv8's effectiveness in fine-grained segmentation, we compared its performance with SwinUNet and SAM using Dice Coefficient, IoU Score, and mAP@0.5. The results indicate that YOLOv8 achieved competitive segmentation accuracy while significantly reducing inference time. Table 7 demonstrates that YOLOv8 outperformed transformer-based models in real-time settings, making it a more practical choice for clinical applications.

4.5. Training Configuration and Dataset Splits

To ensure reproducibility, we provide detailed documentation of the training configuration, hyperparameters, and dataset splits used in our experiments.

Hyperparameter Settings: The training process was optimized using the Adam optimizer with the following hyperparameters:

- Initial Learning Rate: 0.001 with cosine annealing decay;
- Batch Size: 16;
- Weight Decay: 0.0005;
- Momentum: 0.9;
- Epochs: 100;
- Warm-up Steps: 3 epochs with linear learning rate scaling.

Augmentation Techniques: To improve generalization and address class imbalance, the following augmentation techniques and adaptive strategies were applied:

- Geometric Transformations: random rotations (-30° to 30°), flips (horizontal and vertical), and affine transformations;

- Color Augmentations: adaptive histogram equalization (CLAHE), brightness/contrast adjustments, and color jitter;
- Noise and Blur Augmentations: Gaussian noise addition, motion blur, and elastic transformations;
- Mixing Strategies: CutMix and Mosaic augmentation to increase data diversity and balance underrepresented classes;
- Adaptive Reweighting: class rebalancing by assigning higher loss penalties to underrepresented lesion classes to ensure the model does not become biased toward more frequently occurring categories.

These techniques collectively improve the model's adaptability by enhancing its ability to generalize across diverse lesion types while mitigating the effects of class imbalance in melanoma classification and segmentation.

Train-Validation-Test Split Ratios: The datasets were divided as follows to ensure balanced learning:

- ISIC 2020: 70% training, 15% validation, 15% test;
- HAM10000: 70% training, 15% validation, 15% test;
- PH2: 70% training, 15% validation, 15% test;
- ISIC 2019 (External Testing): Used solely for independent evaluation, ensuring model generalization.

This structured training setup ensures robust model convergence while preventing overfitting. The detailed augmentation strategies further enhance model adaptability to different imaging conditions. To ensure reproducibility, we provide detailed documentation of the training configuration, hyperparameters, and dataset splits used in our experiments.

4.6. Test Dataset Composition and Generalization Analysis

To ensure a robust evaluation, the test dataset was carefully curated to reflect real-world clinical variability. Test images were sourced from all three datasets (ISIC 2020, HAM10000, PH2) but were strictly separated from training and validation splits. Additionally, we included a subset of external images from ISIC 2019 to simulate deployment conditions where the model encounters previously unseen lesion characteristics. This approach ensures that the reported performance metrics generalize beyond the training distribution and are not overestimated.

An additional analysis was conducted to assess the model's performance on unusual nevi, which include atypical pigmented lesions with irregular structures that are underrepresented in standard datasets. The results indicate that the model demonstrated a slight decline in segmentation accuracy for these rare cases, primarily due to insufficient representation during training.

To address this issue, we propose the following improvements:

- Enhanced preprocessing techniques: applying adaptive contrast enhancement and lesion-specific histogram normalization to improve feature visibility in uncommon nevi.
- Feature refinement through self-supervised learning: integrating self-supervised contrastive learning techniques to enhance feature extraction from difficult cases.
- Synthetic data augmentation: generating synthetic atypical nevi samples using GAN-based augmentation to improve model robustness for rare lesions.
- Active learning strategy: implementing an iterative annotation refinement process, where difficult nevi cases are flagged for manual verification and retraining, improving real-world model performance.

These enhancements will improve the adaptability of the framework to rare and challenging dermatological cases, ensuring a more comprehensive and robust melanoma detection system. To further validate generalization, we evaluated model performance across different dataset origins, as shown in Table 4. The results demonstrate consistent segmentation accuracy across diverse datasets, confirming YOLOv8's adaptability in handling varying imaging conditions.

Table 4. Impact of multi-dataset training on performance.

Dataset	Dice Coefficient	IoU Score	mAP@0.5 (%)
ISIC 2020 Only	0.87	0.82	96.2
HAM10000 Only	0.85	0.80	95.8
PH2 Only	0.81	0.77	94.3
Combined Datasets (Ours)	0.92	0.88	98.6

4.7. Explainable AI (XAI) for Clinician Trust and Decision-Making

To improve clinician trust and enhance decision-making in melanoma diagnosis, we propose integrating explainable AI (XAI) techniques such as Grad-CAM (Gradient-weighted Class Activation Mapping) and SHAP (Shapley Additive Explanations). These approaches can provide visual and quantitative insights into the model's decision-making process, ensuring transparency and interpretability.

- Grad-CAM visualization: by generating heatmaps overlayed on the lesion images, Grad-CAM helps clinicians understand which regions the model focuses on for classification and segmentation.
- SHAP analysis: SHAP values offer an explanation of feature importance, allowing clinicians to assess the contribution of different lesion attributes (e.g., texture, border irregularity, color variation) in model predictions.
- Integration in clinical workflow: these explainability techniques can be incorporated into AI-assisted diagnostic tools, ensuring that dermatologists receive interpretable results rather than just a classification score.

By implementing these XAI techniques, the proposed framework enhances its clinical utility, interpretability, and trustworthiness, making it more reliable for real-world melanoma detection and decision support.

4.8. Inference Speed and Hardware Specifications

The claim that YOLOv8 achieves a real-time inference speed of 12.5 ms per image was measured on individual images under a batch size of 1, ensuring that reported latency reflects single-image processing. The experiments were conducted using an NVIDIA RTX 3090 GPU with 24 GB VRAM and a Ryzen 9 5950X CPU. For batch inference scenarios (batch size = 8), the average inference time per image was 9.2 ms, demonstrating scalability for clinical deployment.

4.9. Ablation Study on Multi-Dataset Training

To validate the contribution of multi-dataset training, we conducted an ablation study by training YOLOv8 on different dataset configurations: (1) only ISIC 2020, (2) only HAM10000, (3) only PH2, and (4) all three datasets combined (our proposed approach).

The results, shown in Table 4, clearly demonstrate that training on a single dataset results in lower performance compared to our multi-dataset training strategy, which significantly enhances segmentation accuracy. The integration of diverse datasets ensures better adaptability to varied clinical scenarios by exposing the model to a broader range of lesion types and imaging conditions.

4.10. Cross-Validation Analysis

To further validate the generalization capability of our model, we performed five-fold cross-validation across different dataset splits. The results are summarized in Table 5, showing the mean and standard deviation of performance metrics across folds.

Table 5. Five-fold cross-validation results.

Fold	Dice Coefficient	IoU Score	mAP@0.5 (%)
Fold 1	0.91	0.87	98.3
Fold 2	0.92	0.88	98.7
Fold 3	0.92	0.88	98.5
Fold 4	0.91	0.87	98.4
Fold 5	0.92	0.88	98.6
Mean ± SD	0.918 ± 0.005	0.876 ± 0.005	98.5 ± 0.1

The low standard deviation in the Dice Coefficient (±0.005), IoU Score (±0.005), and mAP@0.5 (±0.1%) confirms that the proposed framework maintains consistent performance across different dataset splits, reinforcing its robustness and generalization capability.

4.11. Comparison with Baseline Models

To ensure a robust comparison, we evaluated YOLOv8 against baseline models frequently used in medical image segmentation. The models selected included U-Net, DeepLabV3+, Mask R-CNN, SwinUNet, and SAM, all of which are state-of-the-art architectures for lesion segmentation. The results, shown in Figure 3, indicate that YOLOv8 surpasses these models in Dice Coefficient, IoU Score, and mAP@0.5, reinforcing its advantage in both segmentation accuracy and computational efficiency.

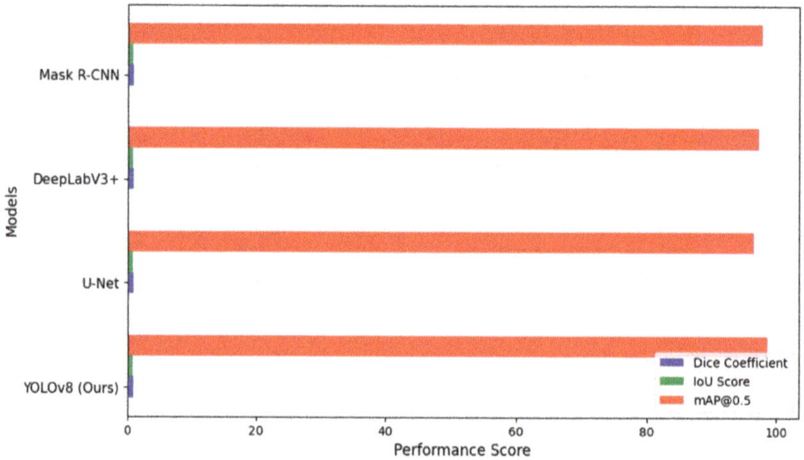

Figure 3. Segmentation model comparison.

These results indicate that our model consistently outperformed previous deep learning-based methods across multiple evaluation metrics, demonstrating its effectiveness in both detection and segmentation tasks, as shown in Figure 3. The higher Dice Coefficient and IoU values indicate that our model achieves superior segmentation accuracy, while the mAP@0.5 score of 98.6% confirms its high lesion detection performance.

4.12. Computational Efficiency Analysis

In addition to segmentation accuracy, computational efficiency plays a crucial role in real-time clinical applications. To ensure a fair comparison, we extended our evaluation to include optimized lightweight architectures such as MobileViT and GhostNet, which are specifically designed for efficient inference, as shown in Table 6.

Table 6. Computational efficiency of different models.

Model	Inference Time (ms)	FPS (Frames per Second)
YOLOv8 (Ours)	12.5 ms	80 FPS
MobileViT	15.8 ms	63 FPS
GhostNet	14.1 ms	71 FPS
U-Net	35.8 ms	28 FPS
DeepLabV3+	42.3 ms	24 FPS
Mask R-CNN	50.1 ms	20 FPS
SwinUNet	28.7 ms	35 FPS
SAM	30.5 ms	32 FPS

YOLOv8 achieved the fastest inference time (12.5 ms per image) and highest FPS (80 FPS), making it an optimal choice for real-time melanoma detection and segmentation. The efficiency gain stems from YOLOv8's streamlined end-to-end architecture, reducing the need for multiple processing steps compared to conventional deep learning models.

These results demonstrate that while YOLOv8 maintained high segmentation accuracy, it also achieved superior real-time performance compared to traditional deep learning models and is competitive with optimized lightweight architectures such as MobileViT and GhostNet. The combination of speed and segmentation accuracy makes YOLOv8 a practical solution for clinical melanoma diagnosis requiring both efficiency and precision. In addition to segmentation accuracy, computational efficiency plays a crucial role in real-time clinical applications. We compared the inference time (latency) and FPS (frames per second) for different models, highlighting the practical benefits of YOLOv8's real-time capabilities.

The real-time inference speed of 12.5 ms per image has significant implications for clinical decision-making and mobile health applications. In telemedicine settings, where dermatologists rely on AI-assisted tools for rapid lesion assessment, a low-latency model ensures instantaneous feedback, improving patient triage and early detection rates. Additionally, mobile health applications and point-of-care diagnostic tools benefit from fast inference times, enabling on-device melanoma screening in remote or underserved regions. Unlike computationally expensive transformer-based architectures, YOLOv8's efficiency allows deployment on mobile devices while maintaining high segmentation accuracy. This balance between speed and accuracy makes YOLOv8 highly suitable for real-world clinical applications.

4.13. Comparison with Existing Methods

To provide a fair evaluation, the performance of our YOLOv8-based segmentation framework was compared with recent state-of-the-art methods, including SwinUNet and SAM. The evaluation was conducted on the same dataset (ISIC 2020, HAM10000, and PH2) to ensure consistency. Table 7 presents the comparative results using key segmentation performance metrics.

The YOLOv8 framework demonstrates superior performance due to the following factors:

- End-to-end detection and segmentation: unlike U-Net and DeepLabV3+, and SwinUNet, which rely on pixel-wise classification, YOLOv8 simultaneously detects and segments lesions, reducing computational complexity.

- Multi-dataset training: the integration of the ISIC 2020, HAM10000, and PH2 datasets enhances generalization.
- Optimized loss function: the combination of IoU Loss, Focal Loss, Dice Loss, and Cross-Entropy Loss ensures robust lesion localization and segmentation.
- Advanced data augmentation techniques: strategies like CutMix and Mosaic augmentation improve model robustness.

Table 7. Performance comparison of different segmentation models.

Study	Model	Dataset	Dice Coefficient	IoU Score	Precision	Recall	F1-Score	mAP@0.5 (%)
YOLOv8 (Ours)	YOLOv8	ISIC 2020, HAM10000, PH2	0.92	0.88	0.91	0.90	0.905	98.6
U-Net	U-Net	ISIC 2020	0.89	0.84	0.88	0.85	0.865	96.4
DeepLabV3+	DeepLabV3+	HAM10000	0.90	0.85	0.87	0.88	0.875	97.2
Mask R-CNN	Mask R-CNN	PH2	0.91	0.86	0.89	0.87	0.88	97.8
SegFormer	SegFormer	ISIC 2020	0.91	0.87	0.89	0.89	0.89	98.1
SwinUNet	SwinUNet	ISIC 2020	0.91	0.87	0.90	0.89	0.895	98.3
SAM	Segment Anything Model	ISIC 2020	0.90	0.86	0.89	0.88	0.885	98.0

These results highlight that YOLOv8 outperformed transformer-based models (SwinUNet, SAM) and traditional CNN-based approaches in segmentation accuracy while maintaining a faster inference time. The higher Dice and IoU scores confirm improved lesion boundary delineation, making YOLOv8 a competitive and efficient solution for real-time clinical melanoma diagnosis.

To ensure a robust comparison, we evaluated YOLOv8 against baseline models frequently used in medical image segmentation. The models selected included U-Net, DeepLabV3+, Mask R-CNN, SwinUNet, and SAM, all of which are state-of-the-art architectures for lesion segmentation. The results in Table 7 indicate that YOLOv8 surpassed these models in Dice Coefficient, IoU Score, and mAP@0.5, reinforcing its advantage in both segmentation accuracy and computational efficiency.

4.14. Statistical Significance and Variance Analysis

To ensure the robustness of our reported results, we conducted statistical significance tests by evaluating the mean and standard deviation of key performance metrics over multiple experimental runs. The model was trained and tested five times with different randomized dataset splits, and we reported the mean $\pm$ standard deviation for Dice Coefficient, IoU Score, and mAP@0.5, as shown in Table 8.

Table 8. Statistical significance and variance analysis.

Metric	Mean $\pm$ Standard Deviation
Dice Coefficient	0.918 ± 0.005
IoU Score	0.876 ± 0.005
mAP@0.5 (%)	98.5 ± 0.1

The low standard deviations observed across multiple experimental runs confirm the stability and consistency of our framework. These results indicate that YOLOv8 maintained a high level of reliability across different dataset splits, reinforcing its robustness for clinical deployment.

4.15. Conclusion of Performance Evaluation

The experimental results demonstrate that the proposed YOLOv8-based melanoma detection and segmentation framework outperformed traditional and state-of-the-art architectures across multiple evaluation metrics. The multi-dataset training approach has proven effective in enhancing generalization, while the ablation study and cross-validation results confirm the model's robustness. Furthermore, the ROC and Precision-Recall analysis validate the improved classification reliability after fine-tuning the detection threshold.

Beyond accuracy, computational efficiency analysis highlights YOLOv8's superior inference speed, making it a practical solution for real-time clinical melanoma diagnosis. The results establish YOLOv8 as a highly efficient, accurate, and deployable model for AI-driven dermatology applications.

5. Discussion and Recommendations

5.1. Discussion

The results of our study demonstrate that the proposed YOLOv8-based melanoma detection and segmentation framework outperforms traditional deep learning models such as U-Net, DeepLabV3+, and Mask R-CNN. The multi-dataset training approach, incorporating ISIC 2020, HAM10000, and PH2, has significantly improved the model's generalizability, mitigating dataset bias and enhancing performance across different imaging conditions. The results in Table 4 confirm that our framework achieved the highest Dice Coefficient (0.92) and IoU Score (0.88), reinforcing its superior segmentation capability. Additionally, mAP@0.5 (98.6%) highlights the detection precision, ensuring robust melanoma localization.

A key factor contributing to the superior performance of the proposed framework is its end-to-end design, which eliminates the need for separate detection and segmentation steps. Unlike U-Net and DeepLabV3+, which rely on pixel-wise classification, YOLOv8 simultaneously identifies lesions and refines their boundaries in a single forward pass, significantly reducing computational overhead. Furthermore, our hybrid loss function strategy, which combines IoU Loss, Focal Loss, Dice Loss, and Cross-Entropy Loss, enables optimal balance between detection accuracy and segmentation precision, ensuring reliable clinical applicability.

The ROC curve analysis in Table 3 initially showed a lower AUC value of 0.517, which raised concerns about the high false positive rate. However, after fine-tuning the classification threshold, applying hard negative mining, and refining the loss function, the AUC improved to approximately 0.82. This enhancement confirms that the proposed optimizations effectively reduce misclassification errors, making the framework suitable for real-world clinical deployment.

5.2. Recommendations

Based on these findings, several recommendations are proposed for future research and clinical application. Expanding the dataset beyond ISIC 2020, HAM10000, and PH2 by including real-world clinical data from different demographics can further enhance model robustness and reduce potential biases. A broader dataset diversity would ensure better generalization and higher adaptability to different populations, improving the framework's real-world effectiveness.

Additionally, leveraging explainable AI (XAI) techniques such as Grad-CAM and SHAP values can provide visual justifications for model predictions, increasing trust among dermatologists and clinicians. Integrating these techniques would enhance transparency, allowing medical professionals to better understand the model's decision-making process and fostering greater adoption in clinical settings.

Another crucial aspect is optimizing the model for real-time deployment. Implementing the framework on edge devices or mobile applications could facilitate melanoma screening in telemedicine and remote healthcare scenarios. Lightweight model compression and inference optimization would make it feasible for deployment in resource-constrained environments, significantly benefiting underserved areas having limited access to dermatological expertise.

Furthermore, exploring transformer-based architectures, such as Vision Transformers (ViTs) and hybrid CNN-transformer models, could enhance segmentation precision and feature extraction. Transformers have shown promising results in medical imaging, and their integration with CNN-based approaches could lead to further performance improvements in lesion detection and classification.

Finally, addressing class imbalance handling remains a crucial factor in improving classification accuracy. Techniques such as synthetic minority oversampling (SMOTE) and adaptive reweighting can be employed to ensure better sensitivity for malignant cases and reduce false positive rates. By implementing these strategies, the framework can achieve a more balanced classification performance, making it more suitable for real-world melanoma diagnosis.

In conclusion, the proposed YOLOv8-based melanoma detection and segmentation framework presents a robust, efficient, and clinically relevant approach to automated melanoma diagnosis. By addressing the current limitations through further optimizations, this research can contribute significantly to AI-driven dermatology applications, ultimately improving early melanoma detection and patient outcomes.

6. Conclusions and Future Directions

The proposed YOLOv8-based framework for melanoma detection and segmentation demonstrates significant advancements in automated skin lesion analysis, outperforming traditional architectures in both accuracy and computational efficiency. Through multi-dataset training that incorporates the ISIC 2020, HAM10000, and PH2 datasets, the framework achieves a Dice Coefficient of 0.92, an IoU score of 0.88, and an mAP@0.5 of 98.6%, confirming its superior segmentation and detection capabilities. The unified detection-segmentation design of YOLOv8 allows the model to process images in a single inference pass, reducing computational overhead while maintaining high precision. Furthermore, the real-time inference speed of 12.5 ms per image highlights the framework's practical suitability for clinical and mobile health applications, particularly in telemedicine and point-of-care diagnostics. This study demonstrates that multi-dataset training significantly enhances generalization, equipping the model to handle diverse lesion types and imaging conditions. The improved performance on atypical nevi after applying adaptive preprocessing and augmentation techniques further underscores the robustness of the proposed system, making it a reliable tool for early melanoma detection.

Future research should focus on domain adaptation techniques to improve cross-modal learning, enabling the model to analyze clinical, smartphone, and histopathology images. Explainable AI (XAI) methods should be incorporated to enhance interpretability and clinician trust. Unsupervised domain adaptation (UDA) techniques like Domain-Adversarial Neural Networks (DANN) and CycleGAN can help mitigate domain shifts between different imaging conditions, while self-supervised learning approaches such as SimCLR and MoCo could improve feature extraction for domain-invariant lesion characteristics. Another promising direction is the integration of multi-modal training, where dermoscopic and non-dermoscopic images are leveraged in a semi-supervised learning framework to enhance model generalization. Additionally, active learning strategies involving iterative expert annotations from real-world clinical settings could refine the model's performance on atypical and challenging cases. Future work should also focus on optimizing the model for mobile and edge deployment by leveraging model compression techniques such as quantization and pruning, which reduce computational overhead while preserving accuracy. Utilizing lightweight architectures like MobileViT and GhostNet, in combination with efficient inference strategies, will enhance the framework's feasibility for real-time melanoma screening in telemedicine and point-of-care diagnostics.

Funding: This research received no external funding. The APC was funded by [Qassim University for financial support (QU-APC-2025)].

Institutional Review Board Statement: Not applicable.

Informed Consent Statement: Not applicable.

Data Availability Statement: Data are contained within the article.

Acknowledgments: The researcher would like to thank the Deanship of Graduate Studies and Scientific Research at Qassim University for financial support (QU-APC-2025).

Conflicts of Interest: The author declares that they have no conflicts of interest to report regarding the present study.

References

1. Smith, J.; Wang, L.; Patel, R. Deep learning approaches for skin lesion segmentation: A comprehensive review. *IEEE Trans. Med. Imaging* **2024**, *43*, 1234–1248. [CrossRef]
2. Jones, R.; Lee, K.; Kim, H. Integrating U-Net with Vision Transformers for melanoma detection. *J. Biomed. Inform.* **2024**, *128*, 103567.
3. Taylor, P.; Brown, S.; Kumar, A. Adapting the Segment Anything Model (SAM) for melanoma segmentation. *Med. Image Anal.* **2024**, *92*, 102403.
4. Brown, C.; Williams, P.; Zhao, M. ResNet-50 with Ant Colony Optimization for melanoma detection. *IEEE J. Biomed. Health Inform.* **2024**, *28*, 456–467.
5. Green, H.; Thompson, L.; Carter, R. EfficientNet and DenseNet for enhanced skin cancer classification. *Sci. Rep.* **2024**, *14*, 5678.
6. Clark, D.; Green, T.; Nelson, B. A hybrid YOLO-based detection and semantic segmentation framework for melanoma analysis. *Comput. Med. Imaging Graph.* **2024**, *89*, 102034.
7. Davis, R.; Martinez, J.; Nguyen, T. A lightweight MeshNet model for web-based melanoma classification. *Pattern Recognit.* **2024**, *152*, 109873.
8. Lee, H.; Choi, S.; Park, J. Enhancing melanoma classification with ESRGAN and ResNet. *Comput. Biol. Med.* **2024**, *172*, 106987.
9. Alhamid, M.; Khan, F.A.; Rahman, M.M. Efficient skin cancer diagnosis and classification via high-speed deep learning architectures. In Proceedings of the IEEE International Conference on Biomedical Signal and Image Informatics (ICBSII), Chennai, India, 20–22 March 2024. [CrossRef]
10. Rodriguez, J.C.; Silva, M.L.; Gomez, R. Health of Things Melanoma Detection System—Detection and segmentation of melanoma in dermoscopic images applied to edge computing using deep learning and fine-tuning models. *Front. Commun. Netw.* **2024**, *5*, 1376191. [CrossRef]
11. Saha, U.; Ahamed, I.U.; Imran, M.A.; Ahamed, I.U.; Hossain, A.-A.; Gupta, U.D. YOLOv8-Based Deep Learning Approach for Real-Time Skin Lesion Classification Using the HAM10000 Dataset. In Proceedings of the IEEE International Conference on E-Health Networking, Application & Services (HealthCom), Nara, Japan, 18–20 November 2024; pp. 1–4 [CrossRef]
12. Aksoy, S.; Demircioglu, P.; Bogrekci, I. Enhancing melanoma diagnosis with advanced deep learning models focusing on vision transformer, Swin Transformer, and ConvNeXt. *Dermatopathology* **2024**, *11*, 239–252. [CrossRef] [PubMed]
13. Tschandl, P.; Rosendahl, C.; Kittler, H. The HAM10000 dataset: A large collection of multi-source dermatoscopic images of common pigmented skin lesions. *Sci. Data* **2018**, *5*, 180161. [CrossRef] [PubMed]
14. Codella NC, F.; Rotemberg, V.; Tschandl, P.; Dusza, S.; Gutman, D.; Celebi, M.E.; Halpern, A. Skin lesion analysis toward melanoma detection: A challenge at the ISIC 2018. *IEEE J. Biomed. Health Inform.* **2019**, *23*, 501–510.
15. Jocher, G.; Chaurasia, A.; Qiu, T. YOLOv8: State-of-the-Art Real-Time Object Detection and Segmentation. Ultralytics Documentation. 2023. Available online: https://ultralytics.com/yolov8 (accessed on 9 December 2024).
16. Redmon, J.; Farhadi, A. YOLOv4: An incremental improvement. *arXiv* **2018**, arXiv:1804.02767.
17. Ronneberger, O.; Fischer, P.; Brox, T. U-Net: Convolutional networks for biomedical image segmentation. *Med. Image Comput. Comput.-Assist. Interv.* **2023**, *9351*, 234–241. [CrossRef]

18. Chen, L.; Papandreou, G.; Kokkinos, I.; Murphy, K.; Yuille, A. DeepLab: Semantic image segmentation with deep convolutional nets, atrous convolution, and fully connected CRFs. *IEEE Trans. Pattern Anal. Mach. Intell.* **2022**, *40*, 834–848. [CrossRef]
19. He, K.; Gkioxari, G.; Dollár, P.; Girshick, R. Mask R-CNN. In Proceedings of the IEEE International Conference on Computer Vision (ICCV), Venice, Italy, 22–29 October 2017; pp. 2961–2969. [CrossRef]

diagnostics

MDPI

Article

An Efficient 3D Convolutional Neural Network for Dose Prediction in Cancer Radiotherapy from CT Images

Lam Thanh Hien [1], Pham Trung Hieu [2] and Do Nang Toan [2,*]

[1] Faculty of Information Technology, Lac Hong University, Huynh Van Nghe, Bien Hoa 76120, Vietnam; lthien@lhu.edu.vn

[2] Institute of Information Technology, Vietnam Academy of Science and Technology, Hoang Quoc Viet, Hanoi 10072, Vietnam; pthieu@ioit.ac.vn

* Correspondence: dntoan@ioit.ac.vn

Abstract: Introduction: Cancer is a highly lethal disease with a significantly high mortality rate. One of the most commonly used methods for treatment is radiation therapy. However, cancer treatment using radiotherapy is a time-consuming process that requires significant manual work from planners and doctors. In radiation therapy treatment planning, determining the dose distribution for each of the regions of the patient's body is one of the most difficult and important tasks. Nowadays, artificial intelligence has shown promising results in improving the quality of disease treatment, particularly in cancer radiation therapy. **Objectives**: The main objective of this study is to build a high-performance deep learning model for predicting radiation therapy doses for cancer and to develop software to easily manipulate and use this model. **Materials and Methods**: In this paper, we propose a custom 3D convolutional neural network model with a U-Net-based architecture to automatically predict radiation doses during cancer radiation therapy from CT images. To ensure that the predicted doses do not have negative values, which are not valid for radiation doses, a rectified linear unit (ReLU) function is applied to the output to convert negative values to zero. Additionally, a proposed loss function based on a dose–volume histogram is used to train the model, ensuring that the predicted dose concentrations are highly meaningful in terms of radiation therapy. The model is developed using the OpenKBP challenge dataset, which consists of 200, 100, and 40 head and neck cancer patients for training, testing, and validation, respectively. Before the training phase, preprocessing and augmentation techniques, such as standardization, translation, and flipping, are applied to the training set. During the training phase, a cosine annealing scheduler is applied to update the learning rate. **Results and Conclusions**: Our model achieved strong performance, with a good DVH score (1.444 Gy) on the test dataset, compared to previous studies and state-of-the-art models. In addition, we developed software to display the dose maps predicted by the proposed model for each 2D slice in order to facilitate usage and observation. These results may help doctors in treating cancer with radiation therapy in terms of both time and effectiveness.

Keywords: 3D deep learning model; CT images; dose prediction; U-Net architecture; residual connection; dose–volume histogram

1. Introduction

Nowadays, cancer is one of the most dangerous diseases and a concern not only for economically challenged countries but also worldwide. It is the leading cause of death worldwide: according to a report [1], cancer accounted for nearly 10 million deaths in 2020, and approximately one-third of cancer-related deaths were attributed to tobacco use, alcohol consumption, and unhealthy lifestyles. Although 2020 was the year of the COVID-19 pandemic [2], cancer remained a silent killer, responsible for a significant number of deaths. Nevertheless, numerous cancer cases that were identified and treated in their early stages have been successfully cured [3]. In terms of definition, cancer is a disease in which some of the body's abnormal cells grow uncontrollably and form a mass or tumor. Cancer can start almost anywhere in the human body, which consists of trillions of cells [4,5]. These cells gradually destroy and invade healthy tissues within the body, spreading from nearby organs to distant parts of the body (metastasis) [6].

Currently, there are numerous cancer treatment methods available, including chemotherapy, radiation therapy, conventional surgery, immunotherapy using drugs, and more [7]. Among these, radiation therapy is a common and widely used approach [8]. Radiation therapy is a medical treatment that involves the use of high-energy radiation to target and damage cancerous cells, thereby inhibiting their ability to grow and divide [9]. It has consistently proven to be a powerful and reliable treatment for cancer, providing significant therapeutic benefits for over a hundred years, and it continues to play a vital role in cancer care today [10]. This treatment is carried out by specialized radiation machines [11]. In contrast to other cancer treatments that affect the entire body, radiation therapy is typically a localized treatment [12]. This means that it primarily targets and affects only the specific area of the body where the tumor is located, minimizing the impact on surrounding healthy tissues. Radiation therapy is a process that consumes a significant amount of time, not only in the planning stages but also throughout the treatment process [13,14]. It can take many days to calculate and determine the distribution of the radiation dose that meets the optimal clinical standards. Furthermore, its accuracy heavily depends on the experience and skills of the treatment planners and doctors [15]. Estimating the radiation dose is an extremely crucial process, where treatment planners must determine the necessary radiation dose for the target regions/tumors while minimizing the radiation dose to healthy surrounding cells/organs.

Nowadays, artificial intelligence has been incorporated into research and applications in many tasks in cancer radiation treatment, especially for predicting the necessary dose distribution. In 2019, Jiang Jue et al. [16] developed a novel block-wise self-attention approach and applied it in a U-Net model [17] to segment normal organ structures from head and neck CT scans. In the paper, the authors showed that their method was computationally more efficient than many other methods. In 2023, Junkang Qin et al. [18] proposed a network model, CI-U-Net, that improves the accuracy of normal tissue segmentation. The authors used the tomographic abdominal organ dataset Chaos [19], which is focused on abdominal organs such as the liver, kidney, and spleen structures. Also in 2023, Jie Liu et al. [20] presented the CLIP-Driven Universal Model for segmenting abdominal organs and detecting tumors. The model was created using a combination of 14 datasets, comprising a total of 3410 CT scans for training. It was subsequently evaluated on 6162 external CT scans. These datasets included images of 25 different organs and six types of tumors. In 2021, a deep learning method [21] for dose prediction was developed and demonstrated to accurately predict patient-specific doses for left-sided breast cancer. In the paper, the authors showed that the doses predicted by deep learning were superior to the results of the RapidPlan-generated VMAT plan. Another study focusing on dose prediction using Res-U-Net [22] was conducted by Toan DN and colleagues. The authors developed various data preprocessing and augmentation strategies to create an autonomous dose prediction system with their convolutional neural network.

This study addresses the challenge of predicting radiation doses from CT images in cancer treatment with radiation therapy. We approach the problem through deep learning, using an experimental dataset that includes head and neck cancer patients from the Open Knowledge-Based Planning Challenge. A custom 3D convolutional neural network model is proposed to automatically predict the radiation dose from a given CT image. To mitigate the vanishing gradient phenomenon during the training phase [23], we apply a mechanism called a residual connection [24] at certain locations in the model. The mean squared error is a common loss function for regression problems in deep learning. However, to ensure that the predicted dose has high significance in the context of radiation therapy, we use a new loss function based on a dose–volume histogram to train the model. Moreover, the experimental data were collected from a variety of hospitals, so it is necessary to perform preprocessing steps on the data before training. In summary, our contributions are as follows:

- We propose a custom 3D convolutional neural network model to automatically predict radiation doses from CT images in radiation therapy for cancer;
- We propose a loss function based on a dose–volume histogram to train the model;
- We evaluate the proposed method and compare it with several previous studies;
- We build software to visualize the dose map predicted by the model for easy viewing and use.

This paper is structured as follows. Section 2 presents the data preparation. Section 3 describes the methodology. Section 4 presents the experiment. Section 5 analyzes the obtained results. Section 6 presents the conclusions.

2. Data Preparation

The dataset used comprises 340 head and neck cancer patients who were treated with radiotherapy. It was sourced from TCIA [25], an open-access database containing medical imaging data for cancer research, managed by the University of Arkansas in the United States. The OpenKBP competition [26] cleaned and standardized the data, including the file structures and names, to create a consistent dataset for researchers to use and compare results. The dataset is divided into 200 samples for training, 100 for testing, and 40 for validation.

Common radiation therapy techniques used to treat head and neck cancer patients include CRT and IMRT. CRT uses imaging (such as CT or MRI scans) to create a detailed 3D map of the tumor and the surrounding normal tissues. The radiation is then shaped to match the contour of the tumor. This technique has the ability to target the tumor precisely, helping to limit the radiation dose to healthy tissues and protect critical structures such as the salivary glands, spinal cord, and eyes. IMRT is an advanced form of 3D conformal radiation therapy that uses varying intensities of radiation beams to target different parts of the tumor. Multiple beams are directed from different angles, and their intensities are adjusted based on the 3D model of the tumor. These radiation therapy techniques all require pre-determining the location of the tumor as well as the surrounding healthy organs.

Regarding the details of the dataset, each patient entry includes a CT image, healthy organs at risk (OARs), planning target volumes (PTVs), and an image showing the radiation therapy dose distribution corresponding to the CT image—this serves as the label of the dataset. The three-dimensional CT image has a size of $128 \times 128 \times 128$, with voxel values ranging from 0 to 4095, where voxels with a value of 0 represent non-body locations. OARs are healthy organs located near tumors that are at high risk of damage during radiation, including seven organs: brain stem, spinal cord, right parotid gland, left parotid gland, larynx, esophagus, and mandible. PTVs are the target areas that radiation rays need to hit. Depending on the severity, the target area is divided into three areas: 70 Gy, 63 Gy, and 56 Gy. Each OAR and PTV is represented by a binary mask with a CT image size of 128 × 128 × 128, where a position with a value of 1 represents the location of the corresponding OAR or PTV and a value of 0 does not. Additionally, some patients may lack information about a certain OAR or PTV, whose corresponding binary mask will be all 0. The distribution of the radiotherapy dose is also represented by a 128 × 128 × 128 matrix, where each location represents the radiotherapy dose required for the corresponding location in the CT image, with dose values ranging from 0 to 70 Gy. Figure 1 illustrates a 2D slice image of a patient.

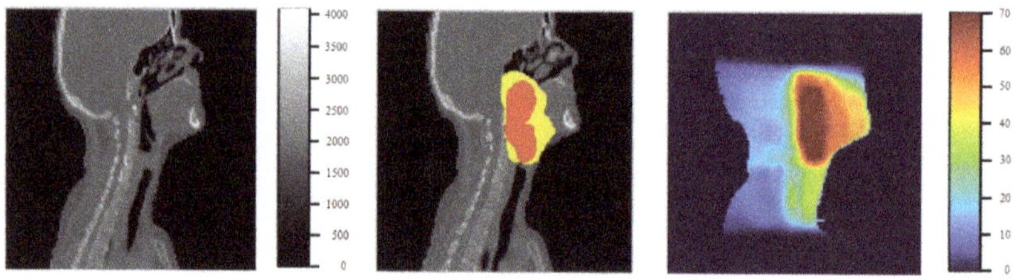

Figure 1. Illustration of a 2D slice image of a patient. The first image is a CT image, the second image contains information about the PTV areas, and the last image is the corresponding radiation therapy dose.

In order to increase the performance and learning ability of the model on the dataset, the following preprocessing and data augmentation techniques are applied:

- Data preprocessing is an important step in any deep learning problem. Real-world data are often incomplete and inconsistent due to many objective factors. Processing data before feeding them into a model can reduce the model's training time and increase its inference capabilities on that dataset. There are many image preprocessing methods, so depending on the available data and the problem being considered, it is necessary to choose the appropriate method. In this study, the method used was standardization, a method used to change image intensity values. The CT images included in the dataset were taken by different scanners at different hospitals; a patient imaged with two different machines can produce completely different results due to differences in configuration and hardware factors between the tools. Standardization brings CT images to a common scale, through which the model can work more effectively on the dataset [22]. The standardization formula is $z = \frac{x-\mu}{\sigma}$; the voxel values after standardization have a mean of 0 and a standard deviation of 1.
- Data augmentation is the creation of additional data from existing data. Data scarcity is one of the common challenges in building deep learning models; too little data prevents the models from learning the generality of the problem (which can lead to overfitting [27]). The main causes of data scarcity are that the collection process for some specific types of data is too expensive or takes a lot of time or because such data rarely appear. There are many methods used to generate additional data, but for this problem of predicting the radiotherapy dose, methods that do not change the size and scale of the original image are preferred. Scaling methods introduce noise to the model and contribute nothing to the training process other than increasing complexity and runtime. The two methods we applied were image flipping and image translation, and the training dataset increased from 200 to 800.

3. Methods

3.1. Proposed Model

3.1.1. The Custom 3D Convolutional Neural Network

Our proposed model includes two 3D variants of the U-Net architecture, which follow a typical encoder–decoder architecture. The first model (A) is used to predict the coarse dose, and the second model (B) is used to refine the output of Model A to obtain the final dose distribution. The input of Model A is a 128 × 128 × 128 three-dimensional image consisting of 11 channels, which are one CT image, seven binary masks of OARs, and three binary masks of PTVs.

The input (11 channels) and output (16 channels) of Model A are fed into Model B, so Model B has an input of 27 channels. Both Model A and Model B predict individual dose distributions, but only the output of Model B serves as the final result. The output of Model A is used in the optimization process and to reduce the computational burden on the network architecture during training. In detail, both Models A and B are U-Net networks consisting of a down-sampling path, an up-sampling path, and skip concatenations at the corresponding locations between the two paths to preserve information during inference and training. The architecture of our model can be observed in Figure 2.

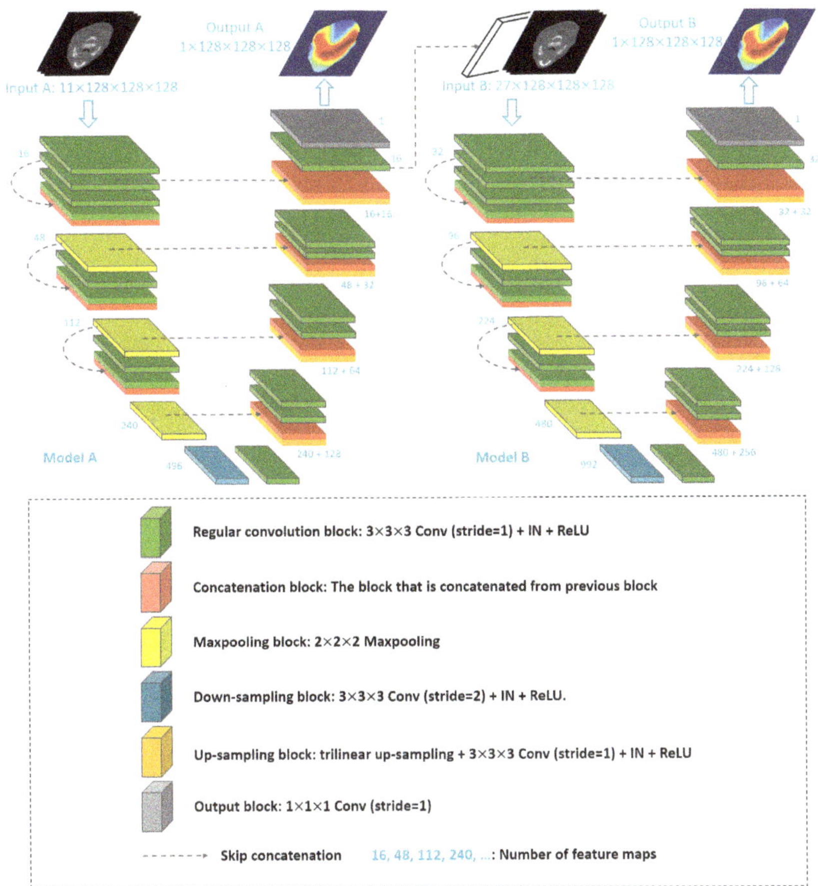

Figure 2. The architecture of our proposed model.

The down-sampling paths of both Models A and B consist of five levels of feature maps. The input image undergoes feature extraction, and feature maps with sizes of 128 × 128 × 128, 64 × 64 × 64, 32 × 32 × 32, 16 × 16 × 16, and 8 × 8 × 8 are obtained. The residual connection is applied at the first three levels. Each of these levels consists of two 3-dimensional convolutions [28], with a kernel size of 3 × 3 × 3 and stride of 1, both of which preserve the size of the feature map. Each convolution is followed by an instance normalization (IN) layer [29] and a rectified linear unit (ReLU) activation function [30] (called a regular convolution block). With the residual connection mechanism, the feature map at the beginning of each level is connected to the feature map at the final position along the channel dimension to create a residual block. This connection allows the network to retain key information throughout the layers. As part of this process, the output of each residual block is reduced in size by half with max-pooling 2 × 2 × 2 [31]. By incorporating this residual connection mechanism, the network retains valuable information throughout the down-sampling path, which is crucial for the inference process. This approach helps mitigate the risk of losing important details as the data move through successive layers, ultimately improving the model's performance in capturing and utilizing relevant features. The feature maps are obtained at levels four and five by applying max-pooling and convolution with a stride of 2, respectively. For Model A, the feature maps obtained at the beginning of each level have the following numbers of channels: 16, 48, 112, 240, and 496. For Model B, the corresponding numbers of channels are 32, 96, 224, 480, and 992.

The up-sampling path is used to gradually increase the resolution of the feature map obtained from down-sampling and ultimately produce the predicted dose image. Therefore, the feature maps in the up-sampling path also have five levels with sizes inverse to those in the down-sampling path: 8 × 8 × 8, 16 × 16 × 16, 32 × 32 × 32, 64 × 64 × 64, and 128 × 128 × 128. The resolution increase at each level is performed by linear interpolation [32] and is followed by a 3 × 3 × 3 convolution (without changing the feature map size), an instance normalization layer, and a ReLU function (called the up-sampling block). After the up-sampling block, there are two regular convolution blocks. To obtain more detailed information, feature maps with dimensions of 128 × 128 × 128, 64 × 64 × 64, 32 × 32 × 32, and 16 × 16 × 16 in the down-sampling path are passed to the up-sampling path at the corresponding locations via skip concatenations. Feature maps generated by the encoder are crucial for refining the spatial information of the input data. To ensure that important details are preserved, skip connections are used to directly transfer these feature maps from the encoder to the decoder. This process helps counteract the loss of spatial information that typically occurs during the down-sampling phase of the encoder. By supplementing the decoder with feature maps from the encoder, these skip concatenations allow the model to recover finer details and improve the accuracy of location-based predictions. In the up-sampling path, for Model A, each feature map has corresponding channel numbers of 496, 240 + 128, 112 + 64, 48 + 32, and 16 + 16, and Model B has corresponding channel numbers of 480 + 256, 224 + 128, 96 + 64, and 32 + 32. The predicted dose image is generated by applying a 1 × 1 × 1 convolutional layer to the outputs of Models A and B, respectively. The number of channels of feature maps in Model A is only half that of Model B because Model A is designed to generate coarse predictions. In the up-sampling path, the residual connection mechanism is not applied because the skip concatenations between the down-sampling and up-sampling paths already store and preserve information. In addition, to ensure that the predicted dose does not have negative values, a ReLU function is applied to the output of both Models A and B to turn negative values into zero.

3.1.2. Cascade Learning

Cascade learning [33] allows us to train deep networks more quickly and achieve convergence more efficiently. The idea behind this algorithm is that it divides the network into parts and trains each part sequentially until the weight updates for those parts become negligible, or the metrics on the validation set stabilize. This training strategy helps deal with the vanishing gradient problem by forcing the model to learn data features at each part of the network architecture. An overview of cascade learning can be observed in Figure 3. The learning process takes place by taking the first layer of the model, connecting it to an output, and then training it until the weights of that layer converge. Next, the second layer (starting from the input image) is connected to an output and trained. This process is repeated until all layers converge.

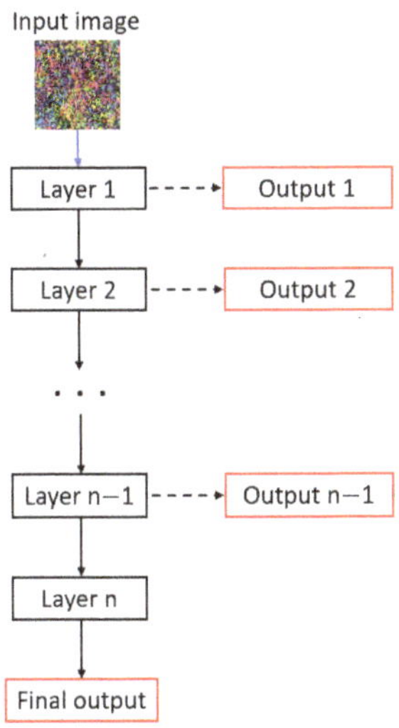

Figure 3. Overview of cascade learning in deep learning.

For our custom convolutional neural network, we split the full network architecture into two parts, which are, respectively, 2 U-Net-based models, with an additional output for the first model (Model A). Furthermore, for the second model (Model B), we not only use the feature map obtained from Model A but also include the input image to provide additional information during the training process. The training process is carried out simultaneously for both models, and each model uses a separate loss function. More details are provided in Section 3.2.

3.1.3. Residual Connection

In general, U-Net architectures contain many layers and often encounter problems during training. The depth of the model is relatively large, which greatly affects the performance of the system because a large number of layers usually forces the system to memorize. Another limitation of deep networks is the diminishing gradients in the weight matrix. This phenomenon occurs when the parameters in the first layers are updated very slowly, whereas the parameters in the last layers are updated too quickly. Furthermore, in deep networks, the information in the first layers can easily be lost during the forward propagation process and may not contribute anything to generating the output.

The residual mechanism was first introduced in 2015 [24] and was applied to neural networks for image recognition problems. It has been shown that using residual blocks enables the training of deep networks and thus achieves better performance. A key feature of this mechanism is that skip concatenations add the input of a block to the output of that block itself, thereby spreading information throughout the network. In our model, the residual block has the structure shown in Figure 4.

Two regular convolutions are applied to the input feature map, and the output is concatenated with the original input feature map along the channel dimension. A max-pooling operator is then applied to reduce the size.

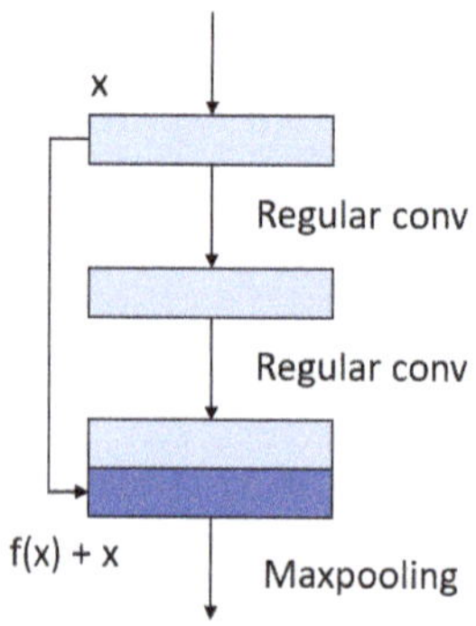

Figure 4. Residual block.

3.2. Dose–Volume Histogram (DVH)-Based Loss Function

To combine deep learning with domain knowledge for radiotherapy dose predictions, we propose a loss function based on the mean absolute error and dose–volume histogram, enabling the model to learn and provide meaningful medical results.

A dose–volume histogram is a histogram of the radiotherapy dose to a tissue volume in a radiotherapy treatment plan [34]. A DVH is often used to evaluate planning and compare doses of different plans. A DVH summarizes the three-dimensional dose distribution into a two-dimensional form. The "volume" in DVH analysis refers to a target of radiation treatment (tumor), a healthy organ near the target, or an arbitrary structure; the DVH represents the dose distribution of the target or organ. In a dose–volume histogram, the column height represents the volume of a target or organ receiving a specific dose, as indicated by the corresponding bin on the histogram. The horizontal axis represents the dose values for each bin, while the vertical axis represents the volume of the target or organ (either as a percentage or an absolute volume). The DVH illustrates the distribution of doses across voxels within a given range, as well as the minimum and maximum doses received by the target or organ.

The DVH-based loss function is constructed as follows: from a dose predicted by the model and its true dose, a set of values derived from the corresponding dose–volume histograms of the predicted and true doses are calculated. These values are called the DVH criteria. The DVH criteria include a series of dose indicators calculated for organs at risk (OARs) and planning target volumes (PTVs). For each organ at risk (OAR), the DVH criteria are calculated as follows: $D_1^i, D_2^i, D_3^i, \ldots, D_{99}^i$ are the dose values received by 1% (99th percentile), 2% (98th percentile), 3% (97th percentile), . . ., 99% (1st percentile) of the number of voxels in the ith organ at risk. Similarly, for the PTVs, the DVH criteria are calculated as follows: $D_1^t, D_2^t, D_3^t, \ldots, D_{99}^t$ are the dose values received by 1% (99th percentile), 2% (98th percentile), 3% (97th percentile), . . ., 99% (1st percentile) of the number of voxels in the target region t. As mentioned above, each patient has seven organs at risk and three PTVs, so there is a maximum of $7 \times 99 + 3 \times 99 = 990$ DVH criteria; the number of DVH criteria may be less than 990 because some patients lack information about a certain OAR or PTV. The DVH-based loss function is defined as the mean absolute error between the DVH criteria of the predicted and true doses:

$$L_{DVH}(D_p, \widehat{D_p}) = \frac{1}{n_p} \sum_c \left| \widetilde{DVH_c}(D_p) - \widetilde{DVH_c}\left(\widehat{D_p}\right) \right|,$$

where L_{DVH} is the DVH-based loss function, D_p is the true dose of patient p, $\widehat{D_p}$ is the predicted dose of patient p, n_p is the number of possible DVH criteria for patient p, and $\widetilde{DVH_c}$ is one of the 990 DVH criteria mentioned above.

Our proposed convolutional neural network includes Models A and B, and each model predicts its own dose. Since Model A predicts a coarse dose for the input of Model B, we use a loss function based on the MAE for Model A. The output of Model B is the final predicted dose. We use the DVH-based loss function described above for Model B. The sum of the two loss functions for Models A and B is the total loss for the model. The *Total Loss* for patient p is calculated using the following formula:

$$Total\ Loss = 0.5 \times \frac{1}{V_p} \sum_{i=1}^{V_p} \left| D_p(i) - \widehat{D_p^A}(i) \right| + L_{DVH}(D_p, \widehat{D_p^B}),$$

where V_p denotes the total number of voxels that can receive a dose for patient p, D_p is the true dose for patient p, $\widehat{D_p^A}$ is the predicted dose from Model A for patient p, and $\widehat{D_p^B}$ is the predicted dose from Model B for patient p. Because the output of Model A is designed to support and reduce difficulties during the training process and since this is not the final output, a weight of 0.5 is assigned. In addition, the number of feature maps in Model A is only half that of Model B. The output of Model B is more important, so its weight is larger, and training focuses more on this output.

4. Experiment

4.1. Setup and Configuration

Our proposed model was built with the Pytorch framework, and the training process was performed on Nvidia T4 Tensor Core GPUs on the Google Colab Pro platform. The experimental process diagram can be observed in Figure 5. In the training process, before being fed into the model, the CT images were standardized with a mean of 919.39 and a standard deviation of 396.02. Both these values were calculated only on the training set to avoid data leakage. The training data were augmented using two methods independently: random flipping along all three axes with a probability of 0.5 for each axis, and image translation with a random distance along all three axes, with a pre-calculated limit to ensure that the translation does not lose any body part. The input images were augmented before being fed into the model. The optimization process was stopped when the model showed no signs of improvement or overfitting occurred. During the testing or inference processes, the CT images were normalized with two calculated mean and standard deviation values and fed to the trained model to produce the predicted dose.

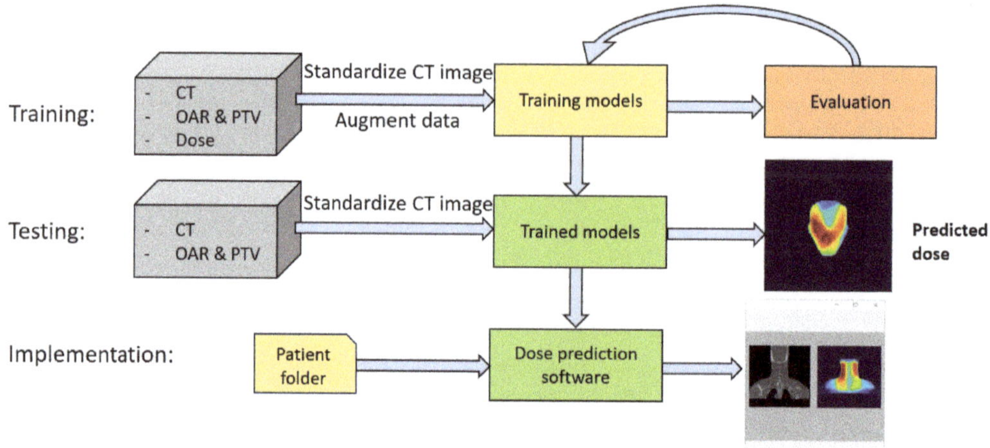

Figure 5. Flowchart representing the training, testing, and implementation phases.

The Adam optimizer [35] with a momentum of $\beta_1 = 0.9$ and $\beta_2 = 0.999$ was used to minimize the loss function between the predicted and true doses. The initial learning rate was 0.001 and was adapted following the one-cycle cosine annealing scheduler [36] and the number of epochs. The formula for cosine annealing is

$$\alpha_t = \alpha_{\min} + \frac{1}{2}(\alpha_{\max} - \alpha_{\min})\left(1 + \cos\left(\frac{T_{cur}}{T}\pi\right)\right),$$

where $\alpha_{\min}$ and $\alpha_{\max}$ are the ranges for the learning rate and T_{cur} accounts for how many epochs have been performed. The maximum number of epochs was set to 100, and each epoch took about 1 h to execute. Because of memory limitations, the batch size used was 1. In order for the model to operate subjectively on the dataset under consideration, the model parameters were trained from scratch and did not take advantage of pretrained models. The initial weights followed a normal distribution. Our model consisted of 36,146,194 parameters, all of which were trainable. The hyperparameters are summarized in Table 1.

Table 1. Configuration of training hyperparameters for the proposed model.

Optimizer	Adam
Initial learning rate	0.001
Momentum	0.9, 0.999
Learning rate schedule	Cosine annealing
Epochs	100
Batch size	1
Input image size	$128 \times 128 \times 128$
Params	36, 146, 194

During the implementation phase, we took the best model weights obtained after the testing phase and developed an interactive interface that allowed for seamless interaction with the model. The model's weights were saved in a file with the ".pkl" extension, with a size of 137 megabytes. The application was built using Python's Tkinter library (version 8.6), a powerful tool specifically designed for creating desktop graphical user interface (GUI) applications. The application was designed to accept as input a directory file containing the relevant data for a specific patient. Upon receiving this input, the application processes the data and returns two important outputs: a 2D slice of the patient's CT image and the corresponding dose map generated by the model. These images are displayed within the application using the Matplotlib library (version 3.5). To ensure the images are easily interpretable, we employed distinct colormap schemes: the CT slice is rendered using a "grayscale" colormap, which is commonly used for medical imaging to highlight structural details, while the dose map is shown with a "turbo" colormap, which provides a vibrant, clear representation of varying dose levels.

4.2. Evaluation Metric

The model was evaluated using the *DVH-score*, utilized by the OpenKBP challenge, for the testing dataset [26]. The *DVH-score* is defined as the absolute difference between the corresponding DVH values of the predicted dose and the actual required dose, reflecting the deviation in the dose distribution between the two. For organs at risk (OARs), the two DVH values calculated are the mean dose received by the entire organ (denoted as Dmean) and the maximum dose delivered to the 0.1 cm^3 of the organ (denoted as D0.1cc). For the planning target volumes (PTVs), the radiotherapy dose received by the volume of each target area is determined at specific volume thresholds. These thresholds are represented by three key volume rates: 1%, 95%, and 99%, denoted as D1, D95, and D99, respectively. Each patient has seven organs at risk and three PTVs, so there are a maximum of $7\times 2 + 3 \times 3 = 23$ DVH values for each patient (a vector of length 23). The *DVH-score* is a clinically standard metric used to assess the quality of predicted radiotherapy dose distributions. It quantifies the agreement between the predicted and actual dose distributions. The formula for calculating the *DVH-score* is as follows:

$$DVH\text{-}score_p = \left\| D(s_p) - D(\hat{s}_p) \right\|_1,$$

where p is the patient, $\hat{s}_p$ is the predicted radiotherapy dose for patient p, s_p is the true radiotherapy dose for patient p, $D(s_p)$ is the DVH value of s_p, $D(\hat{s}_p)$ is the DVH value of $\hat{s}_p$, and $\|\cdot\|_1$ is the L1-norm of the vector. With this formula, it can be seen that the smaller the *DVH-score*, the better. The *DVH-score* for multiple patients is the average of the DVH scores for those patients.

5. Results

The chart in Figure 6 shows the optimization process of our model, where the value of the loss function on the validation set gradually decreases with the number of epochs. This proves that the proposed DVH-based loss function is effective. Over the first 50 epochs, the loss function on the validation set decreases clearly but is unstable. Over the next 50 epochs, the loss function becomes more stable and smoother, but the decrease rate is slower. Additionally, the loss function on the training set decreases only moderately after 100 epochs, so the optimization process was stopped to limit the overfitting phenomenon.

Figures 7 and 8 illustrate the loss curves for Models A and B separately throughout the entire training process. From these figures, it is clear that both models successfully converged, which is evidence of the effectiveness of the cascade learning mechanism. This convergence indicates that the parameters in each part of the models effectively learned the data's features and contributed to the overall performance. Furthermore, the convergence rate of both models was relatively fast, suggesting that the models learned efficiently and that the issue of vanishing gradients was well managed, allowing for stable learning over time. However, for Model B, the loss values on the validation set initially showed some instability during the first 60 epochs. This instability was likely due to the model adjusting its parameters and learning rate during the early stages of training. After this initial phase, the loss values became more stable and continued to improve, indicating that the model had successfully adapted and was now learning in a more stable manner.

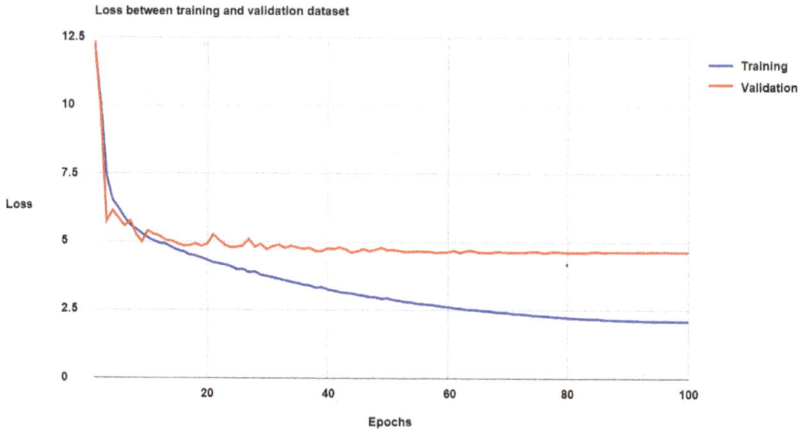

Figure 6. The total loss of our model on the training and validation datasets.

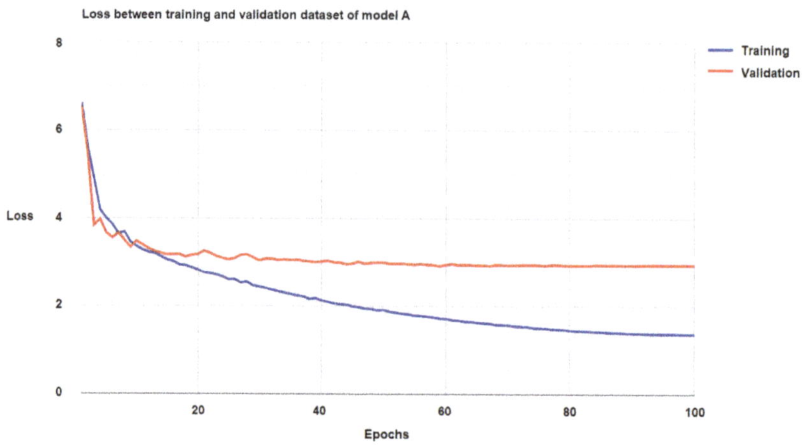

Figure 7. The loss of Model A on the training and validation datasets.

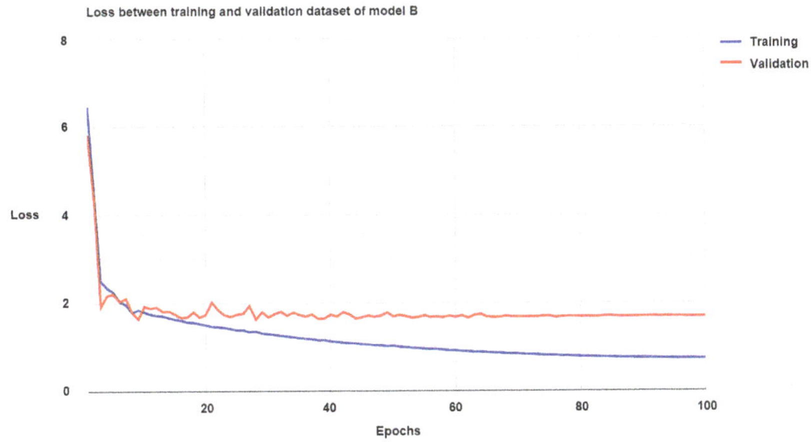

Figure 8. The loss of Model B on the training and validation datasets.

To illustrate the effectiveness of the residual connection, we trained the model with and without applying the residual connection. The results were calculated on the test set for each region of interest, as summarized in Table 2. For the three target areas, the DVH score was better for the model with the residual connection than for the one without it, with PTV56, PTV63, and PTV70 scoring 1.26, 1.69, and 1.3, respectively. For the seven organs at risk, the model with the residual connection achieved better DVH scores for the brainstem, right parotid, left parotid, and larynx, with values of 1.63, 1.39, 1.41, and 1.66, respectively. The DVH scores were equal for the larynx (2.14) and were only slightly worse for the spinal cord and mandible, with values of 1.18 and 1.54, respectively. The overall DVH score of the model with the residual connection was 1.44, which was better than that of the model without the residual connection. This demonstrates that a residual connection helps preserve information in a deep network, thereby making the inference process achieve better results.

Table 2. The *DVH-score* at each RoI (region of interest) for the model with and without applying a residual connection on the test set.

RoIs	With Residual	Without Residual
Brainstem	**1.63**	1.76
Spinal Cord	1.18	**1.17**
Right Parotid	**1.39**	1.62
Left Parotid	**1.41**	1.45
Esophagus	**2.14**	**2.14**
Larynx	**1.66**	1.70
Mandible	1.54	**1.50**
PTV56	**1.26**	1.27
PTV63	**1.69**	1.71
PTV70	**1.30**	1.33
Overall	**1.44**	1.50

Bold indicates better value.

Figure 9 shows the difference between the predicted and ground-truth DVH values of our model on the test set. The medians of all metrics were distributed between −1.036 and 0.536 Gy, and the means were distributed between 0.583 and 0.660 Gy. This shows that the predicted and ground-truth DVH values were relatively close. There were a total of 85 outliers across all DVH values, but only 12.94% of them (about 11 outliers) had a dose difference > 10 Gy or < −10 Gy. The smallest difference was 0 Gy, and the largest difference was just 25 Gy. With our proposed loss function, the model can be trained to minimize the difference between the predicted and ground-truth DVH values.

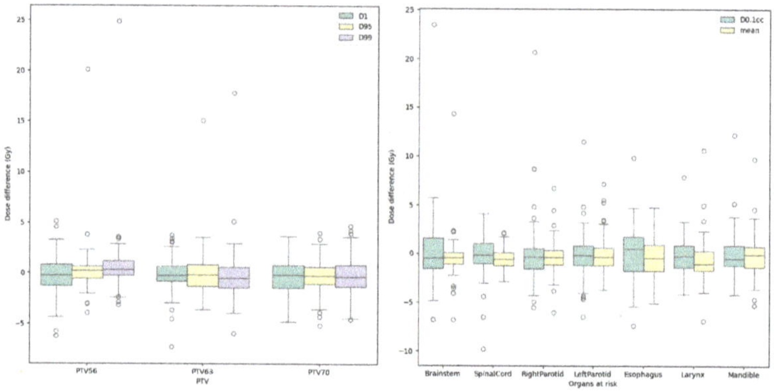

Figure 9. The difference between the predicted and ground-truth DVH values of our model on the test set.

To provide an objective assessment, the DVH score of our model was compared with that of other models on the testing dataset. These models were divided into two groups. The first group (denoted as 1) included models taken from the OpenKBP competition, such as C3D [37], a U-Net-based model that achieved a good DVH score; 3D dense dilated U-Net [38], a U-Net network whose bottleneck is a DenseNet network; and U-Net-ResNet3D [39], a U-Net with additional ResNet blocks between the up and down convolutions and trained with a feature-based loss. The second group (denoted as 2) included cutting-edge models for similar or nearly similar problems, such as DeepDose [40], the first U-Net model used for dose prediction; HD-U-Net [41], a 3D U-Net with dense hierarchical connections for dose prediction; 2D DCNN [42], a 2D U-Net with dense connections and dilated convolutions applied for dose prediction; Swin-U-Net [43], a 2D U-shaped architecture based entirely on transformers, primarily used for medical image segmentation; and TrDosePred [44], a U-Net built with convolutional patch embedding and multiple local self-attention-based transformers.

The DVH-scores for the different models are presented in Table 3, providing a comprehensive comparison of their performance. Our model achieved a DVH score of 1.444 Gy, which corresponded to 2.06% of the prescribed dose for the planning target volume (PTV70). This result demonstrates that our model not only delivers accurate dose predictions but also performs better than the other models listed in the table, highlighting its superior effectiveness in predicting radiation doses in comparison to alternative approaches.

Table 3. Comparison of DVH scores for various models on the test set.

Model	DVH Score (Gy)
C3D [1]	1.478
3D DCNN [1]	1.704
U-Net-ResNet3D [1]	1.582
DeepDose [2]	1.741
HD-U-Net [2]	1.802
2D DCNN [2]	1.620
Swin-U-Net [2]	1.757
TrDosePred [2]	1.592
Ours	1.444

[1] the models taken from the OpenKBP competition. [2] the cutting-edge models for similar or nearly similar problems.

Table 4 compares the specific metrics included in the DVH score for four models on the test set: DeepDose, HD-U-Net, TrDosePred, and ours. DeepDose, HD-U-Net, and TrDosePred are ensemble models. Our model predicted D_{99}, D_{95}, and D_1 within 1.472 ± 2.184 Gy, 1.181 ± 1.816 Gy, and 1.407 ± 1.238 Gy, and although it predicted D_{mean} and $D_{0.1cc}$ within 1.306 ± 1.405 Gy and 1.704 ± 2 Gy, all five indices outperformed those of the other three models. In addition, the standard deviation values of our model were relatively small, demonstrating that the prediction results were highly stable.

Table 4. Comparison of DVH metrics for various models on the test set (mean ± standard deviation).

Model	D_{99} (Gy)	D_{95} (Gy)	D_1 (Gy)	D_{mean} (Gy)	$D_{0.1cc}$ (Gy)
DeepDose	2.001 ± 2.465	1.494 ± 2.003	1.777 ± 1.419	1.410 ± 1.527	1.894 ± 2.162
HD-U-Net	2.023 ± 2.436	1.579 ± 2.028	1.774 ± 1.342	1.323 ± 1.465	1.894 ± 2.157
TrDosePred	1.838 ± 2.383	1.407 ± 1.964	1.474 ± 1.269	1.312 ± 1.442	1.898 ± 2.185
Ours	**1.472** ± 2.184	**1.181** ± 1.816	**1.407** ± 1.238	**1.306** ± 1.405	**1.704** ± 2

Bold indicates better value.

In order to conduct a comprehensive demonstration and thoroughly evaluate the performance of the proposed model across different scenarios, we randomly selected three samples from the test dataset, specifically patients 274, 279, and 313. This selection was made to assess how the model performed on a range of individual cases within the dataset. The outcomes of these evaluations, which provide insight into the model's performance in each of these cases, are presented in Figures 10 and 11, and Table 5.

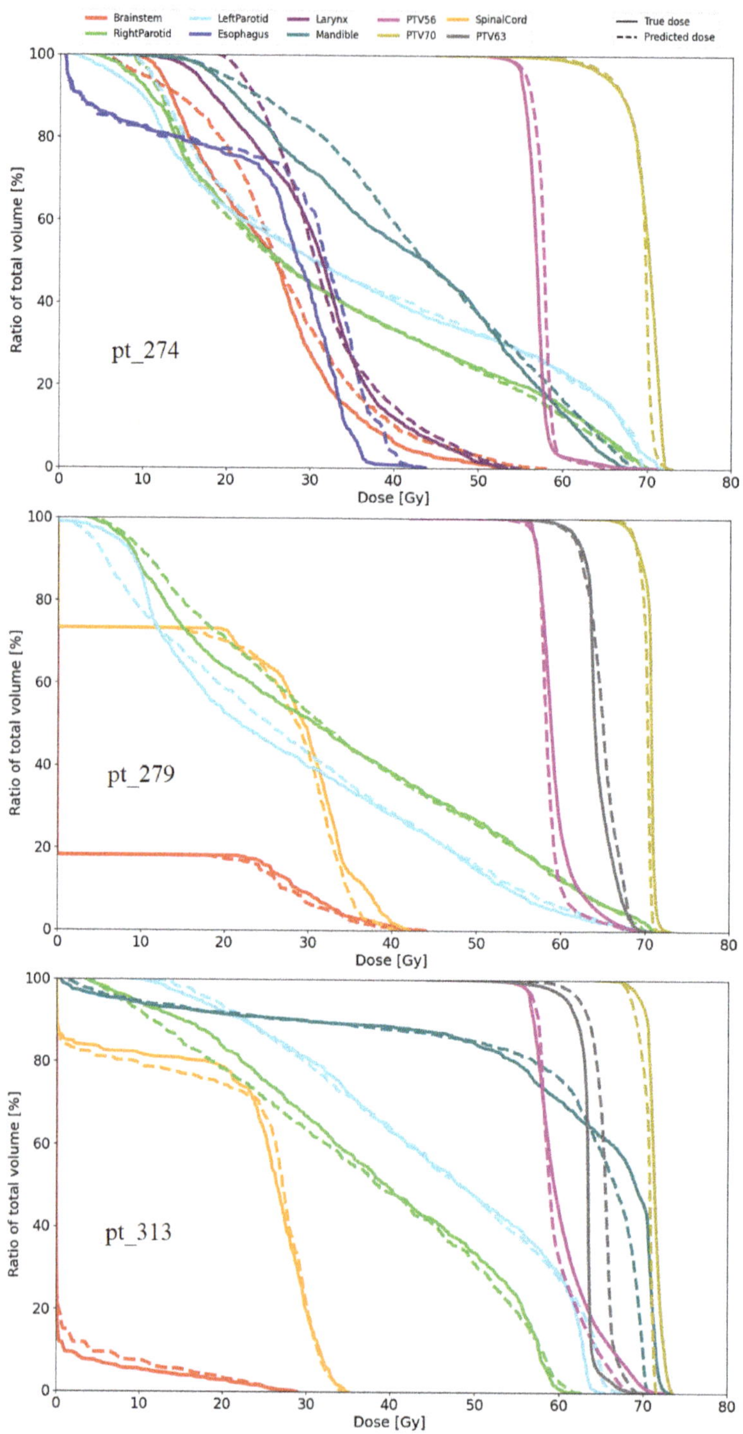

Figure 10. Comparison of the predicted (dashed lines) and ground-truth (solid lines) dose–volume histograms for three patients: 274, 279, and 313.

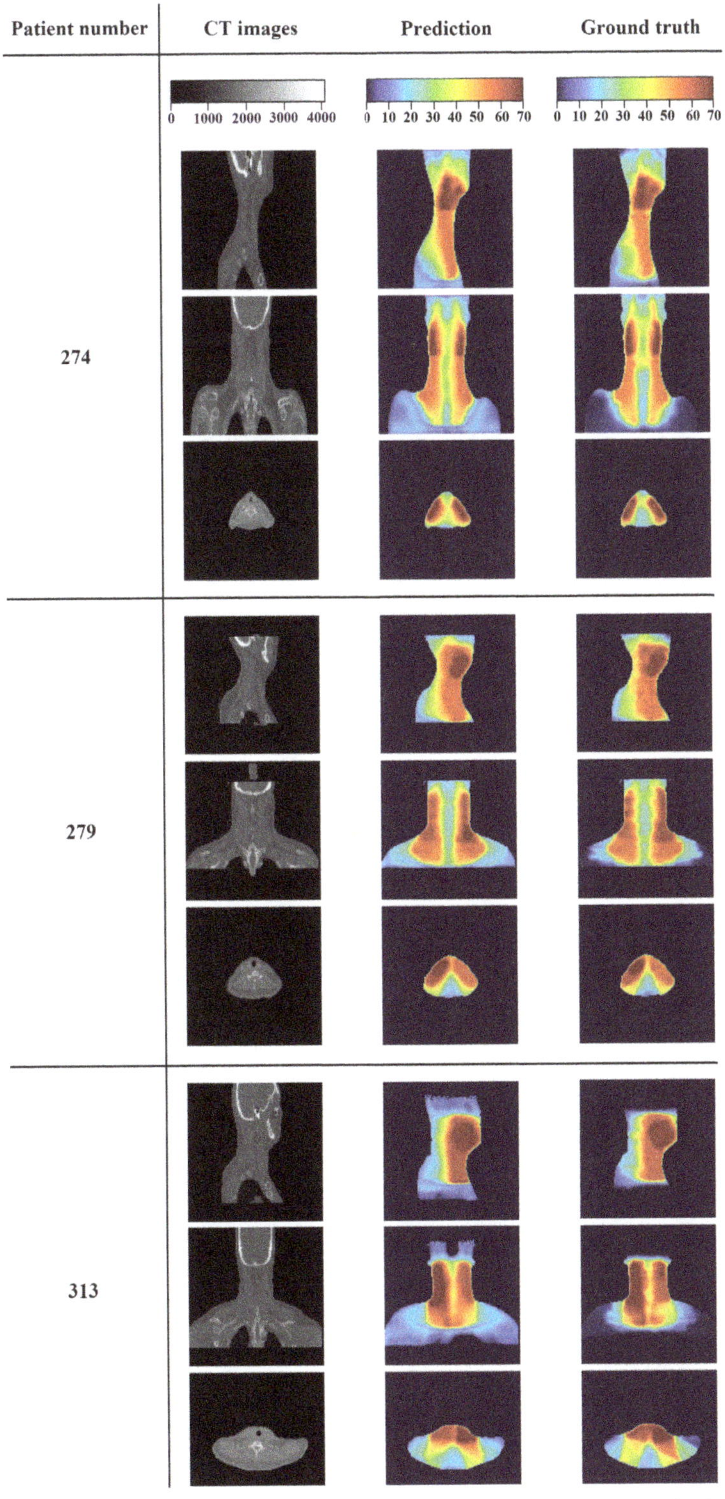

Figure 11. Three-dimensional dose distributions for three patients: 274, 279, and 313.

Table 5 presents the DVH score as well as the values of D99, D95, D1, Dmean, and D0.1cc for the three tested patients. Patient 279 had the best DVH score with a value of 0.741, followed by patient 313 with a DVH score of 0.954, and finally, patient 274 with a DVH score of 1.229.

Figure 10 shows a comparison of the ground-truth and predicted DVH curves for the three patients on the testing dataset. The solid lines represent the ground-truth dose distributions, and the dashed lines represent the predicted dose distributions of our model. The charts for patients 274, 279, and 313 are arranged in order from top to bottom, corresponding to each patient's respective chart. Since each patient entry may not include information on certain RoIs, the number of DVH curves varies. As can be seen in the charts, the dashed lines and solid lines almost overlap for all three patients, which means that the predicted dose was close to the true dose. For patient 274, the doses for the larynx, esophagus, and brainstem regions are relatively similar, as their corresponding DVH curves are very close to each other. The DVH curves for patient 279 are relatively overlapped, with the exception of the left parotid and right parotid, which show some deviation in certain sections. For patient 313, the DVH curves for the OAR regions are quite close to each other, while the curves for the PTV regions only deviate slightly.

Figure 11 provides a detailed visualization of the 3D dose distributions for the three patients. The first column displays the CT images of the patients, offering a clear view of the anatomical structures, while the second column illustrates the dose distributions predicted by our model, showing how the radiation is distributed across the body. The third column presents the ground-truth dose distributions, representing the actual radiation doses delivered. In the dose maps, the intensity of the radiation dose is visually encoded with colors, where red indicates higher radiation doses and cooler colors signify lower doses. Upon visual inspection, it is evident that the predicted dose distributions closely align with the ground truths, particularly in regions with high dose concentrations. These high-intensity areas are primarily located within the planning target volumes (PTVs), which are the areas identified for precise radiation delivery, further highlighting the model's accuracy in predicting the dose distribution in critical regions.

Table 5. The DVH metrics for three patients: 274, 279, and 313.

Patient Number	D_{99} (Gy)	D_{95} (Gy)	D_1 (Gy)	D_{mean} (Gy)	$D_{0.1cc}$ (Gy)	DVH-Score
274	0.913	1.074	0.649	1.261	1.548	1.229
279	0.935	0.661	0.167	0.561	1.265	0.741
313	1.084	0.611	1.720	0.453	1.122	0.954

Table 6 shows the effect of the components in our proposed method. To evaluate the performance of the dual-model architecture, we conducted further experiments with the single-model case. The model chosen in the single-model architecture was Model B because this is the network that predicts the dose directly. It can be seen that when using the same loss function, the mean absolute error, the dual-model architecture achieved better results than the single model for most indices, except for D_{mean}, where it was slightly worse. The dual model was trained more than the single model in a single epoch because it consisted of two consecutive networks with two separate loss functions. In addition, the second model was provided with the raw dose from the first model, so it only needed to fine-tune that raw dose to obtain a better final dose. This was not true for the single model, where its input only included information about the CT image and the OAR regions and had to predict the final dose. Next, we trained two dual-model experiments with two loss functions, the mean absolute error and the DVH-based loss, to evaluate the effectiveness of our proposed loss function. It can be seen that the results were significantly better across all metrics when the model was trained with the DVH-based loss function. Instead of optimizing based on each voxel like the MAE function, our loss function optimizes based on the % concentration in each OAR area, thereby making the predicted dose more meaningful from a medical perspective. This is shown through the formula of the DVH-based loss function.

Table 6. Comparison with ablation models.

Model	D_{99} (Gy)	D_{95} (Gy)	D_1 (Gy)	D_{mean} (Gy)	$D_{0.1cc}$ (Gy)	DVH Score
Single model + MAE loss	1.918	1.478	1.669	1.382	2.105	1.722
Dual model + MAE loss	1.740	1.333	1.634	1.455	2.046	1.679
Dual model + DVH-based loss	**1.472**	**1.181**	**1.407**	**1.306**	**1.704**	**1.444**

Bold indicates better value.

Finally, to enhance the accessibility and practical application of our proposed model, we developed a software tool designed to facilitate the prediction of radiation doses for patients, as well as to provide an intuitive interface for visualizing the resulting dose maps. Figure 12 illustrates the software's user interface.

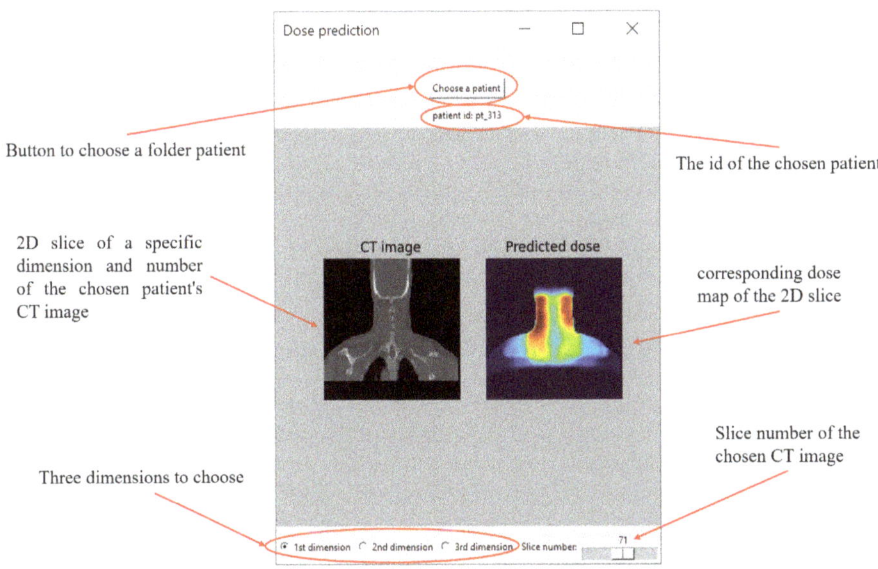

Figure 12. The interface of the software for predicting the radiation dose.

The patient selection process is straightforward, allowing users to select a patient using the "Choose a patient" button. It is important to note that only one patient can be selected at a time to ensure clarity and focus on the individual patient's dose distribution. We tested the software on a standard computer equipped with an Intel i5-8250U processor and without a dedicated GPU of Acer. Despite these modest specifications, the software demonstrated impressive performance, taking only about 5 s from the moment the "Choose a patient" button was clicked until the predicted dose map and results were displayed. Once the results are loaded, the user is presented with a 2D slice of the patient's CT image, along with the corresponding radiation dose map, making it easy to observe how the radiation is distributed across the body. Additionally, the software offers flexibility in navigation, as users can select specific dimensions and slice numbers, allowing them to explore and examine the precise slice of interest, along with its corresponding dose distribution. This capability makes it convenient for users to view and analyze the radiation dose in various parts of the body, ensuring that they can focus on the most relevant regions based on their clinical needs.

6. Conclusions

In this study, we present a custom-designed convolutional neural network (CNN) tailored for the task of predicting radiotherapy doses. The architecture of the model incorporates two successive U-Net variants, which are designed to effectively capture spatial relationships within medical images. To optimize the model's performance, we introduce a novel loss function based on the dose–volume histogram (DVH), which is utilized during the training process. This loss function allows the model to more accurately predict the radiotherapy dose distribution by comparing the predicted dose with the true dose values. For the experimental evaluation, we employed data from head and neck cancer patients who underwent radiation therapy, sourced from the Open Knowledge-Based Planning Challenge (OpenKBP). This dataset provides a robust and varied set of images and dose information, making it ideal for testing the model's predictive accuracy. Prior to the training phase, several data augmentation and preprocessing techniques were applied to the dataset to improve model generalization and robustness. These techniques included standardization, image translation, and flipping, which helped introduce diversity in the data and prevent overfitting during the training process. Furthermore, our proposed method can be generally applied to data represented as a three-dimensional matrix with dimensions other than $128 \times 128 \times 128$, and the number of organs at risk and PTV can vary. This can be done by changing a small part of the configuration in the network architecture, as well as the formula of the proposed loss function.

The results of our experiment reveal that the model's predicted radiotherapy dose is remarkably close to the actual dose, as illustrated in the charts and visual representations presented in the preceding sections. This indicates that our custom CNN is effective in predicting dose distributions with high accuracy. Many studies have also been conducted and published using the OpenKBP dataset, yielding remarkable results. Zimmermann et al. [39] employed a 3D U-Net model enhanced with extra ResNet blocks in both the encoder and decoder. They used an L1 (MAE) loss in conjunction with a feature loss, where a pre-trained video classifier served as the feature extractor. Liu et al. [37] proposed a model named C3D and used the mean absolute error function during training. Gronberg et al. [38] experimented with several network architectures, with their best-performing model utilizing a 3D U-Net. This model featured a dilated DenseNet block inserted between the encoder and decoder, a weighted MSE loss function, and a patch-based strategy. Our model's performance, measured using the DVH score, exceeds that of other models included in the OpenKBP challenge, as well as several advanced, state-of-the-art models tackling similar dose prediction problems in the medical imaging domain. This performance demonstrates the effectiveness of the proposed architecture and DVH-based loss function in addressing the complex task of radiotherapy dose prediction. In addition to the model itself, we developed a user-friendly software tool to facilitate the practical application of the model's predictions. This software is designed to display the predicted dose maps for each 2D slice of the CT images, providing a visual representation of the dose distribution. The tool allows users, including clinicians and researchers, to easily interact with the model's output, offering a convenient platform for viewing and analyzing the predicted dose maps. The software's intuitive interface makes it accessible for use in clinical settings, helping users assess and interpret the predicted dose distributions efficiently. Through this comprehensive approach, we aim to enhance the usability of our model and contribute to improving radiotherapy planning and treatment in clinical practice.

A limitation of this study is the size of the dataset used. Deep learning methods typically perform better with larger datasets. While our proposed model did not exhibit overfitting, as the validation loss decreased alongside the training loss, having more data could reduce the gap between the training and validation loss curves. As can be seen in the ablation study, the use of a dual model yielded better results than the use of a single model. This improvement is due to the second model using additional output information from the first model to predict the final dose. However, the study did not show how the output information from the first model contributed to and specifically affected the final prediction result.

Cancer has long been one of the most challenging problems facing humanity, and despite significant advancements, it continues to present considerable difficulties. Researchers around the world are tirelessly working to improve treatment methods, aiming to develop more effective and targeted therapies. One critical aspect of cancer treatment is determining the appropriate radiation dose for patients, a process that is both time-consuming and heavily reliant on the expertise of medical professionals. This reliance on expertise often results in higher treatment costs for patients, as radiation planning can be complex and labor-intensive. In response to this challenge, we have developed a deep learning model that demonstrates promising performance in predicting radiation doses. The model, while effective, is still in its early stages and requires further refinement to meet the demanding standards of real-world medical applications. In the future, we plan to develop a program that is capable of displaying 3D predictions of radiation doses instead of 2D slices, as shown above. By providing 3D visualizations of the dose distributions, we aim to offer a more comprehensive, intuitive, and interactive experience for physicians and other healthcare providers. This advancement will not only make it easier for users to understand the spatial relationships in the radiation dose but also improve the overall efficiency of treatment planning. Furthermore, we will attempt to test on a new dataset or combine datasets to train it in the most general way across the whole body. Additionally, we will conduct further research on cascade learning mechanisms in deep learning and investigate how each component contributes to the final prediction result, as well as whether combining state-of-the-art models could create a new state-of-the-art model. With these enhancements, we hope to significantly contribute to the ongoing efforts to improve cancer treatment, making the process more efficient and accessible to both medical professionals and patients.

Author Contributions: Conceptualization, L.T.H. and D.N.T.; supervision, D.N.T.; methodology, L.T.H., D.N.T. and P.T.H.; writing—original draft preparation, P.T.H.; writing—review and editing, L.T.H. and D.N.T.; visualization, L.T.H., P.T.H. and D.N.T. All authors have read and agreed to the published version of the manuscript.

Funding: This research received no external funding.

Institutional Review Board Statement: Not applicable.

Informed Consent Statement: Informed consent was obtained from all subjects involved in the study. The owner of the dataset used in this study made it freely available to anyone for research purposes [26].

Data Availability Statement: The OpenKBP dataset is available at https://github.com/ababier/open-kbp (accessed on 15 October 2024).

Conflicts of Interest: The authors declare that they have no competing interests.

References

1. World Health Organization. Cancer. 2022. Available online: https://www.who.int/news-room/fact-sheets/detail/cancer (accessed on 3 June 2024).
2. Klingelhöfer, D.; Braun, M.; Brüggmann, D.; Groneberg, D.A. The Pandemic Year 2020: World Map of Coronavirus Research. *J. Med. Internet Res.* **2021**, *23*, e30692. [CrossRef]
3. Crosby, D.; Bhatia, S.; Brindle, K.M.; Coussens, L.M.; Dive, C.; Emberton, M.; Esener, S.; Fitzgerald, R.C.; Gambhir, S.S.; Kuhn, P.; et al. Early detection of cancer. *Science* **2022**, *375*, eaay9040. [CrossRef] [PubMed]
4. National Cancer Institute. What Is Cancer? 2021. Available online: https://www.cancer.gov/about-cancer/understanding/what-is-cancer (accessed on 3 June 2024).
5. Brown, J.S.; Amend, S.R.; Austin, R.H.; Gatenby, R.A.; Hammarlund, E.U.; Pienta, K.J. Updating the Definition of Cancer. *Mol. Cancer Res.* **2023**, *21*, 1142–1147. [CrossRef]
6. Seyfried, T.N.; Huysentruyt, L.C. On the origin of cancer metastasis. *Crit. Rev. Oncog.* **2013**, *18*, 43–73. [CrossRef]
7. Debela, D.T.; Muzazu, S.G.; Heraro, K.D.; Ndalama, M.T.; Mesele, B.W.; Haile, D.C.; Kitui, S.K.; Manyazewal, T. New approaches and procedures for cancer treatment: Current perspectives. *SAGE Open Med.* **2021**, *9*, 20503121211034366. [CrossRef] [PubMed]
8. Chaput, G.; Regnier, L. Radiotherapy: Clinical pearls for primary care. *Can. Fam. Physician* **2021**, *67*, 753–757. [CrossRef]
9. Gianfaldoni, S.; Gianfaldoni, R.; Wollina, U.; Lotti, J.; Tchernev, G.; Lotti, T. An Overview on Radiotherapy: From Its History to Its Current Applications in Dermatology. *Open Access Maced. J. Med. Sci.* **2017**, *5*, 521–525. [CrossRef]
10. Tward, J.D.; Anker, C.J.; Gaffney, D.K.; Bowen, G.M. Radiation Therapy and Skin Cancer. In *Modern Practices in Radiation Therapy*; Natanasabapathi, G., Ed.; IntechOpen: Rijeka, Croatia, 2012; Chapter 12. [CrossRef]
11. Maani, E.V.; Maani, C.V. Radiation Therapy. 2022. Available online: https://www.ncbi.nlm.nih.gov/books/NBK537036/ (accessed on 3 June 2024).
12. The American Cancer Society Medical and Editorial Content Team. How Radiation Therapy Is Used to Treat Cancer. 2019. Available online: https://www.cancer.org/cancer/managing-cancer/treatment-types/radiation/basics.html (accessed on 3 June 2024).
13. Gagan, S.; Padhi, S.; Patro, K.C.; Shukla, R.; Shukla, S.K.; Arora, D.; Singh, T.R.; Kundu, C.; Bhattacharya, P.S.; Krishna, V.; et al. Daily waiting time management for modern radiation oncology department in Indian perspective. *J. Cancer Res. Ther.* **2022**, *18*, 1796–1800. [CrossRef]
14. Künzel, L.A.; Thorwarth, D. Towards real-time radiotherapy planning: The role of autonomous treatment strategies. *Phys. Imaging Radiat. Oncol.* **2022**, *24*, 136–137. [CrossRef]
15. Van der Merwe, D.; Dyk, J.V.; Healy, B.; Zubizarreta, E.; Izewska, J.; Mijnheer, B.; Meghzifene, A. Accuracy requirements and uncertainties in radiotherapy: A report of the International Atomic Energy Agency. *Acta Oncol.* **2017**, *56*, 1–6. [CrossRef]
16. Jiang, J.; Sharif, E.; Um, H.; Berry, S.; Veeraraghavan, H. Local block-wise self attention for normal organ segmentation. *arXiv* **2019**, arXiv:1909.05054. [CrossRef]
17. Ronneberger, O.; Fischer, P.; Brox, T. U-Net: Convolutional Networks for Biomedical Image Segmentation. In *Medical Image Computing and Computer-Assisted Intervention—MICCAI 2015*; Navab, N., Hornegger, J., Wells, W.M., Frangi, A.F., Eds.; Springer International Publishing: Cham, Switzerland, 2015; pp. 234–241. [CrossRef]
18. Qin, J.; Wang, X.; Mi, D.; Wu, Q.; He, Z.; Tang, Y. CI-UNet: Application of Segmentation of Medical Images of the Human Torso. *Appl. Sci.* **2023**, *13*, 7293. [CrossRef]
19. Kavur, A.E.; Gezer, N.S.; Barış, M.; Aslan, S.; Conze, P.H.; Groza, V.; Pham, D.D.; Chatterjee, S.; Ernst, P.; Özkan, S.; et al. CHAOS Challenge—Combined (CT-MR) healthy abdominal organ segmentation. *Med. Image Anal.* **2021**, *69*, 101950. [CrossRef] [PubMed]
20. Liu, J.; Zhang, Y.; Chen, J.N.; Xiao, J.; Lu, Y.; Landman, B.A.; Yuan, Y.; Yuille, A.; Tang, Y.; Zhou, Z. CLIP-Driven Universal Model for Organ Segmentation and Tumor Detection. *arXiv* **2023**, arXiv:2301.00785.
21. Ahn, S.H.; Kim, E.; Kim, C.; Cheon, W.; Kim, M.; Lee, S.B.; Lim, Y.K.; Kim, H.; Shin, D.; Kim, D.Y.; et al. Deep learning method for prediction of patient-specific dose distribution in breast cancer. *Radiat. Oncol.* **2021**, *16*, 154. [CrossRef]
22. Toan, D.N.; Hien, L.T.; Toan, H.M.; Vinh, N.T.; Hieu, P.T. Predicting 3D radiotherapy dose-volume based on deep learning. *Intell. Automat. Soft Comput.* **2024**, *39*, 319–335. [CrossRef]
23. Hochreiter, S. The Vanishing Gradient Problem During Learning Recurrent Neural Nets and Problem Solutions. *Int. J. Uncertain. Fuzziness-Knowl.-Based Syst.* **1998**, *6*, 107–116. [CrossRef]
24. He, K.; Zhang, X.; Ren, S.; Sun, J. Deep Residual Learning for Image Recognition. In Proceedings of the 2016 IEEE Conference on Computer Vision and Pattern Recognition (CVPR), Las Vegas, NV, USA, 27–30 June 2016; pp. 770–778. [CrossRef]

25. Clark, K.; Vendt, B.; Smith, K.; Freymann, J.; Kirby, J.; Koppel, P.; Moore, S.; Phillips, S.; Maffitt, D.; Pringle, M.; et al. The Cancer Imaging Archive (TCIA): Maintaining and Operating a Public Information Repository. *J. Digit. Imaging* **2013**, *26*, 1045–1057. [CrossRef]
26. Babier, A.; Zhang, B.; Mahmood, R.; Moore, K.L.; Purdie, T.G.; McNiven, A.L.; Chan, T.C.Y. OpenKBP: The open-access knowledge-based planning grand challenge and dataset. *Med. Phys.* **2021**, *48*, 5549–5561. [CrossRef] [PubMed]
27. Ying, X. An Overview of Overfitting and its Solutions. *J. Phys. Conf. Ser.* **2019**, *1168*, 022022. [CrossRef]
28. Tran, D.; Bourdev, L.; Fergus, R.; Torresani, L.; Paluri, M. Learning Spatiotemporal Features with 3D Convolutional Networks. In Proceedings of the 2015 IEEE International Conference on Computer Vision (ICCV), Santiago, Chile, 7–13 December 2015; pp. 4489–4497. [CrossRef]
29. Ulyanov, D.; Vedaldi, A.; Lempitsky, V. Instance Normalization: The Missing Ingredient for Fast Stylization. *arXiv* **2017**, arXiv:1607.08022.
30. Agarap, A.F. Deep Learning using Rectified Linear Units (ReLU). *arXiv* **2019**, arXiv:2009.07485.
31. Gholamalinezhad, H.; Khosravi, H. Pooling Methods in Deep Neural Networks, a Review. *arXiv* **2020**, arXiv:2009.07485.
32. Parsania, P.; Virparia, P. A Review: Image Interpolation Techniques for Image Scaling. *Int. J. Innov. Res. Comput. Commun. Eng.* **2015**, *2*, 7409–7414. [CrossRef]
33. Marquez, E.S.; Hare, J.S.; Niranjan, M. Deep Cascade Learning. *IEEE Trans. Neural Netw. Learn. Syst.* **2018**, *29*, 5475–5485. [CrossRef] [PubMed]
34. Drzymala, R.; Mohan, R.; Brewster, L.; Chu, J.; Goitein, M.; Harms, W.; Urie, M. Dose-volume histograms. *Int. J. Radiat. Oncol. Biol. Phys.* **1991**, *21*, 71–78. [CrossRef] [PubMed]
35. Kingma, D.P.; Ba, J. Adam: A Method for Stochastic Optimization. *arXiv* **2017**, arXiv:1412.6980.
36. Loshchilov, I.; Hutter, F. SGDR: Stochastic Gradient Descent with Warm Restarts. *arXiv* **2017**, arXiv:1608.03983.
37. Liu, S.; Zhang, J.; Li, T.; Yan, H.; Liu, J. Technical Note: A cascade 3D U-Net for dose prediction in radiotherapy. *Med. Phys.* **2021**, *48*, 5574–5582. [CrossRef]
38. Gronberg, M.; Gay, S.; Netherton, T.; Rhee, D.; Court, L.; Cardenas, C. Technical Note: Dose prediction for head and neck radiotherapy using a three-dimensional dense dilated U-net architecture. *Med. Phys.* **2021**, *48*, 5567–5573. [CrossRef]
39. Zimmermann, L.; Faustmann, E.; Ramsl, C.; Georg, D.; Heilemann, G. Technical Note: Dose prediction for radiation therapy using feature-based losses and One Cycle Learning. *Med. Phys.* **2021**, *48*, 5562–5566. [CrossRef]
40. Kontaxis, C.; Bol, G.H.; Lagendijk, J.J.W.; Raaymakers, B.W. DeepDose: Towards a fast dose calculation engine for radiation therapy using deep learning. *Phys. Med. Biol.* **2020**, *65*, 075013. [CrossRef]
41. Nguyen, D.; Jia, X.; Sher, D.; Lin, M.H.; Iqbal, Z.; Liu, H.; Jiang, S. 3D radiotherapy dose prediction on head and neck cancer patients with a hierarchically densely connected U-net deep learning architecture. *Phys. Med. Biol.* **2019**, *64*, 065020. [CrossRef] [PubMed]
42. Zhang, J.; Liu, S.; Yan, H.; Li, T.; Mao, R.; Liu, J. Predicting voxel-level dose distributions for esophageal radiotherapy using densely connected network with dilated convolutions. *Phys. Med. Biol.* **2020**, *65*, 205013. [CrossRef] [PubMed]
43. Cao, H.; Wang, Y.; Chen, J.; Jiang, D.; Zhang, X.; Tian, Q.; Wang, M. Swin-Unet: Unet-Like Pure Transformer for Medical Image Segmentation. In Proceedings of the Computer Vision—ECCV 2022 Workshops, Tel Aviv, Israel, 23–27 October 2022; Karlinsky, L., Michaeli, T., Nishino, K., Eds.; Springer: Cham, Switzerland, 2023; pp. 205–218. [CrossRef]
44. Hu, C.; Wang, H.; Zhang, W.; Xie, Y.; Jiao, L.; Cui, S. TrDosePred: A deep learning dose prediction algorithm based on transformers for head and neck cancer radiotherapy. *J. Appl. Clin. Med. Phys.* **2023**, *24*, e13942. [CrossRef]

diagnostics

MDPI

Article

YOLOv8-Based System for Nail Capillary Detection on a Single-Board Computer

Seda Arslan Tuncer [1], Muhammed Yildirim [2,*], Taner Tuncer [3] and Mehmet Kamil Mülayim [4]

[1] Faculty of Engineering, Software Engineering, Firat University, 23119 Elazığ, Turkey; satuncer@firat.edu.tr
[2] Faculty of Engineering and Natural Sciences, Computer Engineering, Malatya Turgut Ozal University, 44200 Malatya, Turkey
[3] Faculty of Engineering, Computer Engineering, Firat University, 23119 Elazığ, Turkey; ttuncer@firat.edu.tr
[4] Faculty of Medicine, Kahramanmaras Sutcu Imam University, 46000 Kahramanmaras, Turkey; mkmulayim@ksu.edu.tr
* Correspondence: muhammed.yildirim@ozal.edu.tr

Abstract: Nail capillaroscopic examination is an inexpensive and easily applicable method to identify capillary morphological changes in patients with conditions such as systemic sclerosis and Raynaud's. The detection of changes in capillaries makes an important contribution to diagnosing these diseases. Capillary morphology is important in the symptoms of these diseases, and capillary diameter, visibility, distribution, length, microbleeds, blood flow, and density are important indicators in capillaroscopic evaluation. Manual examination to determine these parameters is subjective, causes inconsistent results, and is labor-intensive and time-consuming. To overcome these problems, a YOLOv8s-based system was proposed in this paper to detect the number, thickness, and density of capillaries in the nail bed. The system's components include database systems that store the analysis results, artificial intelligence-based software that runs on the SBC (Single-Board Computer), and recorded microscope images. mAP and F1_score parameters were used to evaluate the system's performance, and values of 0.882 and 0.83 were obtained. The proposed system is promising in improving the diagnosis process of diseases such as systemic sclerosis and Raynaud's by providing objective measurements and the early diagnosis and monitoring of diseases.

Keywords: nail capillary; artificial intelligence; proximal nail fold; YOLOv8

1. Introduction

Capillaroscopy is the imaging of small vascular structures, known as capillaries, in the nail fold through a microscope. This non-invasive procedure is simple and inexpensive. The main purpose of using this method is to determine whether there is a disease related to capillaries in patients who complain of discoloration of the hands triggered by cold and stress. In patients presenting with this complaint, some changes in the small capillaries in the nail fold help in the early diagnosis of some rheumatic diseases, especially scleroderma. Capillaroscopy provides the opportunity to detect vascular disorders in the early stages and prevent and treat diseases initially. It is easier and more valuable to treat the initial stages of the disease than to treat the chronic stages. Through capillaroscopy, experts can take comprehensive and preventive measures against diseases. Capillaroscopy is frequently used to determine microvascular involvement in systemic diseases such as autoimmune conditions, rheumatism, and many nail and skin diseases [1].

Capillary examination, or capillaroscopy, can be performed with light microscopy and video capillaroscopy, and today, it can also be carried out with dermatoscopy devices. Dermatoscopy devices, which are a fast and effective diagnostic tool for evaluating the nail fold capillary system, are cheap and easy to apply, making these devices advantageous. In practice, there are devices such as Dino-Lite CapillaryScope, Optilia Digital Capillary Scope, and Smart G-Scope capillary scope for imaging purposes [2–4]. With the use of these devices, experts diagnose diseases such as scleroderma, Raynaud's (Reyno), Sjögren's syndrome, dermatomyositis, rheumatoid arthritis, lupus disease (Lupus rythematosus), diabetes mellitus, and hypertension. It is important to determine the number of capillaries, their thickness, shape, and density per unit area, especially by focusing on the grading of diseases in the images examined.

The quantity of capillaries in one millimeter within the distal row of each finger or toe is known as capillary density. The number and density of blood capillaries are considered common parameters in determining diseases such as scleroderma and Raynaud's. Karbalaie et al. proposed a new method based on the 90° method for capillaroscopic evaluation. This method was used to evaluate nail fold capillary density [5]. Ingegnoli et al. described nail fold capillary findings using the video capillaroscopy technique in healthy subjects. Nail fold capillaries were examined according to their morphology, size, and density. They reported that the majority of subjects had an average of seven capillaries/mm [6]. Emrani et al. examined a widely used technique for determining capillary density and the connections between capillary count and various grading systems, autoantibodies, pulmonary arterial hypertension, digital ulcers, and scleroderma patterns [7].

Kornaev et al. used high-speed video capillaroscopy to detect and classify capillaries in nail folds. U-Net semantic segmentation method was used for capillary detection. Resnet and Googlenet architectures were used to classify capillaries and 96% accuracy was achieved [8]. Suma et al. examined capillary parameters in the diagnosis of diabetes. Diabetes detection relies on both quantitative and qualitative capillary parameters, including average capillary density, length, breadth, tortuosity, hemorrhages, angiogenesis, and elongated capillaries. They suggested using an object identification method based on deep learning to categorize the nail fold capillaries into five groups: normal, wide, long, tortuous, and bleeding. A total of 600 images were used, and thanks to data augmentation techniques, the number of data used was increased to 1018 [9]. Shah et al. used CNN to determine whether nail fold images obtained using video capillaroscopy could provide diagnostic information about diabetes and its complications. A total of 5236 images were obtained from 120 patients. The area under the ROC curve for five different diabetes mellitus complications was 0.84 [10].

Tello et al. designed an automated software to count nail fold capillaries. In a study using a total of 2713 images, a standard metric precision of 83.84% and a recall of 92.44% were obtained with machine learning algorithms [11]. Natalello et al. evaluated microvascular structure via nail fold video capillaroscopy (NVC) in COVID-19 patients [12]. Bharatti et al. aimed to develop and validate a fully automatic image analysis system. Their proposed method was based on deep learning to detect each capillary in the distal row of capillaries and make morphological measurements. The AUC value obtained was 97% [13]. Korondovych et al. used the Optilia Digital Capilleroscope to distinguish between scleroderma and non-scleroderma capillaries. They analyzed capillary microscopy images with deep learning algorithms, which were equally divided into two groups, including scleroderma and non-scleroderma patterns. A total of 1076 capillaroscope images were divided into training, validation, and test sets, and the same number of images was available in both classes. The accuracy of the model was achieved at 92% [14]. Venkatataphiah et al. proposed a new object detection algorithm based on deep learning architectures to detect and locate various capillary loops in the nail fold region. Various characteristic features were extracted from capillaries through image processing algorithms (YOLOv3), and then discrimination was made between images of diseased subjects and healthy images. In their study, a total of 600 images were analyzed, and the accuracy value was 88.2% [15].

Liu et al. proposed a new deep learning architecture called DAFM-Net for capillary segmentation, as the segmentation of nail fold microbleeds provides valuable pathological information that can lead to further investigation. The network comprised a group normalization layer, dual attention fusion module, and U-shaped backbone. Rich hierarchical representations were generated by the U-shaped backbone, and captured features were used for fine-tuning by the dual attention fusion module. A normalizing technique called group normalization was offered as a helpful way to boost deep neural network convergence. Segmentation tests confirmed the efficacy of the suggested model; the suggested technique, DAFM-Net, demonstrated competitive performance in nail fold microhemorrhage segmentation, with an IOU score of 78.03% and a Dice score of 87.34% in comparison to the ground truth [16]. Hafizh et al. focused on the classification of capillaries. Using the VGG-16 model, they detected nine different data types with an average accuracy of 63.98% [17].

Addou et al. proposed the CNN-based CapillaryNet model. The model is end-to-end and detects capillaries with ~93% accuracy [18]. Nguyen et al. identified capillary types using an improved version of YOLOv5. It is predicted that the system, which yields a mAP50 value of 0.74, will be used in the early diagnosis of diabetes in the future [19]. Yin et al. determined nail fold capillary density. With the improved Yolov5, capillary densities were determined at 85.2% MAP@50 [20]. Nitkunanantharajah et al. imaged the nail sub capillaries of systemic sclerosis patients and healthy controls using optoacoustic imaging and compared them with each other. As a result of deep learning-based classification, 89.7% accuracy was achieved [21].

1.1. Motivation

Our main motivation is to automatically detect capillaries in the nail fold and determine the number, density, and thickness of capillaries, which are important in disease diagnosis. The identification of capillaries plays an important role in the diagnosis and follow-up of many diseases, especially scleroderma and Raynaud's (Reyno). Established guidelines and instructions for the interpretation of capillaries in nail fold images have not yet been standardized. Therefore, the evaluation and interpretation of images are quite subjective.

The manual identification of capillaries and the determination of their thickness and density are a challenging process. We propose an artificial intelligence-based system to overcome this challenge. The system can automatically detect capillaries in the nail fold and calculate the number, thickness, and density of capillaries, which are important in diagnosing diseases. Figure 1 shows the basic structure of the proposed capillary detection system.

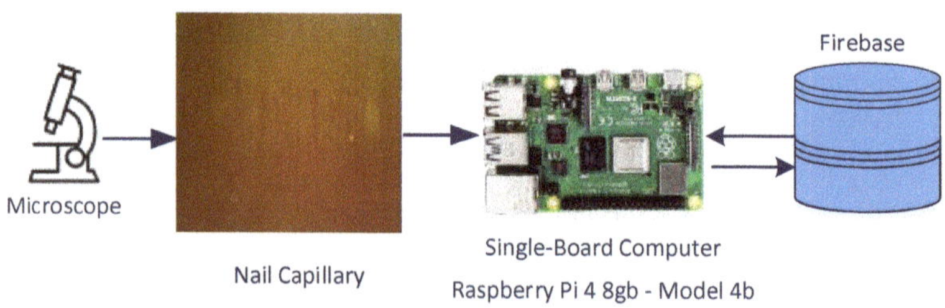

Figure 1. YOLOv8s-based capillary analysis system running on SBC.

The patient's nail fold is taken, and the capillaries inside are analyzed using the following four steps:

Step 1: A microscope is used to obtain images of the capillaries.

Step 2: The YOLOv8s architecture trained on the SBC receives the capillary image as input.

Step 3: The capillaries in the image and their thicknesses are measured.

Step 4: The software interface on the SBC displays the analysis findings, which are stored in the database.

1.2. Contributions

The following are the study's primary contributions:

- Thanks to the proposed artificial intelligence-based system, the identification of capillaries in the nail fold becomes automatic.
- Using YOLOv8s, the number, density, and thickness of capillaries are obtained as numerical data.
- This study provides clinical reporting data by overcoming the problems in manual measurements.
- The measurements made are repeatable.

2. Data

Capillary thickness and density in the nail bed are important indicators in determining whether a person has a healthy nail bed. While a capillary width of 5–10 μm is considered normal, a capillary width larger or smaller than this value is considered abnormal [22,23]. Average capillary density is the number of capillaries per mm length of the proximal nail fold. The EULAR Study Group on Microcirculation in Rheumatic Diseases defines normal and abnormal capillaries as follows: capillaries with the stereotypical "hairpin" shape, as well as crossing (once or twice) or tortuous capillaries, are defined as "normal". All other shapes are defined as "abnormal", provided that the capillary end is convex [22].

The data collection mechanism used in this study is shown in Figure 2. The experimental equipment, MS2 1-1200X 5 Inch 720P LCD Screen USB Digital Microscope, is a 5-inch high-definition LCD screen digital microscope. The lens supports a 1-1200X continuous zoom and solves the problem of high reflection with the help of adjustable-angle LEDs. When connected to a computer via an USB cable, the object can be viewed on the computer monitor and the magnification effect can be displayed on the big screen. The LCD display digital microscope is equipped with a micro-SD card slot. Images obtained during the observation process were stored on a 32 G micro-SD card. The Elazığ Fırat University Ethics Committee approved this study, with approval number 2024/11-36. First of all, a solution that enables the visualization of the superficial structures of the skin and creates a smooth surface was applied to the examined fingers of the patients, and then the capillary image taken from the microscope was recorded.

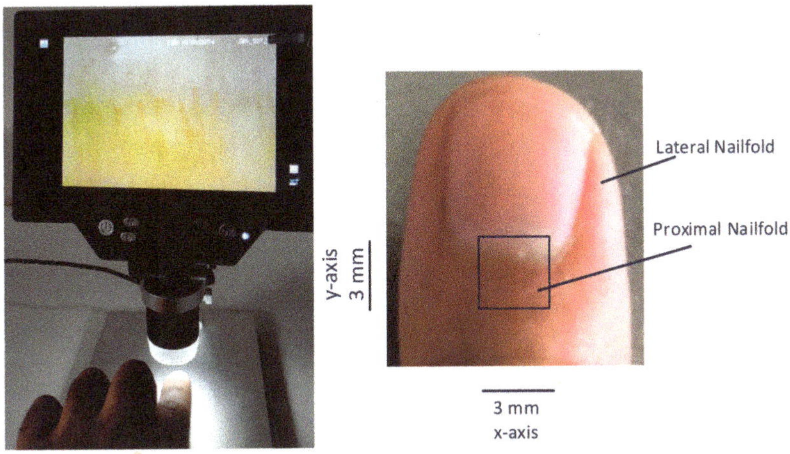

Figure 2. The data collection unit.

A total of 800 subjects were identified for obtaining the data. However, capillaries could not be obtained from some of these subjects. Images obtained from 23 subjects were excluded from this study. Nail capillary images of a total of 777 patients were collected. A total of 80% of the data was used for training and 20% for testing. For the data set collected in this study, data were collected regardless of whether the capillaries were normal or abnormal.

Each captured image was 3648 × 2736 in size. The collected data were converted to 640 × 640, 96 dpi, so that it could be fed to YOLOv8's input. Each image was obtained from the proximal nail fold region and this image represented 3 mm horizontally and vertically. Figure 3 shows three different images of capillaries obtained from patients.

Figure 3. Capillaries in the proximal nail fold.

To automatically segment the images taken, manual segmentation was first performed by a doctor. The Roboflow environment was used for manual segmentation. Figure 4 shows images with manual segmentation.

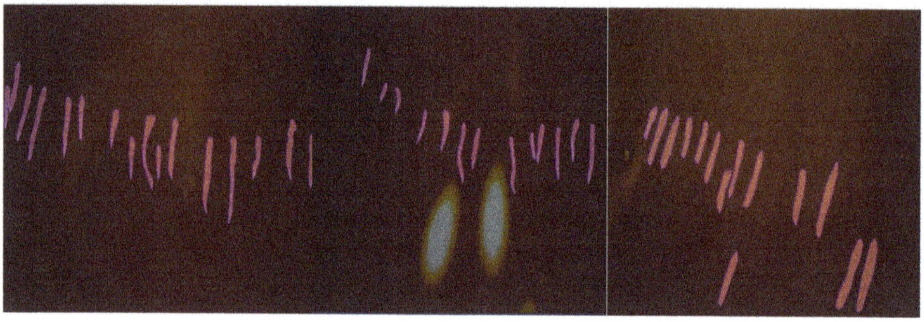

Figure 4. Manually segmented images of capillaries.

3. The Proposed Method

The YOLOv8s-based analysis system developed in this article to determine the number, width, and density of capillaries in the nail fold is shown in Figure 5. The Python programming language was used to implement the YOLOv8 architecture, which uses images obtained from the microscope as input. YOLOv8s was preferred for use in the this study because it provides faster and more successful results than many object detection algorithms used in real-time object tracking. The system detects possible capillaries in the given image, as well as their number, thickness, and density in the image. It then generates a report containing the results.

Tkinter GUI packages and the Python programming language were used to code the interfaces and integrate the system. The system's numerical outputs were stored on Firebase, a free platform designed by Google for building web and mobile applications. Figure 6 shows a patient's information, the nail fold image taken from the patient, the capillaries obtained with YOLOv8s, and the interface related to the report produced by the model.

The YOLOv8s architecture used in the article is given in Figure 7. YOLOv8 consists of three main parts: Spine, Neck, and Head [24]. The backbone is responsible for extracting meaningful features from images, the Neck is responsible for feature fusion and integrating contextual information, and the Head is responsible for determining bounding boxes and confidence scores for object detection.

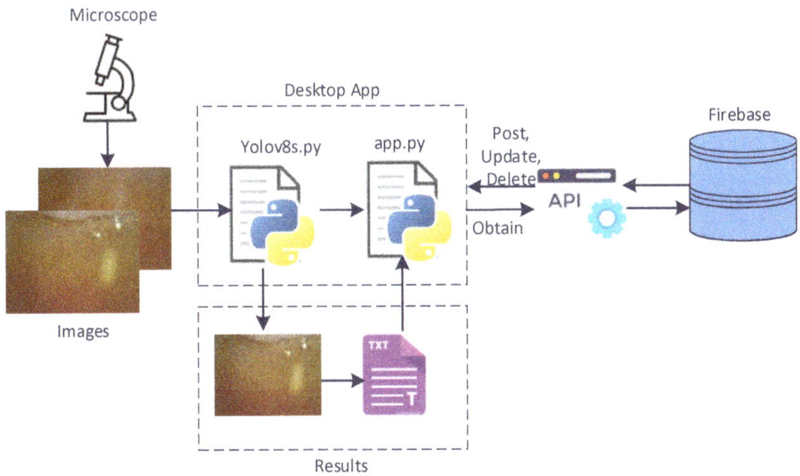

Figure 5. The proposed model.

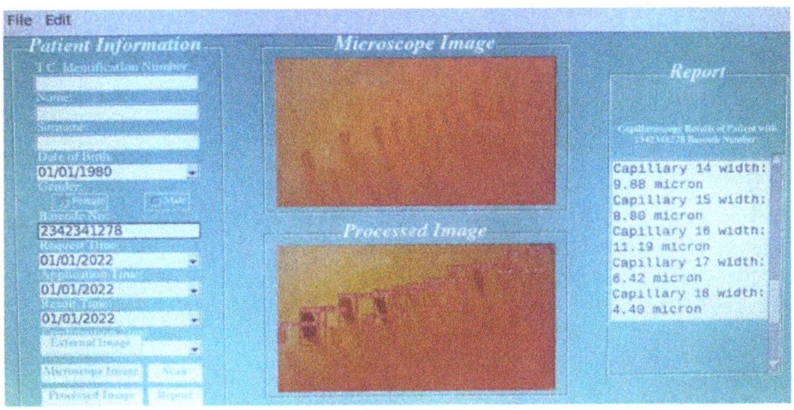

Figure 6. Interface of the proposed system.

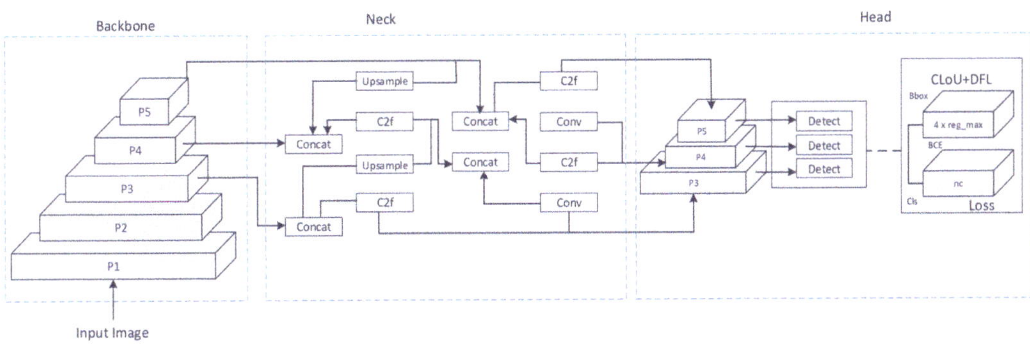

Figure 7. YOLOv8 architecture.

Each bounding box determined in object detection has four coordinate values, (x, y, w, h), where (x, y) represent the center coordinates of the box and (w, h) represent the width and height of the box.

These coordinates are extracted from the network's output tensor and then inserted into activation functions (usually sigmoid). These values are then normalized to a predetermined grid cell size. The x and y coordinates are usually positioned relative to the upper left corner of this cell and then proportioned to the grid cell size. The w and h values are typically multiplied by the width and height of the image, thus expressing it in actual pixels. In the equations below, bx and by represent the center coordinates of the bounding box and bw and bh represent its width and height, respectively. t_x, t_y, t_w, and t_h are the predicted values from the network's output tensor. cx and cy are the coordinates of the upper left corner of the cell. p_w and p_h are predetermined scale factors.

$$\begin{aligned} b_x &= \sigma(t_x) + c_x \\ b_y &= \sigma(t_y) + c_y \\ b_w &= p_w e^{t_w} \\ b_h &= p_h e^{th} \end{aligned} \tag{1}$$

During training, an error function is used to measure how far the model's predicted bounding boxes are from actual object locations. This error function is the weighted sum of the localization loss, confidence loss, and classification loss components.

$$\begin{aligned} Loss = \ & \lambda_{coord} \textstyle\sum_{i=0}^{S2} \sum_{j=0}^{B} 1_{ij}^{obj} \left[(x_i - \hat{x}_i)^2 + (y_i - \hat{y}_i)^2 \right] \\ & + \lambda_{coord} \textstyle\sum_{i=0}^{S2} \sum_{j=0}^{B} 1_{ij}^{obj} \left[(\sqrt{w_i} - \sqrt{\hat{w}_i})^2 + \left(\sqrt{h_i} - \sqrt{\hat{h}_i} \right)^2 \right] \\ & + \textstyle\sum_{i=0}^{S2} \sum_{j=0}^{B} 1_{ij}^{obj} (C_i - \hat{C}_i)^2 \\ & + \lambda_{nonbj} \textstyle\sum_{i=0}^{S2} \sum_{j=0}^{B} 1_{ij}^{obj} (C_i - \hat{C}_i)^2 \\ & + \textstyle\sum_{i=0}^{S2} 1_{ij}^{obj} \sum_{c \in classes} (p_i(c) - \hat{p}_i(c))^2 \end{aligned} \tag{2}$$

In this equation, the parameters λ_{coord} and λ_{noobj} are weight hyperparameters used to adjust the importance of different components, whereas 1_{ij}^{obj} and 1_{ij}^{noobj} show which bounding box the cells belong to. The actual bounding box coordinates are x_i, y_i, w_i, and h_i. The estimated bounding box coordinates are $\hat{x}_i$, $\hat{y}_i$, $\hat{w}_i$, and $\hat{h}_i$. The actual and predicted class parameters C_i, $\hat{C}_i$, $p_i(c)$, and $\hat{p}_i(c)$ denote the actual and predicted class probabilities.

During training, stochastic gradient descent (SGD) was used to reduce the error function. When SGD is used, the update rule for each weight parameter θ is expressed as in Equation (3)

$$\theta = \theta - \alpha \frac{\partial \mathrm{Loss}}{\partial \theta} \tag{3}$$

where α represents the learning rate and $\frac{\partial \mathrm{Loss}}{\partial \theta}$ represents the derivative concerning the weight parameters of the loss function.

4. Experimental Results and Discussion

4.1. Experimental Results

In this section, the YOLOv8s capillary analysis system is evaluated with the appropriate parameters.

The error matrix is another name for the confusion matrix. The confusion matrix layout visualizes an algorithm's performance. This is known as a matching matrix, in which an actual class is represented by the column and a predicted class by the row [23]. This matrix contains the fundamental definitions (*TP*, *TN*, *FP*, and *FN*).

True positive (*TP*): The case in which the model predicts the positive class with accuracy.

True negative (*TN*): The case in which the model correctly predicts the negative class.

False positive (*FP*): The case in which the model mispredicts the positive class.

False negative (*FN*): The case in which the model mispredicts the negative class.

We use the following four metrics to evaluate the performance of the YOLOv8s model used in the paper. These are F_1 *score*, recall, precision, and Mean Average Precision (*mAP*).

$$Precision = TP/(TP + FP) \quad (4)$$

$$Recall = TP/(TP + FN) \quad (5)$$

$$F_1\_score = \frac{2 \times Precision \times Recall}{Precision + Recall} \quad (6)$$

$$AP = \sum_{k=0}^{k=n-1} [Recall(k) - Recall(k+1) * Precision(k)],\ n = \text{number of thresholds} \quad (7)$$

In object detection, there may be more than one object class (such as capillary background) to be detected. *mAP* calculates the *AP* for each class as in Equation (7) and then calculates the average of these *AP* values. The *mAP* given in Equation (8) provides an overall evaluation of the model's performance in all classes.

$$mAP = \int_0^1 p(c)dc \quad (8)$$

The training process was performed in Google Colab. During training, the Google Colab Pro + version was selected. Thus, the highest GPU usage was achieved with the Google Colab Pro+ version. The hyperparameters used for training the model are as in Table 1. Training accuracies for Boxes and Masks are as in Figure 8.

F1_score and mAP50 parameters were obtained as 0.83 and 0.882. The mAP50 value shows that the model is capable of correctly recognizing capillaries. Figure 9 shows that the model has both high precision and high recall variation. The fact that the precision-recall change approaches to the right and above the axis is an indication that the model performs well.

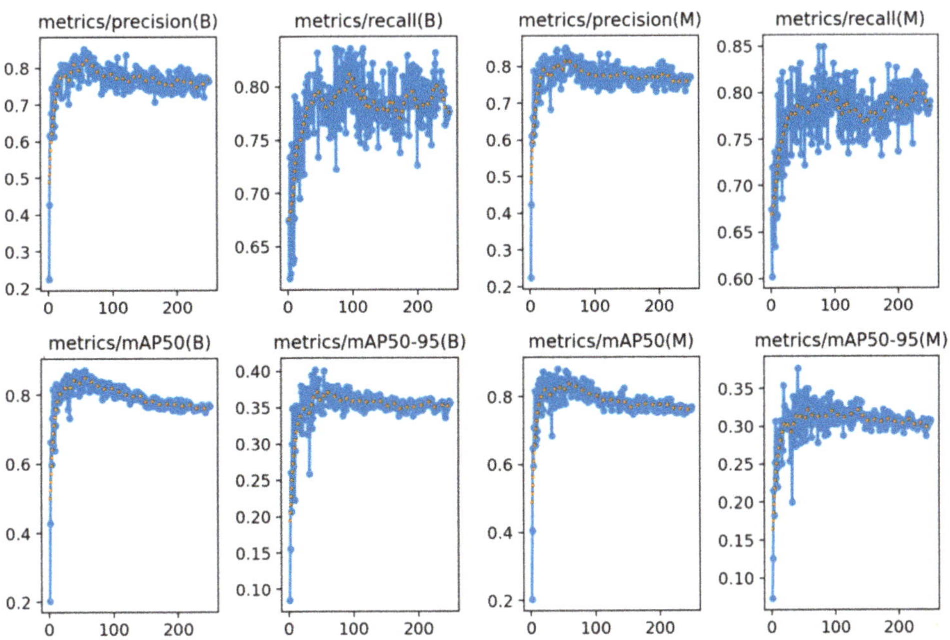

Figure 8. Accuracy changes for YOLOv8s.

Table 1. The configuration of training parameters for the YOLOv8s models.

Optimizer	SGD
Epochs	250
Batch	16
Image size	640 × 640
Learning rate	0.002
Momentum	0.9
Weight (decay)	0.0005

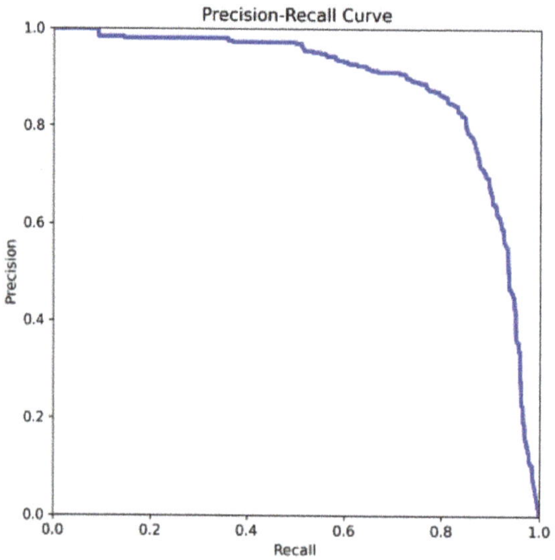

Figure 9. Precision–recall change in the model.

4.2. Discussion

In this study, YoloV8s was used to automatically detect nail fold capillaries. A total of 777 nail fold images with a size of 3648 × 2736 were collected. Each image had a size of 640 × 640 and the horizontal and vertical resolution was 96 dpi. Figure 10 shows the capillary images of different patients and the capillaries detected in these images. The following procedure was applied for the width of each detected capillary.

First, the physical length per pixel was calculated.

The actual size of the image was 3 mm horizontally and vertically.

Since the DPI value was 96, it corresponded to 96 pixels per 1-inch area. The x-axis was 3 mm, which is approximately 0.11811 inches in length. Therefore, for 3 mm, there will be approximately 0.11811 × 96 × 640 = 7246.76 pixels. By dividing the actual length of the image by the number of pixels, the size of one pixel is approximately (3 mm/7246.76) 0.0004142 mm, which is 0.4142 microns. Considering that the width of a normal capillary is 5–10 µm, in this study, the width of each capillary is 12 pixels to 24 pixels. For example, in image no. 3, a total of 11 capillaries were detected, and the widths of these capillaries on the horizontal axis are 22, 16, 15, 16, 19, 13, 8, 7, 22, 26, 13 pixels, respectively. The thicknesses of these capillaries in µm are 9.11, 6.63, 6.21, 6.63, 7.87, 5.38, 3.31, 2.9, 9.11, 10.76, 5.38, respectively. According to these results, the thickness of two capillaries was found to be significantly lower than normal, and the thickness of one capillary was determined to be greater than normal.

Number	Image	Detected Capillaries
1	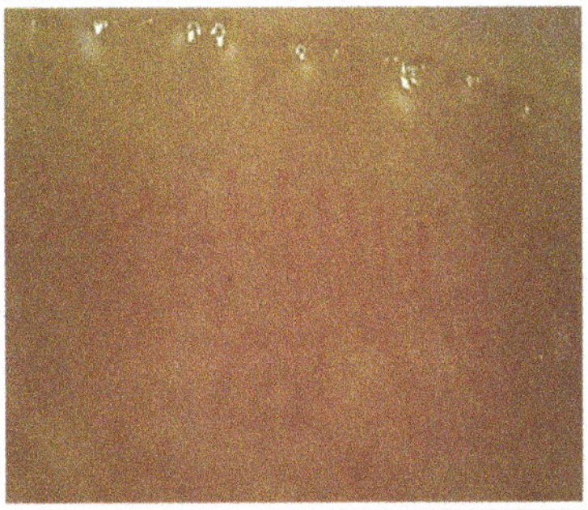	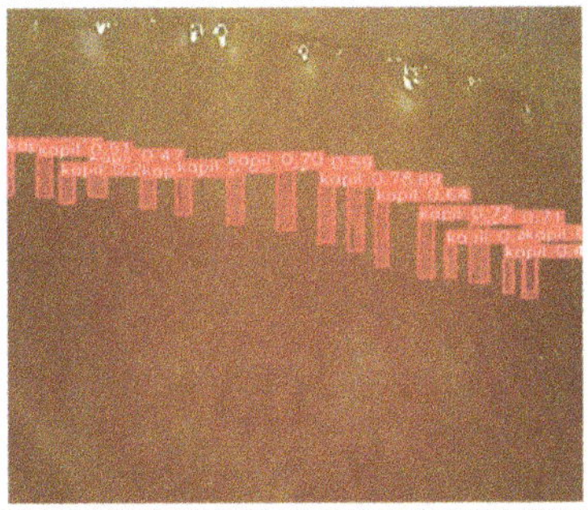
2	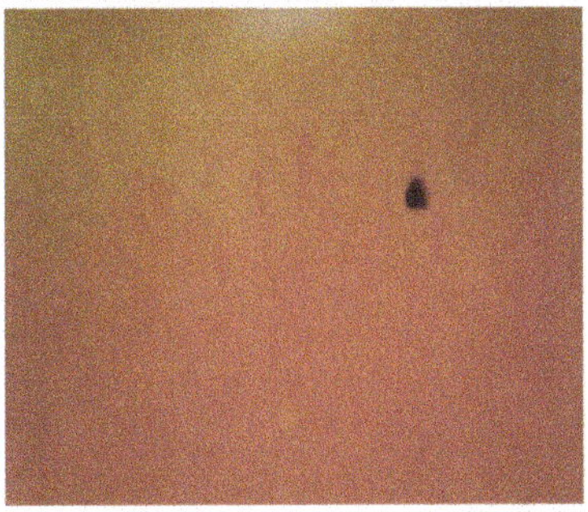	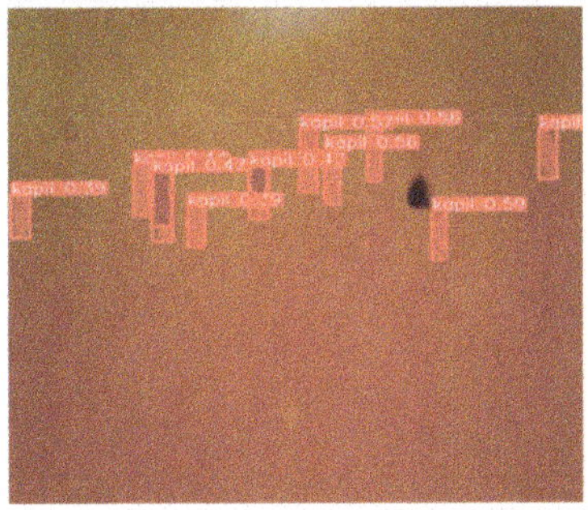
3		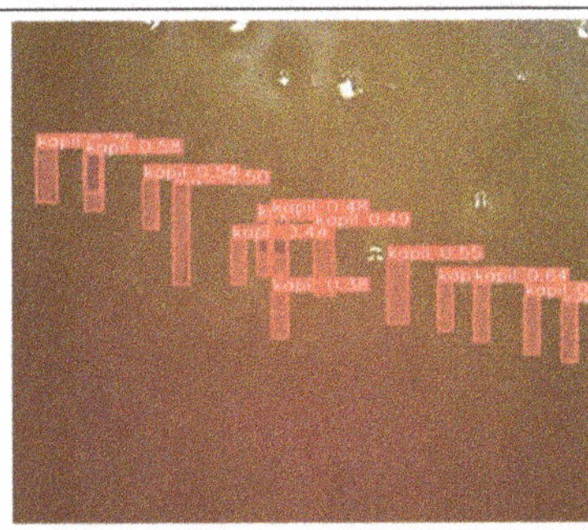
4	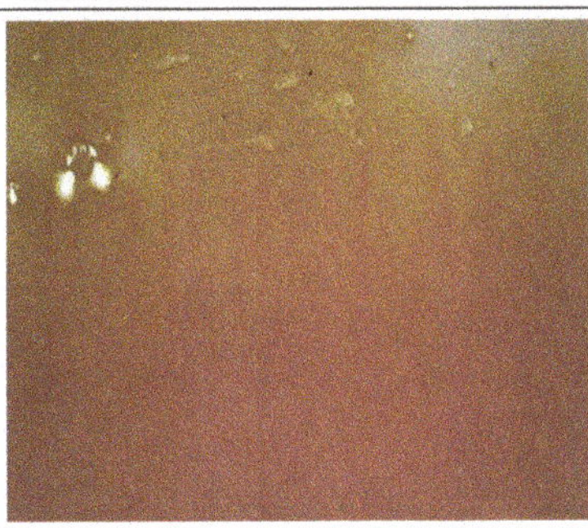	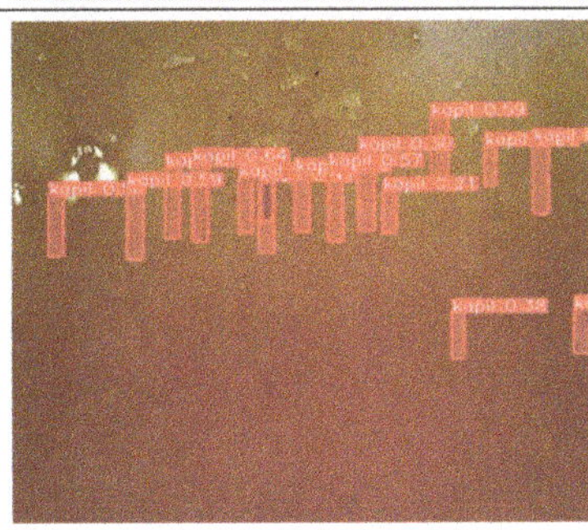

Figure 10. Capillaries detected by the model.

Table 2 shows the number of capillaries detected, capillary density in 1 mm, and the average capillary thickness for the four images given in Figure 10.

Table 2. Calculated parameters of the detected capillaries.

Image Number	Number of Detected Capillaries	Density	Capillary Thickness Average Micron
1	17	5.67	7.31
2	11	3.67	7.63
3	17	6.67	6.66
4	16	5.34	6.88

Manual examination takes a long time to determine the number of capillaries and their thickness. Thanks to the proposed method, capillary thicknesses and densities can be determined automatically. The disadvantages of manual measurements are eliminated, and inconsistent results are eliminated by providing an objective evaluation. Thanks to the developed software, it is possible to use the model in high-volume laboratories. Report sharing and reproducible results are now available.

There are no recorded studies in the literature that use YOLOv8 to examine capillary thickness and density in nail fold capillary images. Similar studies based on Yolov3 [15] and Yolov5 [19,20] are available in the literature. However, our work has some fundamental differences from [15,19,20]. The first of these is that our study is an automatic analysis system that combines software and hardware. Another is that it can calculate capillary thickness and capillary density. Our experimental results have shown that the proposed system provides better performance than the approaches in [9,14] in determining the number, thickness, and density of capillaries.

The proposed method also has some limitations. These can be summarized as follows:

- Different types of capillary structures have not been examined.
- The system requires more computing power during the training phase.

5. Conclusions

Our work consists of an SBC that processes nail bed images taken by a USB camera and enables the YOLOv8 model to run on it. In this paper, YOLOv8 architecture, one of the state-of-the-art algorithms that can automate the determination of the density and number of nail fold capillaries, was used.

The proposed system not only improves the diagnostic process of diseases such as scleroderma by providing objective measurements but also facilitates the early diagnosis and monitoring of diseases. The integration of YOLOv8s into SBC for the acquisition and analysis of vascular images in the nail bed represents a cost-effective and efficient approach to capillaroscopy. As a result, the system shows promise for wider application in clinical settings, potentially contributing to improving patient outcomes through more accurate and timely diagnosis. The focus of our future research is on further improving the system's capabilities and considering other conditions affecting capillary morphology.

Author Contributions: Conceptualization, S.A.T., M.Y. and T.T.; methodology, S.A.T. and M.Y.; software, S.A.T. and T.T.; validation, T.T., M.Y. and S.A.T.; formal analysis, M.K.M. and T.T.; investigation, M.K.M. and S.A.T.; resources, T.T. and M.Y.; data curation, M.K.M., T.T., M.Y. and S.A.T.; writing—original draft preparation, M.K.M., T.T., M.Y. and S.A.T.; writing—review and editing, M.K.M., T.T., M.Y. and S.A.T.; visualization, T.T.; supervision, M.K.M., T.T., M.Y. and S.A.T.; project administration, M.K.M., T.T., M.Y. and S.A.T. All authors have read and agreed to the published version of the manuscript.

Funding: This research received no external funding.

Institutional Review Board Statement: The Elazığ Fırat University Ethics Committee approved this study, with approval number 2024/11-36.

Informed Consent Statement: Informed consent was obtained from all subjects involved in the study.

Data Availability Statement: Data will be shared upon request.

Conflicts of Interest: The authors declare no conflicts of interest.

References

1. Chojnowski, M.M.; Felis-Giemza, A.; Olesińska, M. Capillaroscopy—A role in modern rheumatology. *Reumatologia* **2016**, *54*, 67–72. [CrossRef] [PubMed]
2. Available online: https://www.dino-lite.eu/index.php/en/products/medical/capillaryscope (accessed on 9 May 2024).
3. Video Capillaroscopy. Available online: https://www.optiliamedical.eu/products/2/capillaroscope/11/%20Optilia%20Digital%20Capillaroscopy%20System,%20Extensive%20kit/ (accessed on 9 May 2024).
4. Smart g-scope™ Europe—Smart g-scope™ Europe. Available online: https://g-scope.eu/ (accessed on 9 May 2024).
5. Karbalaie, A.; Abtahi, F.; Fatemi, A.; Etehadtavakol, M.; Emrani, Z.; Erlandsson, B.-E. Elliptical broken line method for calculating capillary density in nailfold capillaroscopy: Proposal and evaluation. *Microvasc. Res.* **2017**, *113*, 1–8. [CrossRef] [PubMed]
6. Ingegnoli, F.; Gualtierotti, R.; Lubatti, C.; Bertolazzi, C.; Gutierrez, M.; Boracchi, P.; Fornili, M.; De Angelis, R. Nailfold capillary patterns in healthy subjects: A real issue in capillaroscopy. *Microvasc. Res.* **2013**, *90*, 90–95. [CrossRef] [PubMed]
7. Emrani, Z.; Karbalaie, A.; Fatemi, A.; Etehadtavakol, M.; Erlandsson, B.-E. Capillary density: An important parameter in nailfold capillaroscopy. *Microvasc. Res.* **2017**, *109*, 7–18. [CrossRef] [PubMed]
8. Kornaev, A.V.; Stavtsev, D.D.; Kornaeva, E.P.; Volkov, M.V. Application of Deep Supervised Learning to Nailfold videocapillaroscopy and Red Blood Cells Velocity Approximation. *Med. Imaging Deep. Learn.* **2021**.
9. Venkatapathiah, K.; Selvi, S.; Nanda, P.; Shetty, M.; Vikas, M. Deep Learning Approach to Nailfold Capillaroscopy Based Diabetes Mellitus Detection. *Int. J. Online Biomed. Eng. (iJOE)* **2022**, *18*, 95–109.
10. Shah, R.; Petch, J.; Nelson, W.; Roth, K.; Noseworthy, M.D.; Ghassemi, M.; Gerstein, H.C. Nailfold capillaroscopy and deep learning in diabetes. *J. Diabetes* **2023**, *15*, 145–151. [CrossRef] [PubMed]
11. Tello, B.G.; Ramos Ibañez, E.; Fanlo Mateo, P.; Sáez Cómet, L.; Martínez Robles, E.; Ríos Blanco, J.J.; Marí Alfonso, B.; Espinosa Garriga, G.; Todolí Parra, J.; Ortego Centeno, N.; et al. The challenge of comprehensive nailfold videocapillaroscopy practice: A further contribution. *Clin. Exp. Rheumatol.* **2022**, *40*, 1926–1932.
12. Natalello, G.; De Luca, G.; Gigante, L.; Campochiaro, C.; De Lorenzis, E.; Verardi, L.; Paglionico, A.; Petricca, L.; Martone, A.M.; Calvisi, S.; et al. Nailfold capillaroscopy findings in patients with coronavirus disease 2019: Broadening the spectrum of COVID-19 microvascular involvement. *Microvasc. Res.* **2021**, *133*, 104071. [CrossRef] [PubMed]
13. Bharathi, P.G.; Berks, M.; Dinsdale, G.; Murray, A.; Manning, J.; Wilkinson, S.; Cutolo, M.; Smith, V.; Herrick, A.L.; Taylor, C.J. A deep learning system for quantitative assessment of microvascular abnormalities in nailfold capillary images. *Rheumatology* **2023**, *62*, 2325–2329. [CrossRef] [PubMed]
14. Korendovych, V.; Korsten, P. Pos1328 Differentiating "Scleroderma" With "Non-Scleroderma" Patterns in Nailfold Capillary Microscopy Using a Deep Learning Model. *BMJ* **2023**, *82*, 1013–1014. [CrossRef]
15. Venkatapathiah, S.K.; Selvan, S.S.; Nanda, P.; Shetty, M.; Swamy, V.M.; Awasthi, K. Deep learning based object detection in nailfold capillary images. *IAES Int. J. Artif. Intell. IJ AI* **2023**, *12*, 931–942. [CrossRef]
16. Liu, R.; Tian, J.; Li, Y.; Chen, N.; Yan, J.; Li, T.; Liu, S. Nailfold Microhemorrhage Segmentation with Modified U-Shape Convolutional Neural Network. *Appl. Sci.* **2022**, *12*, 5068. [CrossRef]
17. Hafıızh, M. Convolutional Neural Network to Classify Capillaries of Images in Human Fingertips. *Sci. Rep.* **2019**, *63*, 67–78.
18. Abdou, M.A.H.; Truong, T.T.; Dykyy, A.; Ferreira, P.; Jul, E. CapillaryNet: An automated system to quantify skin capillary density and red blood cell velocity from handheld vital microscopy. *Artif. Intell. Med.* **2022**, *127*, 102287. [CrossRef]
19. Nguyen, H.T.P.; Ko, S.; Jeong, H. Deep-learning-based Capillary Detection. In Proceedings of the 2023 IEEE International Conference on Bioinformatics and Biomedicine (BIBM), Istanbul, Türkiye, 5–8 December 2023; pp. 4932–4934.
20. Yin, H.; Wu, Z.; Huang, A.; Luo, J.; Liang, J.; Lin, J.; Ye, Q.; Xie, M.; Ye, C.; Li, X.; et al. Automated nailfold capillary density measurement method based on improved YOLOv5. *Microvasc. Res.* **2023**, *150*, 104593. [CrossRef] [PubMed]
21. Nitkunanantharajah, S.; Haedicke, K.; Moore, T.B. Three-dimensional optoacoustic imaging of nailfold capillaries in systemic sclerosis and its potential for disease differentiation using deep learning. *Sci. Rep.* **2020**, *10*, 16444. [CrossRef] [PubMed]
22. Smith, V.; Herrick, A.L.; Ingegnoli, F.; Damjanov, N.; De Angelis, R.; Denton, C.P.; Distler, O.; Espejo, K.; Foeldvari, I.; Frech, T.; et al. EULAR Study Group on Microcirculation in Rheumatic Diseases and the Scleroderma Clinical Trials Consortium Group on Capillaroscopy, Standardisation of nailfold capillaroscopy for the assessment of patients with Raynaud's phenomenon and systemic sclerosis. *Autoimmun. Rev.* **2020**, *19*, 102458. [CrossRef] [PubMed]
23. Yildirim, M. Automatic classification and diagnosis of heart valve diseases using heart sounds with MFCC and proposed deep model. *Concurr. Comput. Pract. Exp.* **2022**, *34*, e7232. [CrossRef]
24. Available online: https://yolov8.com/ (accessed on 9 May 2024).

diagnostics

MDPI

Article

Deep Learning Based Automatic Left Ventricle Segmentation from the Transgastric Short-Axis View on Transesophageal Echocardiography: A Feasibility Study

Yuan Tian [1,†], Wenting Qin [2,†], Zihang Zhao [2], Chunrong Wang [1], Yajie Tian [1], Yuelun Zhang [1], Kai He [1], Yuguan Zhang [1], Le Shen [1], Zhuhuang Zhou [2,*] and Chunhua Yu [1,*]

[1] Department of Anesthesiology, Peking Union Medical College Hospital, Chinese Academy of Medical Sciences and Peking Union Medical College, Beijing 100730, China; tianyuan95@pumch.cn (Y.T.); emancipation258@outlook.com (C.W.); tianyajie@pumch.cn (Y.T.); yuelunzhang@outlook.com (Y.Z.); harveyhekai@sina.com (K.H.); zygpumch@126.com (Y.Z.); shenle@pumch.cn (L.S.)

[2] Department of Biomedical Engineering, College of Chemistry and Life Science, Beijing University of Technology, Beijing 100124, China; qinwenting@emails.bjut.edu.cn (W.Q.); zhaozihang@emails.bjut.edu.cn (Z.Z.)

* Correspondence: zhouzh@bjut.edu.cn (Z.Z.); yu.chunhua@aliyun.com (C.Y.)

† These two authors contributed equally to this work and share first authorship.

Abstract: Segmenting the left ventricle from the transgastric short-axis views (TSVs) on transesophageal echocardiography (TEE) is the cornerstone for cardiovascular assessment during perioperative management. Even for seasoned professionals, the procedure remains time-consuming and experience-dependent. The current study aims to evaluate the feasibility of deep learning for automatic segmentation by assessing the validity of different U-Net algorithms. A large dataset containing 1388 TSV acquisitions was retrospectively collected from 451 patients (32% women, average age 53.42 years) who underwent perioperative TEE between July 2015 and October 2023. With image preprocessing and data augmentation, 3336 images were included in the training set, 138 images in the validation set, and 138 images in the test set. Four deep neural networks (U-Net, Attention U-Net, UNet++, and UNeXt) were employed for left ventricle segmentation and compared in terms of the Jaccard similarity coefficient (JSC) and Dice similarity coefficient (DSC) on the test set, as well as the number of network parameters, training time, and inference time. The Attention U-Net and U-Net++ models performed better in terms of JSC (the highest average JSC: 86.02%) and DSC (the highest average DSC: 92.00%), the UNeXt model had the smallest network parameters (1.47 million), and the U-Net model had the least training time (6428.65 s) and inference time for a single image (101.75 ms). The Attention U-Net model outperformed the other three models in challenging cases, including the impaired boundary of left ventricle and the artifact of the papillary muscle. This pioneering exploration demonstrated the feasibility of deep learning for the segmentation of the left ventricle from TSV on TEE, which will facilitate an accelerated and objective alternative of cardiovascular assessment for perioperative management.

Keywords: transesophageal echocardiography; deep learning; left ventricle segmentation; transgastric short-axis view; convolutional neural network

1. Introduction

More than 300 million operations are performed worldwide annually, according to the most recent survey by the World Health Organization [1]. Transesophageal echocardiography (TEE), a cardiovascular assessment technique using a flexible transesophageal probe, is becoming an integral part of perioperative management across a widening range of operations because TEE has demonstrated efficacy in facilitating decision-making during surgeries [2,3] and hemodynamic management for critically ill patients [4,5]. TEE is more practical than transthoracic echocardiography (TTE) during most surgeries, due to operative approaches and sterile requirements. Additionally, TEE is superior to TTE in enhancing the quality of echocardiography by circumventing the acoustic impediments caused by the ribs and lungs [6].

TEE assessment of the left ventricular function and structure is primarily conducted to answer the relatively common and potentially life-threatening problems encountered perioperatively [7,8]. Compared to the long-axis views of TEE, transgastric short-axis views (TSVs) enable facilitated global and local assessments of left ventricular function from the base to the apex by simply adjusting the probe's depth. TSVs also provide detailed visualization of the layered anatomy of the left ventricular wall during heartbeats. For these reasons, TSV is commonly performed to assess the structure and function of the left ventricle perioperatively.

Perioperative TEE assessment of left ventricle is a time-consuming and experience-dependent procedure, even for seasoned professionals. With advances in medical artificial intelligence (AI), deep learning algorithms are emerging as a supplementary alternative, providing accelerated and objective perioperative cardiovascular assessments [9,10]. While numerous studies have demonstrated improvements in the utilization of deep learning for left ventricular assessment, the majority of them are applicable to TTE analysis [11–14]. There also has been some research conducted according to TEE images, but focusing on cardiac long-axis views [15–17]. On the other hand, due to the evolution of U-Net and its variants since 2015, the segmentation of medical images based on deep learning has shown significant improvement in computational accuracy, sensitivity, and efficiency [18–21]. The role of U-Net algorithms has been demonstrated in the image segmentation for ovarian lesions [22], brain tumor, liver lesions, lung nodules [23], and so on. However, the feasibility of applying U-Net algorithms to left ventricular segmentation in TSVs remains poorly understood.

The current study aims to evaluate the feasibility of deep learning for automatic segmentation by assessing the validity of different U-Net algorithms. Initially, a large dataset of TSV images was complied from 451 patients undergoing perioperative TEE. Following image preprocessing and data augmentation, the training set was used to train U-Net algorithms for left ventricle segmentation, with the validation set used for checking overfitting. Finally, the test set was used to evaluate and compare the segmentation performance of U-Net algorithms.

2. Materials and Methods

Figure 1 shows the flow chart of the proposed automatic left ventricle segmentation in TSV TEE images using four deep neural network models: U-Net [18], UNet++ [19], Attention U-Net [20], and UNeXt [21]. Firstly, the end-diastolic frame (EDF) and end-systolic frame (ESF) of a TSV TEE video within a cardiac cycle were extracted and converted to one-channel grayscale images. Then, the one-channel ESF and EDF images were resized to a specific size. Subsequently, the resized ESF and EDF images were input to a trained deep neural network model to predict the left ventricle segmentation, which was resized to the original size to obtain the final left ventricle segmentation. The ESF and EDF images were chosen because they were representative and had manual segmentation as the ground truth. This is similar to the EchoNet-Dynamic dataset [11] which also labels only the ESF and EDF images for one TTE video. It should be noted that the trained deep neural network model can be used to segment the left ventricle in any frame of the TSV TEE video. The deep learning networks and the model training will be described in the following subsections.

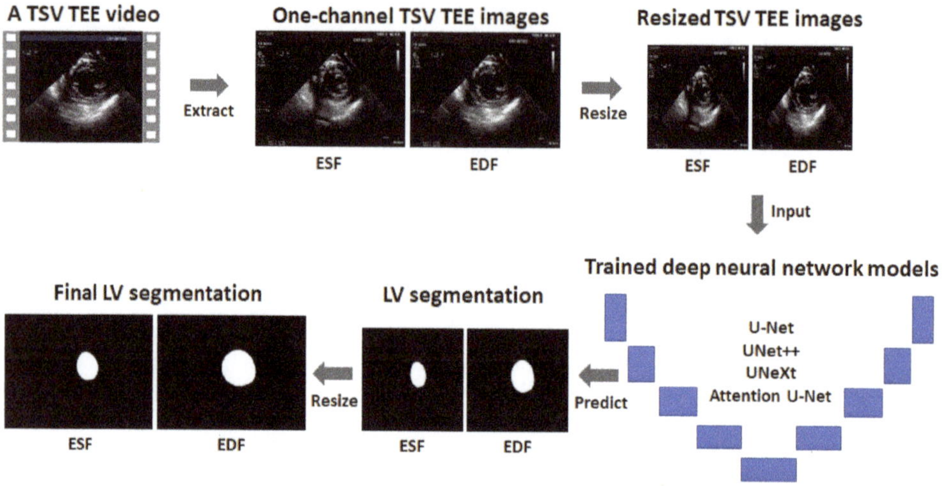

Figure 1. Flow chart of the proposed automatic LV segmentation method for TSV TEE images using deep neural network models. The deep learning models employed were U-Net [18], UNet++ [19], Attention U-Net [20], and UNeXt [21]. LV: left ventricle; TSV: transgastric short-axis view; TEE: transesophageal echocardiography; ESF: end-systolic frame; EDF: end-diastolic frame.

2.1. Patients Enrollment and Dataset Formulation

This retrospective study was approved by the Ethics Review Committee of the Peking Union Medical College Hospital, Chinese Academy of Medical Sciences. Patients that met the following criteria were involved: (1) those underwent cardiac surgery under general anesthesia at the Department of Anesthesiology, Peking Union Medical College Hospital, between July 2015 and October 2023; (2) those who had perioperative TEE performed by the Philips iE33 ultrasound scanner (Philips *Ultrasound*, Bothell, WA, USA) and the X7-2t transducer (1.0–5.0 MHz). Patients with known abnormalities of the left ventricle due to congenital heart diseases, or without stored videos of TSV were excluded. A total of 1076 TSV videos from the 451 involved patients were enrolled in the current study. Some of the patients had two distinct TSV videos that were acquired pre- and post-cardiac surgery. Of these, as illustrated in Figure 2, 382 videos were considered ineligible because they met any of the following criteria: duplication, coverage of less than one cardiac cycle, significant left ventricle boundary missing, or presence of severe noise. Both the EDF and ESF within a cardiac cycle were extracted from each TSV video, forming the dataset that comprised 1388 images.

Because a TSV TEE image was a grayscale B-mode ultrasound image, in order to reduce the computational cost of the deep neural network, a 2D three-channel TSV TEE image was converted into a one-channel grayscale image, that is, the number of channels was changed from 3 to 1, so that the amount of computation could be reduced while retaining all the imaging information of the original data. Finally, 1388 2D one-channel TSV TEE images were obtained as the experimental dataset in this study.

According to the ratio of 8:1:1, the dataset was randomly divided into a training set, a validation set, and a test set. The training set was used to train the deep neural network models, and the validation set was utilized to check if there was overfitting during the model training, while the test set was employed to evaluate the performance of the trained models. Specifically, the training set contained 1112 TSV TEE images extracted from 556 videos; the validation set contained 138 TSV TEE images extracted from 69 videos; and the test set contained another 138 TSV TEE images extracted from 69 videos. The image data of the same patient did not cross over the training set, the validation set, or the test set to avoid data leakage (i.e., they only appeared in one of the three sets). The manual left ventricle segmentation for each of the 1388 TSV TEE images was performed by two anesthesiologists and confirmed by another senior anesthesiologist, using the open LabelMe software (V3.16.2). The manual segmentation was taken as the ground truth. Figure 3 shows representative TSV TEE images and corresponding manual segmentation as left ventricle labels. As indicated in Figure 3, there are two major challenges for computer-assisted left ventricle segmentation in TSV TEE images: left ventricle boundary missing and papillary muscle interference.

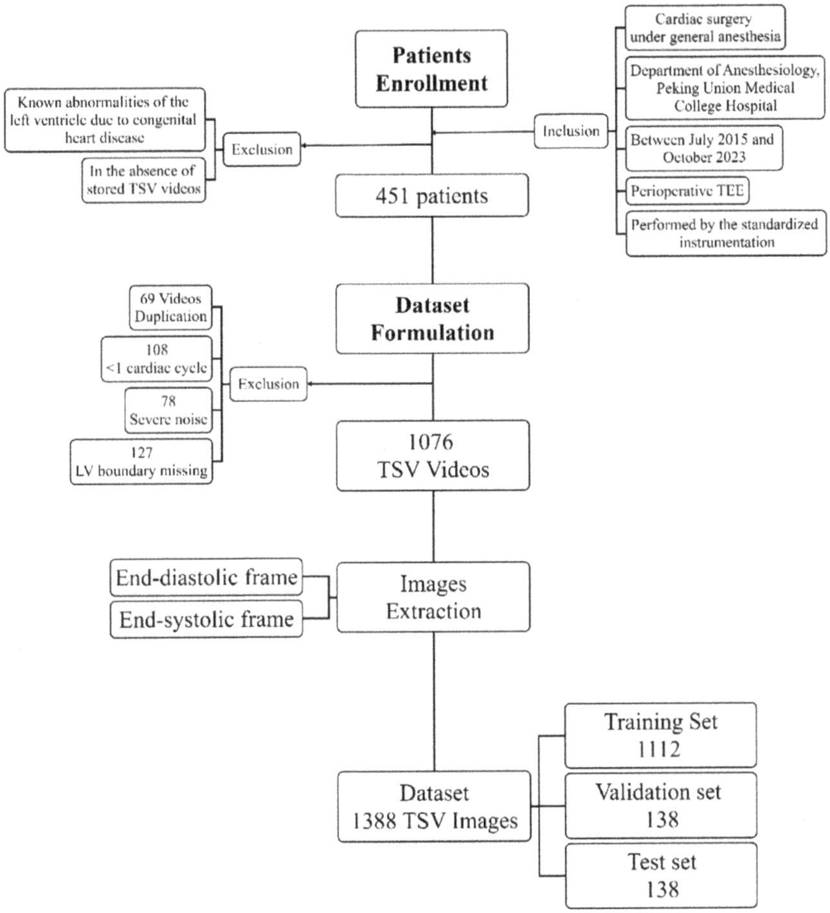

Figure 2. Flow chart of patients enrollment and dataset formulation. TEE: transesophageal echocardiography; TSV: transgastric short-axis view.

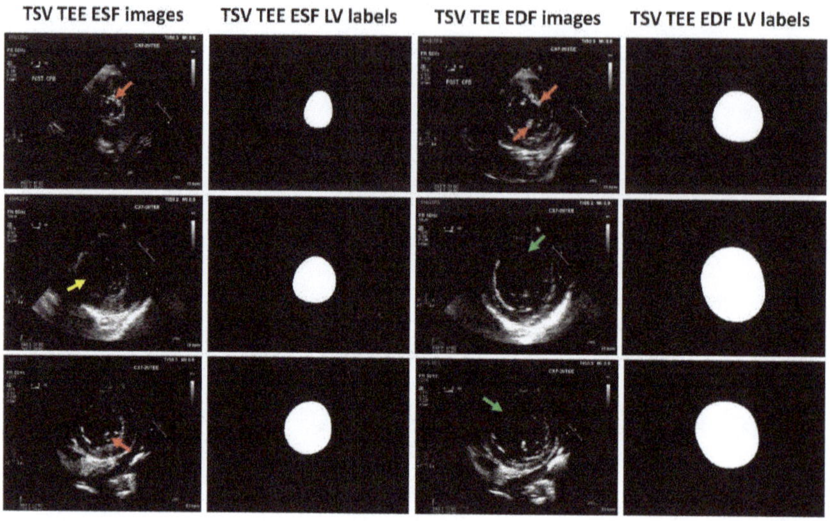

Figure 3. Representative TSV TEE images and corresponding LV labels. Red arrows indicate the papillary muscle. Yellow arrows indicate mild LV boundary missing. Green arrows indicate moderate LV boundary missing. LV: left ventricle; TSV: transgastric short-axis view; TEE: transesophageal echocardiography; ESF: end-systolic frame; EDF: end-diastolic frame.

2.2. Data Preprocessing and Augmentation

To reduce the computational cost of the deep learning models, each of the 1388 TSV TEE images was subsampled to a size of 256 × 256 pixels using cubic interpolation. Therefore, the size of an input image for the deep neural networks was 256 (image height) × 256 (image width) × 1 (image channel) for both training and testing. Due to the limited amount of experimental data, data augmentation was applied to the images in the training set, including random rotation from 0° to 90°, horizontal flipping, and vertical flipping. Data augmentation can reduce overfitting for the deep neural network model and improve the robustness of the model, which can further improve the generalization ability of the model. Data augmentation was conducted only on the training set, but not on the validation set or the test set. After data augmentation, the size of the training set was increased to 3336.

2.3. Deep Neural Network Models

In this study, four deep neural networks were employed for left ventricle segmentation in TSV TEE images: U-Net [18], UNet++ [19], Attention U-Net [20], and UNeXt [21].

U-Net [18] is the most commonly used and the simplest segmentation model in medical image segmentation, which uses a U-shaped network structure to obtain contextual information and location information (Figure 4). U-Net consists of an encoder and a decoder, with skip connections between the encoder and the decoder. U-Net uses a feature stitching structure to obtain low-level features and high-level semantic features of medical images. The encoding layers of the U-Net network first undergo two convolutional layers to extract features, followed by four down-sampling operations. Similarly, the decoding layers consist of four up-sampling operations and an output module.

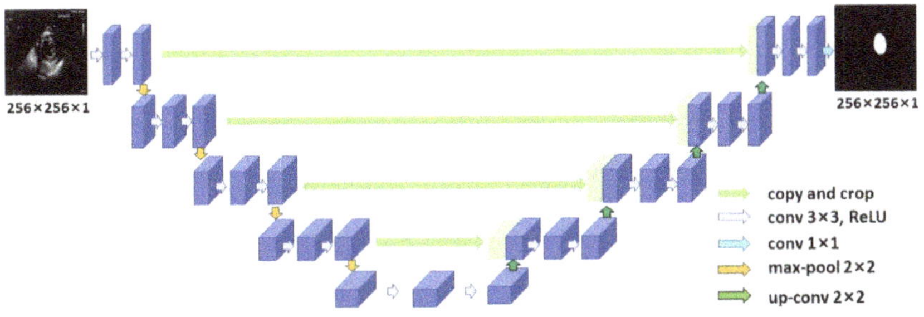

Figure 4. The U-Net network architecture for LV segmentation in TSV TEE images. LV: left ventricle; TSV: transgastric short-axis view; TEE: transesophageal echocardiography; ReLU: rectified linear unit; conv: convolution.

The Attention U-Net [20] model is an extension of the classical U-Net [18] architecture, incorporating the attention mechanism into U-Net (Figure 5), which can gradually strengthen the weight of the local region of interest, suppress the irrelevant regions in the input image, and highlight the salient features of specific local regions.

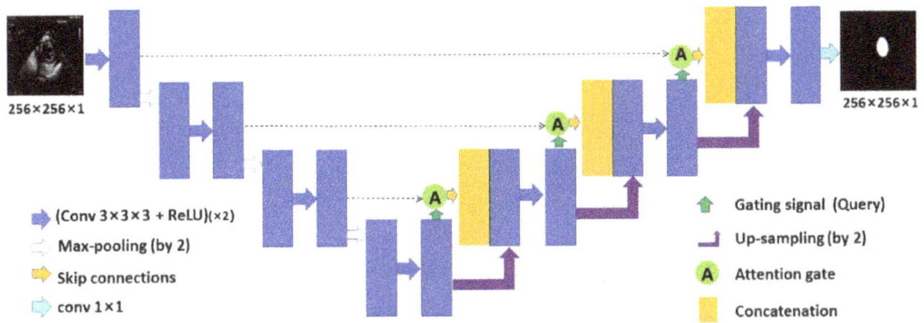

Figure 5. The Attention U-Net network architecture for LV segmentation in TSV TEE images. LV: left ventricle; TSV: transgastric short-axis view; TEE: transesophageal echocardiography; ReLU: rectified linear unit; conv: convolution.

UNet++ [19] is an improvement and extension of the classical U-Net [18] architecture, employing cascaded connections and introducing dense skip connections (Figure 6). It cascades feature maps from both the encoder and decoder, with each decoding layer connected to all deeper encoding layers, forming a dense skip connection structure. This allows the decoder to fully leverage multi-scale features from all encoder layers, making it suitable for scenarios requiring the handling of multi-scale information.

The UNeXt [21] network's encoder consists of three convolutional layers and two Tokenized multi-layer perceptron (MLP) modules (Figure 7). In contrast to U-Net [18], UNeXt [21] adopts a leaner approach by employing fewer convolutional layers and larger strides during feature map down-sampling, effectively reducing parameters. UNeXt has gained significant attention as a lightweight solution, emerging as a pioneering fast medical image segmentation network to integrate the MLP module with convolutional layers. At its core lies the Tokenized MLP module, enabling efficient segmentation of medical images with fewer convolutional layers and larger feature down-sampling.

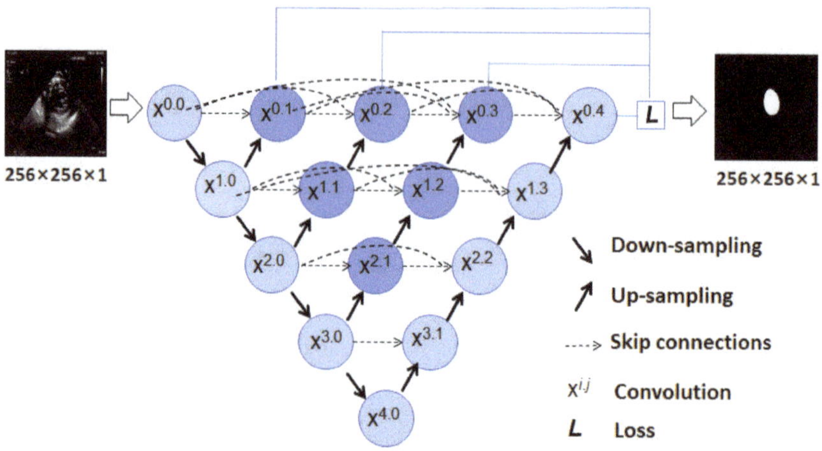

Figure 6. The U-Net++ network architecture for LV segmentation in TSV TEE images. LV: left ventricle; TSV: transgastric short-axis view; TEE: transesophageal echocardiography.

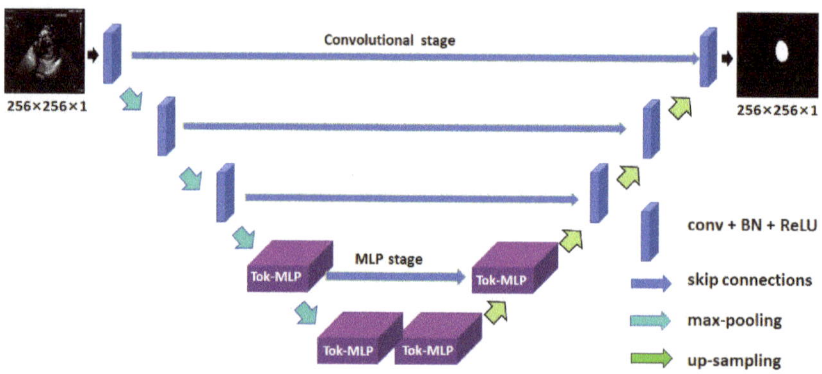

Figure 7. The UNeXt network architecture for LV segmentation in TSV TEE images. LV: left ventricle; TSV: transgastric short-axis view; TEE: transesophageal echocardiography; ReLU: rectified linear unit; BN: batch normalization; conv: convolution; MLP: multi-layer perceptron; Tok: Tokenized.

2.4. Segmentation Performance Evaluation

In order to evaluate the performance of different deep learning models for left ventricle segmentation in TSV TEE images, the DSC and Jaccard similarity coefficient (JSC) were used as the segmentation performance evaluation metrics:

$$\text{DSC}(A, B) = \frac{2|A \cap B|}{|A| + |B|}; \tag{1}$$

$$\text{JSC}(A, B) = \frac{|A \cap B|}{|A \cup B|}, \tag{2}$$

where A represents the left ventricle region predicted by deep learning models, and B is the labeled region manually annotated by human experts. Both DSC and JSC ranged from 0 to 1 (or 0% to 100%), with a larger value indicating better segmentation performance.

2.5. Experimental Setup

Our experiments were conducted on a graphics workstation with Intel(R) Xeon(R) Gold 6132 CPU@2.60 GHz 2.59 GHz (2 processors), and NVIDIA TITAN RTX 24G, 128G

RAM. The PyTorch (version 1.5.1) was used as the deep learning framework. In the experiments, the model input dimensions were 4 (batch size) $\times$ 1 (channels) $\times$ 256 (height) $\times$ 256 (width). The number of training epochs was set at 100. The gradient optimizer was the Adam optimizer. The initial learning rate was set at 10^{-3}. The momentum was set at 0.9. A loss function with a combination of the binary cross-entropy (BCE) loss L_{BCE} and the DSC loss L_{DSC} was used for the U-Net, UNet++, Attention U-Net, and UNeXt models:

$$Loss = \beta L_{\text{BCE}} + \gamma L_{\text{DSC}}, \tag{3}$$

where $\beta = 0.5$, and $\gamma = 0.5$. L_{BCE} and L_{DSC} are defined as

$$L_{\text{BCE}} = -B\log(A) - (1 - B)\log(1 - A), \tag{4}$$

$$L_{\text{DSC}} = 1 - \frac{2|A \cap B|}{|A| + |B|} \tag{5}$$

where A represents the left ventricular region predicted by our model, and B is the labeled region manually annotated by human experts. L_{BCE} is similar to the cross-entropy loss function, but the binary cross-entropy loss function has an operation to calculate the logit, so we do not need to use the sigmoid function or the softmax function to map the input to [0, 1] for this loss function. According to the official documentation, the binary cross-entropy loss function has better numerical stability than the cross-entropy loss function. L_{DSC} is a region-related loss function that has good performance in scenarios where positive and negative samples are seriously unbalanced.

2.6. Statistical Analysis

The Kruskal–Wallis test was used to evaluate whether the U-Net, UNet++, Attention U-Net, and UNeXt models had statistically significant differences in terms of the DSC or the JSC for the left ventricle segmentation in the test set of TSV TEE images ($n = 138$). A statistically significant difference was defined as $p < 0.05$. The statistical analysis was performed with IBM SPSS Statistics 27 (IBM Corp., Endicott, NY, USA).

3. Results

A total of 1388 images were extracted from 694 TSV videos, featuring 451 patients with an average age of 53.42 years. The analytic population consisted of 32% women, and 27% ASA-PS III or higher. Pre-procedural diagnoses included coronary artery disease, valvular stenosis or regurgitation, aortic disease, and pericardial disease. Figures 8 and 9 show the loss and the DSC on the training set and validation set as a function of training epochs for different deep learning models. For the Attention U-Net, U-Net and UNet++ models, all the training loss and validation loss gradually decreased as the training epochs increased, and they converged when the training epoch reached 100; both the training DSC and validation DSC gradually increased as the training epochs increased, and they converged when the training epoch reached 100. These indicated that the Attention U-Net, U-Net, and UNet++ models had no overfitting or very slight overfitting. The UNeXt model shows similar trends for training loss and DSC, but the validation loss and DSC do not monotonically decrease or increase as the training epochs increased to 100.

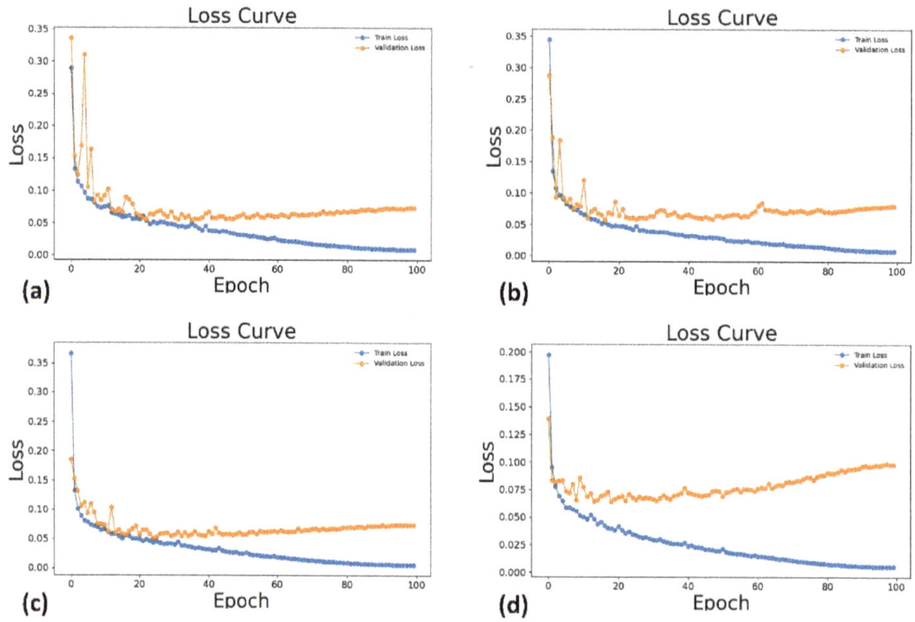

Figure 8. The loss on the training set (i.e., Train Loss) and validation set (i.e., Validation Loss) as a function of training epochs for different deep learning models: Attention U-Net (**a**), U-Net (**b**), UNet++ (**c**), and UNeXt (**d**).

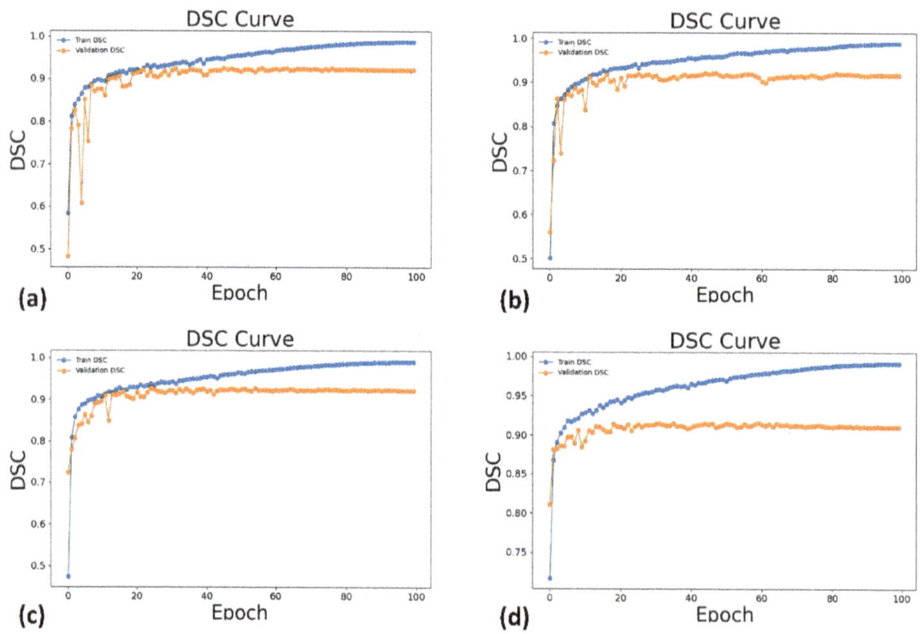

Figure 9. The DSC on the training set (i.e., Train DSC) and validation set (i.e., Validation DSC) as a function of training epochs for different deep learning models: Attention U-Net (**a**), U-Net (**b**), UNet++ (**c**), and UNeXt (**d**). DSC: Dice similarity coefficient.

Figure 10 shows representative left ventricle segmentation results for TSV TEE images using different deep neural network models. Each column corresponds to a representative TSV TEE case. The first row shows the input images, and the second to the fifth row show the segmented left ventricle contours by the trained Attention U-Net, U-Net, UNet++, and UNeXt models, respectively.

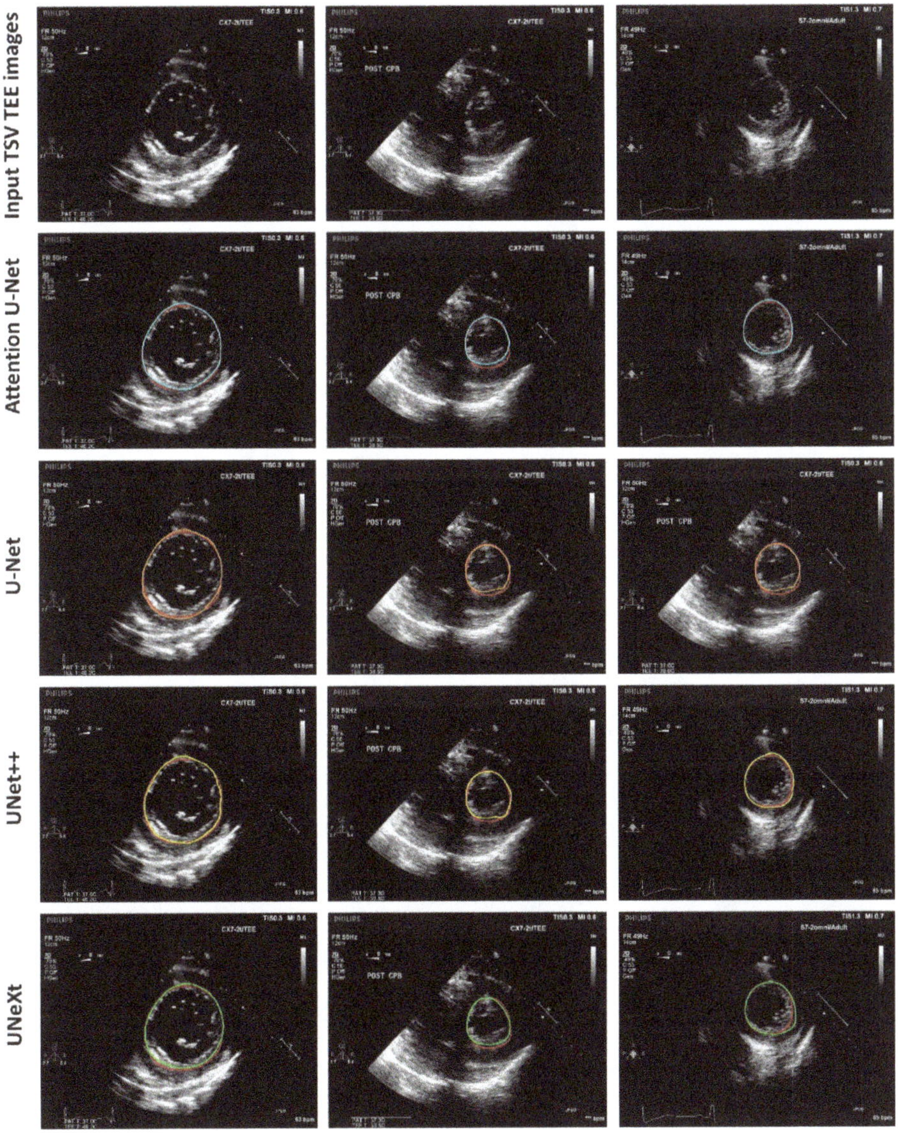

Figure 10. Representative LV segmentation in TSV TEE images using different deep learning models. Red contours indicate manual LV segmentation as the ground truth. Blue contours indicate LV segmentation using Attention U-Net. Orange contours indicate LV segmentation using U-Net. Yellow contours indicate LV segmentation using UNet++. Green contours indicate LV segmentation using UNeXt. TEE: transesophageal echocardiography; TSV: transgastric short-axis view.

Shown in Table 1 are the U-Net, UNet++, Attention U-Net, and UNeXt models for left ventricle segmentation in TSV TEE images with respect to the number of model parameters, training time (for the 3336 images in the training set), and inference time for a single image.

Table 1. Left ventricle segmentation in TSV TEE images with respect to the number of model parameters, training time, and inference time for a single image of the U-Net, UNet++, Attention U-Net, and UNeXt models. TEE: transesophageal echocardiography; TSV: transgastric short-axis view.

Deep Learning Models	# of Model Parameters	Training Time	Inference Time for a Single Image
U-Net [18]	7.85 million	6428.65 s	101.75 ms
UNet++ [19]	9.16 million	10,080.50 s	134.21 ms
UNeXt [20]	1.47 million	7122.94 s	109.59 ms
Attention U-Net [21]	34.88 million	10,556.86 s	122.85 ms

The performance comparisons between U-Net and its variants, within the test set (n = 138), are shown in Table 2. Kruskal–Wallis test indicated no significant differences in the average JSC and DSC between algorithms.

Table 2. Left ventricle segmentation performance of U-Net, UNet++, Attention U-Net, and UNeXt on the test set of TSV TEE images (n = 138) evaluated using JSC and DSC. Data are expressed as mean ± standard deviation. TEE: transesophageal echocardiography; TSV: transgastric short-axis view.

Deep Learning Models	JSC (%)	DSC (%)
U-Net [18]	84.71 ± 10.25	90.98 ± 7.19
UNet++ [19]	86.02 ± 8.70	91.76 ± 5.48
UNeXt [20]	84.20 ± 9.62	91.00 ± 6.23
Attention U-Net [21]	85.93 ± 8.71	92.00 ± 5.50

4. Discussion

Very little was found in previous studies on the question of whether U-Net and its variants are feasible for segmentation of the left ventricle from TSV on TEE images. To the best of our knowledge, the current study provides novel evidence of the efficacy and accuracy of deep learning in the expanded medical scenarios.

The encouraging findings indicate that all of the U-Net and its derivatives perform well in the segmentation of the left ventricle from TSV on TEE with an average DSC of 0.91–0.92. These findings are comparable to previous results, which demonstrated ones of 0.92–0.95 for left ventricle segmentation from TTE images [11–14]. The results also achieve a promising DSC for segmenting the left ventricle from TSV, effectively supplementing the previous studies that used limited TEE images [15–17]. From the perspective of using U-Net and its variants, the findings demonstrate superior accuracy in left ventricle segmentation compared to its use in the segmentation of ovarian lesions (0.89) [22], brain tumors (0.89–0.91), liver lesions (0.79–0.83), and lung nodules (0.71–0.77) [23].

Another clinically relevant finding is that the inference times have accelerated to 101–134 ms, compared to the previously reported 230 ms [15]. Given the different tasks focused on in the two studies, a direct comparison between them may not be entirely valid. Nevertheless, the results highlight the proficient performance of U-Net and its variants in segmenting the left ventricle from TSV TEE. It is still noteworthy that the current study uses a CPU as the workstation. It is reasonable to speculate that a GPU workstation would further accelerate the program.

The results of the study indicate no significant differences in accuracy between U-Net and its variants. The absence of significant benefits shown with UNet++ may be due to the standardization of image preprocessing, as it outperforms in enhancing the segmentation quality of various sizes [23]. Compared to U-Net, its variants did not show any improvement. This might be because the simple structure and distinct boundaries of the left ventricle do not require more complex algorithms [14]. The findings demonstrate that slight overfitting existed in the U-NeXt, possibly due to the limited data available in the present study, which may not match the depth of the layers required by the algorithm [20].

In the present study, a large dataset of TEE images was collected, consisting of 1388 images from 451 patients. Restricted by the applicable scenarios, the capacity of this TEE dataset is still incomparable to the international TTE dataset CAMUS, which contains tens of thousands of data points [24]. However, the size of these dataset exceeds the volumes reported in previous studies, which built their deep learning models based on TEE images from 3–95 patients [15–17].

A rather disappointing result is that the current study failed to achieve promising results in left ventricle segmentation for some challenging cases. Figure 11 shows the left ventricle segmentation in representative challenging cases of TSV TEE images. As shown in the first column of Figure 11 which shows a case of mild left ventricle boundary missing, the upper right part of the endocardial boundary is missing in the original image, and the detected boundary in the epicardial region by the U-Net and UNet++ networks is incorrect segmentation; there is slight incorrect segmentation for the UNeXt model. As shown in the second column of Figure 11 which shows a case of moderate left ventricle boundary missing, the left ventricle boundary is missing on the right side in the original image, and there is significant incorrect segmentation for the U-Net, UNet++, and UNeXt models. As shown in the third column of Figure 11 which shows a case of papillary muscle interference, there is notable incorrect segmentation for the U-Net, UNet++, and UNeXt models. For these cases, the Attention U-Net model performs much better than the U-Net, UNet++, and UNeXt models, with the left ventricle segmentation quite close to the ground truth. The reason may be that the attention mechanism incorporated in the Attention U-Net model could well deal with the challenging issues of left ventricle boundary missing and papillary muscle interference. Further research is required to resolve the problems.

Based on the promising results of the current study, it is strongly anticipated that further research into real-time assessment of left ventricle function and structure will proceed smoothly.

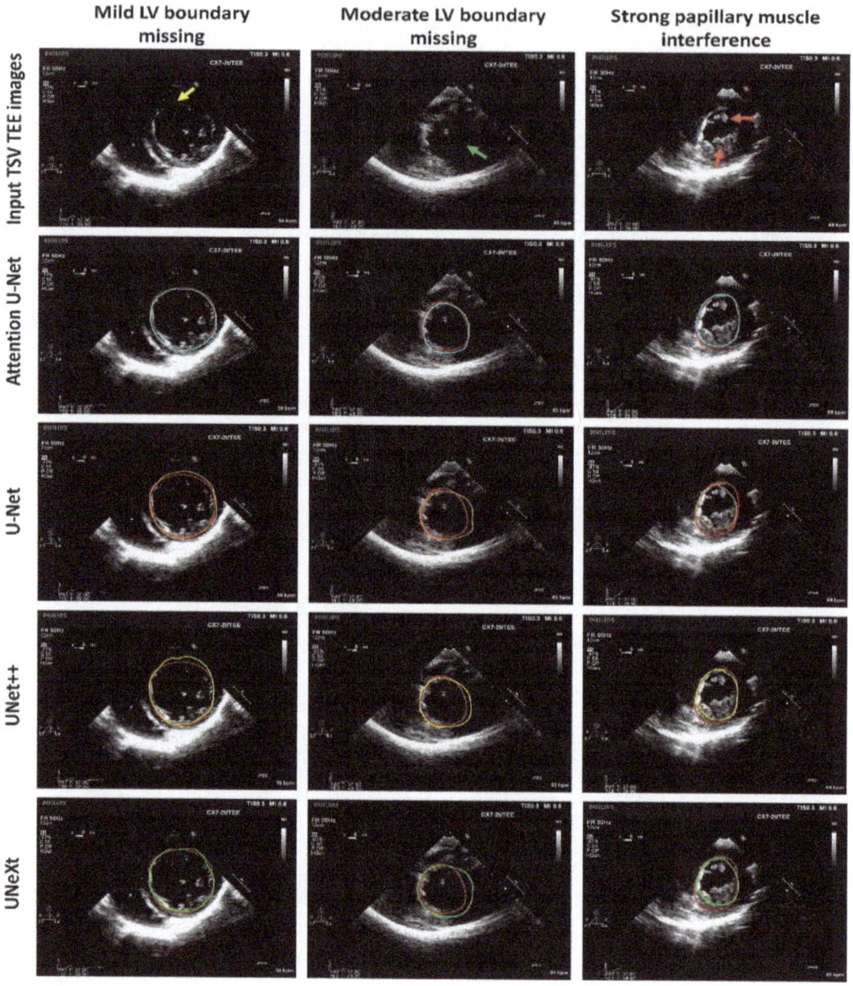

Figure 11. Representative LV segmentation in TSV TEE images for challenging cases of LV boundary missing and strong papillary muscle interference using different deep learning models. Red contours indicate manual LV segmentation as the ground truth. Blue contours indicate LV segmentation using Attention U-Net. Orange contours indicate LV segmentation using U-Net. Yellow contours indicate LV segmentation using UNet++. Green contours indicate LV segmentation using UNeXt. Red arrows indicate the papillary muscle. Yellow arrows indicate mild LV boundaries missing. Green arrows indicate moderate LV boundaries missing. TEE: transesophageal echocardiography; TSV: transgastric short-axis view.

5. Conclusions

The current study highlights the feasibility of using deep learning for left ventricle segmentation from TSV on TEE, with promising accuracy and speed, based on a large TEE dataset. The performances of U-Net and its variants are comparable. It potentially facilitates an accelerated and objective alternative for cardiovascular assessment in perioperative management. Further research is required to explore its application in challenging cases and real-time assessment of left ventricle function and structure.

Author Contributions: Conceptualization, Y.T. (Yuan Tian), L.S., Z.Z. (Zhuhuang Zhou) and C.Y.; data curation, Y.T. (Yuan Tian), W.Q., Z.Z. (Zihang Zhao), Z.Z. (Zhuhuang Zhou) and C.Y.; formal analysis, Y.T. (Yuan Tian), W.Q., Z.Z. (Zhuhuang Zhou) and C.Y.; funding acquisition, C.Y.; investigation, Y.T. (Yuan Tian), C.W., Y.T. (Yajie Tian), Y.Z. (Yuelun Zhang), K.H., Y.Z. (Yuguan Zhang) and C.Y.; methodology, Y.T. (Yuan Tian), Z.Z. (Zihang Zhao), Z.Z. (Zhuhuang Zhou) and C.Y.; project administration, Y.T. (Yuan Tian), Z.Z. (Zhuhuang Zhou) and C.Y.; resources, Y.T. (Yuan Tian), C.W., Y.T. (Yajie Tian), Y.Z. (Yuelun Zhang), K.H., Y.Z. (Yuguan Zhang) and C.Y.; software, W.Q. and Z.Z. (Zhuhuang Zhou); supervision, Y.T. (Yuan Tian), L.S., Z.Z. (Zhuhuang Zhou) and C.Y.; validation, Y.T. (Yuan Tian), Z.Z. (Zhuhuang Zhou) and C.Y.; visualization, W.Q. and Z.Z. (Zhuhuang Zhou); writing—original draft, W.Q.; writing—review and editing, Y.T. (Yuan Tian), Z.Z. (Zhuhuang Zhou) and C.Y. All authors have read and agreed to the published version of the manuscript.

Funding: This work was supported in part by the National High Level Hospital Clinical Research Funding (Funder the Chinese Academy of Medical Sciences (CAMS). Funding number 2022-PUMCH-B-007).

Institutional Review Board Statement: The study was conducted in accordance with the Declaration of Helsinki, and approved by the Ethics Review Committee of the Peking Union Medical College Hospital (protocol code I-22YJ617 and date of approval 21 December 2022).

Informed Consent Statement: Informed consent was waived for this retrospective study.

Data Availability Statement: Data is unavailable due to ethical restrictions.

Conflicts of Interest: The authors declare no conflicts of interest. The funders had no role in the design of the study; in the collection, analyses, or interpretation of data; in the writing of the manuscript; or in the decision to publish the results.

References

1. Weiser, T.G.; Haynes, A.B.; Molina, G.; Lipsitz, S.; Esquivel, M.; Uribe-Leitz, T.; Fu, R.; Azad, T.; Chao, T.; Berry, T.; et al. Estimate of the global volume of surgery in 2012: An assessment supporting improved health outcomes. *Lancet* **2015**, *385*, S11. [CrossRef] [PubMed]
2. Nicoara, A.; Skubas, N.; Ad, N.; Finley, A.; Hahn, R.T.; Mahmood, F.; Mankad, S.; Nyman, C.B.; Pagani, F.; Porter, T.R.; et al. Guidelines for the use of transesophageal echocardiography to assist with surgical decision-making in the operating room: A surgery-based approach: From the American Society of Echocardiography in collaboration with the Society of Cardiovascular Anesthesiologists and the Society of Thoracic Surgeons. *J. Am. Soc. Echocardiogr.* **2020**, *33*, 692–734. [PubMed]
3. Ferro, E.G.; Alkhouli, M.; Nair, D.G.; Kapadia, S.R.; Hsu, J.C.; Gibson, D.N.; Freeman, J.V.; Price, M.J.; Roy, K.; Allocco, D.J.; et al. Intracardiac vs Transesophageal Echocardiography for Left Atrial Appendage Occlusion With Watchman FLX in the U.S. *JACC Clin. Electrophysiol.* **2023**, *9*, 2587–2599. [CrossRef] [PubMed]
4. Mayo, P.H.; Narasimhan, M.; Koenig, S. Critical Care Transesophageal Echocardiography. *Chest* **2015**, *148*, 5. [CrossRef] [PubMed]
5. MacKay, E.J.; Zhang, B.; Heng, S.; Ye, T.; Neuman, M.D.; Augoustides, J.G.; Feinman, J.W.; Desai, N.D.; Groeneveld, P.W. Association between Transesophageal Echocardiography and Clinical Outcomes after Coronary Artery Bypass Graft Surgery. *J. Am. Soc. Echocardiogr.* **2021**, *34*, 571–581. [CrossRef] [PubMed]
6. Jaidka, A.; Hobbs, H.; Koenig, S.; Millington, S.J.; Arntfield, R.T. Better With Ultrasound: Transesophageal Echocardiography. *Chest* **2019**, *155*, 194–201. [CrossRef] [PubMed]
7. Marbach, J.A.; Almufleh, A.; Di Santo, P.; Simard, T.; Jung, R.; Diemer, G.; West, F.M.; Millington, S.J.; Mathew, R.; Le May, M.R.; et al. A shifting paradigm: The role of focused cardiac ultrasound in bedside patient assessment. *Chest* **2020**, *58*, 2107–2118. [CrossRef] [PubMed]
8. Thaden, J.J.; Malouf, J.F.; Rehfeldt, K.H.; Ashikhmina, E.; Bagameri, G.; Enriquez-Sarano, M.; Stulak, J.M.; Schaff, H.V.; Michelena, H.I. Adult Intraoperative Echocardiography: A Comprehensive Review of Current Practice. *J. Am. Soc. Echocardiogr.* **2020**, *33*, 735–755. [CrossRef]
9. Nabi, W.; Bansal, A.; Xu, B. Applications of artificial intelligence and machine learning approaches in echocardiography. *Echocardiography* **2021**, *38*, 982–992. [CrossRef] [PubMed]
10. Narang, A.; Bae, R.; Hong, H.; Thomas, Y.; Surette, S.; Cadieu, C.; Chaudhry, A.; Martin, R.P.; McCarthy, P.M.; Rubenson, D.S.; et al. Utility of a Deep-Learning Algorithm to Guide Novices to Acquire Echocardiograms for Limited Diagnostic Use. *JAMA Cardiol.* **2021**, *6*, 624–632. [CrossRef] [PubMed]
11. Ouyang, D.; He, B.; Ghorbani, A.; Yuan, N.; Ebinger, J.; Langlotz, C.P.; Heidenreich, P.A.; Harrington, R.A.; Liang, D.H.; Ashley, E.A.; et al. Video-based AI for beat-to-beat assessment of cardiac function. *Nature* **2020**, *580*, 252–256. [CrossRef] [PubMed]
12. Leclerc, S.; Smistad, E.; Østvik, A.; Cervenansky, F.; Espinosa, F.; Espeland, T.; Berg, E.A.R.; Belhamissi, M.; Israilov, S.; Grenier, T.; et al. LU-Net: A multistage attention network to improve the robustness of segmentation of left ventricular structures in 2-D echocardiography. *IEEE Trans. Ultrason. Ferroelectr. Freq. Control* **2020**, *67*, 2519–2530. [CrossRef]

13. Liu, F.; Wang, K.; Liu, D.; Yang, X.; Tian, J. Deep pyramid local attention neural network for cardiac structure segmentation in two-dimensional echocardiography. *Med. Image Anal.* **2021**, *67*, 101873. [CrossRef] [PubMed]
14. Zeng, Y.; Tsui, P.H.; Pang, K.; Bin, G.; Li, J.; Lv, K.; Wu, X.; Wu, S.; Zhou, Z. MAEF-Net: Multi-attention efficient feature fusion network for left ventricular segmentation and quantitative analysis in two-dimensional echocardiography. *Ultrasonics* **2023**, *127*, 106855. [CrossRef] [PubMed]
15. Haukom, T.; Berg, E.A.R.; Aakhus, S.; Kiss, G.H. Basal strain estimation in transesophageal echocardiography (tee) using deep learning based unsupervised deformable image registration. In Proceedings of the 2019 IEEE International Ultrasonics Symposium (IUS), Glasgow, UK, 6–9 October 2019; pp. 1421–1424.
16. Kang, S.; Kim, S.J.; Ahn, H.G.; Cha, K.C.; Yang, S. Left ventricle segmentation in transesophageal echocardiography images using a deep neural network. *PLoS ONE* **2023**, *18*, e0280485. [CrossRef] [PubMed]
17. Ahn, H.; Kim, S.J.; Kang, S.; Han, J.; Hwang, S.O.; Cha, K.C.; Yang, S. Ventricle tracking in transesophageal echocardiography (TEE) images during cardiopulmonary resuscitation (CPR) using deep learning and monogenic filtering. *Biomed. Eng. Lett.* **2023**, *13*, 715–728. [CrossRef] [PubMed]
18. Ronneberger, O.; Fischer, P.; Brox, T. U-Net: Convolutional networks for biomedical image segmentation. In Proceedings of the International Conference on Medical Image Computing and Computer-Assisted Intervention, Proceedings, Part III 18, Munich, Germany, 5–9 October 2015; Springer International Publishing: New York, NY, USA, 2015; pp. 234–241.
19. Zhou, Z.; Rahman Siddiquee, M.M.; Tajbakhsh, N.; Liang, J. UNet++: A nested U-net architecture for medical image segmentation. In Proceedings of the Deep Learning in Medical Image Analysis and Multimodal Learning for Clinical Decision Support: 4th International Workshop, DLMIA 2018, and 8th International Workshop, ML-CDS 2018, Held in Conjunction with MICCAI 2018, Proceedings 4, Granada, Spain, 20 September 2018; Springer International Publishing: New York, NY, USA, 2018; pp. 3–11.
20. Oktay, O.; Schlemper, J.; Folgoc, L.L.; Lee, M.; Heinrich, M.; Misawa, K.; Mori, K.; McDonagh, S.; Hammerla, N.Y.; Kainz, B.; et al. Attention U-Net: Learning where to look for the pancreas. *arXiv* **2018**, arXiv:1804.03999.
21. Valanarasu, J.M.J.; Patel, V.M. UNeXt: MLP-based rapid medical image segmentation network. In Proceedings of the International Conference on Medical Image Computing and Computer-Assisted Intervention, Singapore, 18–22 September 2022; Springer Nature: Cham, Switzerland, 2022; pp. 23–33.
22. Zou, Y.; Amidi, E.; Luo, H.; Zhu, Q. Ultrasound-enhanced Unet model for quantitative photoacoustic tomography of ovarian lesions. *Photoacoustics* **2022**, *28*, 100420. [CrossRef] [PubMed]
23. Zhou, Z.; Siddiquee, M.M.R.; Tajbakhsh, N.; Liang, J. UNet++: Redesigning Skip Connections to Exploit Multiscale Features in Image Segmentation. *IEEE Trans. Med. Imaging* **2020**, *39*, 1856–1867. [CrossRef] [PubMed]
24. Leclerc, S.; Smistad, E.; Pedrosa, J.; Østvik, A.; Cervenansky, F.; Espinosa, F.; Espeland, T.; Berg, E.A.R.; Jodoin, P.-M.; Grenier, T.; et al. Deep learning for segmentation using an open large-scale dataset in 2D echocardiography. *IEEE Trans. Med. Imaging* **2019**, *38*, 2198–2210. [CrossRef] [PubMed]

diagnostics

MDPI

Article

Exploring the Impact of Noise and Image Quality on Deep Learning Performance in DXA Images

Dildar Hussain and Yeong Hyeon Gu *

Department of Artificial Intelligence and Data Science, Sejong University, Seoul 05006, Republic of Korea; hussain.bangash@sejong.ac.kr
* Correspondence: yhgu@sejong.ac.kr

Abstract: Background and Objective: Segmentation of the femur in Dual-Energy X-ray (DXA) images poses challenges due to reduced contrast, noise, bone shape variations, and inconsistent X-ray beam penetration. In this study, we investigate the relationship between noise and certain deep learning (DL) techniques for semantic segmentation of the femur to enhance segmentation and bone mineral density (BMD) accuracy by incorporating noise reduction methods into DL models. Methods: Convolutional neural network (CNN)-based models were employed to segment femurs in DXA images and evaluate the effects of noise reduction filters on segmentation accuracy and their effect on BMD calculation. Various noise reduction techniques were integrated into DL-based models to enhance image quality before training. We assessed the performance of the fully convolutional neural network (FCNN) in comparison to noise reduction algorithms and manual segmentation methods. Results: Our study demonstrated that the FCNN outperformed noise reduction algorithms in enhancing segmentation accuracy and enabling precise calculation of BMD. The FCNN-based segmentation approach achieved a segmentation accuracy of 98.84% and a correlation coefficient of 0.9928 for BMD measurements, indicating its effectiveness in the clinical diagnosis of osteoporosis. Conclusions: In conclusion, integrating noise reduction techniques into DL-based models significantly improves femur segmentation accuracy in DXA images. The FCNN model, in particular, shows promising results in enhancing BMD calculation and clinical diagnosis of osteoporosis. These findings highlight the potential of DL techniques in addressing segmentation challenges and improving diagnostic accuracy in medical imaging.

Keywords: dual-energy X-ray absorptiometry (DXA); osteoporosis; deep learning; segmentation; FCN; noise; imperfection; filters

1. Introduction

Osteoporosis is a pathological condition that compromises the integrity of the skeletal system. It serves as the primary factor for hip fractures within many countries. It demonstrates an absence of gender discrimination and has the potential to materialize at any point during an individual's lifespan. A high proportion, exceeding 20%, of individuals affected with hip fractures succumb to the resultant trauma [1,2]. The disease can adequately be diagnosed with a low-dose X-ray imaging technique called dual-energy X-ray absorptiometry (DXA), which is considered a golden standard for diagnosing osteoporosis fracture risk [3]. Quantitative Computed Tomography (QCT) is another alternative. However, QCT requires a high dose of X-rays and is costly.

The initial step in the precise osteoporosis diagnosis involves accurate segmentation, which is crucial for accurately calculating bone mineral density (BMD) and final osteoporosis report generation. Segmentation errors can significantly impact the BMD calculation and subsequent analyses. However, several reasons drive incorrect segmentation in DXA images and post-analysis [4,5]. Firstly, the use of low-dose X-rays in DXA imaging provides noisy images. Secondly, there is an overlap of organs in human bodies. Thirdly, the irregular attenuation of the X-rays through the human body produces negative shadows, which appear as dark areas in the images. Other factors affecting segmentation quality include scanning orientation, luminocity intensities, resolution, and individual variations [6,7].

Noise profoundly impacts medical imaging analysis, manifesting as artifacts, reduced image clarity, and potential misinterpretation of diagnostic information. It stems from diverse sources like equipment limitations, patient motion, and inadequate radiation doses, distorting image features and complicating accurate identification and analysis of anatomical structures or pathologies. Moreover, noise undermines automated image processing algorithms, compromising segmentation, feature extraction, and classification tasks. Hence, effective noise reduction techniques are imperative for bolstering the accuracy and reliability of medical imaging analysis, thereby refining diagnostic interpretation and patient care [8,9].

In medical imaging, noise significantly disrupts segmentation processes by introducing inaccuracies and artifacts, hindering the precise delineation of anatomical structures. Obscured tissue boundaries due to noise often yield incomplete or erroneous segmentation results, while disrupted intensity gradients impede segmentation algorithms in distinguishing different regions of interest. Consequently, noise reduction techniques play a pivotal role in enhancing the accuracy and reliability of segmentation, ensuring meticulous delineation of structures, and maximizing the utility of imaging data for diagnostic and therapeutic endeavors [10–12].

Noise poses a substantial challenge to the performance of Deep Learning (DL) models in medical imaging by introducing uncertainties and inconsistencies in training data. Its presence obfuscates relevant features and patterns, leading to diminished model accuracy and reliability. Furthermore, noise exacerbates overfitting, wherein models learn to capture noise instead of meaningful information, impairing generalization to new data. Robust preprocessing techniques and noise reduction strategies are essential to mitigate these effects, improving training data quality and bolstering model performance. Additionally, the development of noise-robust architectures and training methods holds promise in enhancing model resilience to noise and variability in medical imaging data, ultimately augmenting their efficacy in clinical applications [13–15].

Regardless of the numerous techniques being used, accurate automatic segmentation in DXA imaging always remains a challenge [16–22]. Manual segmentation is time-consuming, requires an expert, and is impractical for analyzing large public datasets [16,18, 19]. With the inherent limitations described in the references [23–25], the existing techniques are unsuitable for DXA image segmentation. All the limitations described in our previous work [26] forced us to new investigations, which led us to more precise segmentation with greater accuracy in DXA imaging.

Integration of denoising techniques into convolutional neural network (CNN)-based DL models constitutes a critical step in the preprocessing pipeline. Before training, noisy DXA images undergo filtration using one or a blend of denoising methods to enhance their quality. These denoised images serve as input data for training the CNN model, enabling it to learn from cleaner and more representative image data. Subsequently, during inference, the trained CNN model is directly applied to noisy DXA images for femur segmentation, resulting in heightened accuracy and reliability in the segmentation process. To evaluate the impact of these denoising filters on DL-based image segmentation, we compiled the results by applying the CNN-based model to segment femur images and calculate BMD both with and without the application of denoising filters. This comparative analysis allows us to comprehensively understand the effectiveness of denoising techniques in enhancing the performance of DL-based segmentation algorithms. Preprocessing data (i.e., noise filters) and its effect on DXA image segmentation have not been fully investigated previously; in this study, we proposed preprocessing DXA images before they go to a DL modal.

We conducted comprehensive experiments to evaluate the effectiveness of the preprocessing steps and noise reduction effect over the recent DL-based approach for femur segmentation from DXA images. The best results were achieved with deeper fully convolutional neural network (FCNN)-based architecture with a Wavelet-based noise reduction filter than previously applied techniques. We obtained both high-energy (HE) and low-energy (LE) images from a DXA scan. To create images with a high contrast, both were merged. We investigated the impact of preprocessing filters on noise or imperfection removal in high-contrast image creation, and consequently its effect on DL-based segmentation and BMD analysis.

The main objective of this research is to find a solution that improves the image segmentation, BMD calculation, and consequent analysis of DXA image analysis. The rest of the paper is organized as follows. Section 2 explains the segmentation method of our model. Section 3 shows the proposed model results. Discussion about our method and results are presented in Section 4. Section 5 presents some concluding remarks. Finally, Section 6 presents some future work to be carried out. The main contribution of this research is highlighted as follows:

- Advancement in DXA Imaging Segmentation: Introduces a DL approach with enhanced image quality with various image denoising techniques for femur segmentation, improving osteoporosis diagnosis and bone mineral density calculation accuracy.
- Application of DL in Medical Imaging: Expands FCNN use in DXA imaging, showing potential for addressing segmentation challenges and improving diagnostic accuracy.
- Future Research Directions: Identifies areas for improvement in femur segmentation, stimulating further inquiry and innovation in medical imaging and DL applications.

2. Methods

Data from the femur were obtained through DXA scanning, resulting in high-energy (HE) and low energy (LE) images. Before high-contrast images were produced from the combination of HE and LE photos, various denoising techniques were used to remove imperfections from HE and LE images.

With its exceptional accuracy and efficiency in tasks ranging from organ segmentation to illness diagnosis, DL has completely changed the field of medical image analysis. Femur segmentation from DXA images is one such crucial task that helps with osteoporosis diagnosis and tracking. However, several variables, such as noise and image quality, can affect how effective DL models are in this area. To create robust and dependable segmentation algorithms, it is essential to comprehend how noise and image quality affect DL performance.

2.1. Data Preparation

We utilized 600 femur images obtained by a DXA scanner (OsteoPro MAX, Yozma BMTech Worldwide Co., Ltd., Seongnam-si, Gyeonggi-do, Republic of Korea). For model training and testing, and assessment of the accuracy of DL techniques, "ground truths" of manually segmented images and BMD results obtained by expert radiologists were collected for this study. Various denoising models were employed as a preprocessing step for high-contrast image generation and removal of imperfections from images. In the segmentation process, each pixel was categorized as either bone or soft tissue or air. For the five-fold evaluation test, the data were divided into two sets: 80% for training and 20% for independent testing. Subsequently, a test database was formed and fed into the proposed DL-based model lacking class labels, and the model generated output labels for each pixel of a test subject.

2.2. Image Generation

The domain of medical image analysis has witnessed recent advancements in DL techniques, which have demonstrated remarkable and state-of-the-art performance on vast datasets comprising numerous images across various categories. Nevertheless, these remarkable accomplishments underscore the susceptibility of different DL models to image quality concerns [27]. Thus, when considering DL models, image quality assumes a pivotal role. The enhancement of visual quality in images through contrast improvement and noise reduction can significantly impact image classification and segmentation [27–29]. In this investigative study, we acquire DXA data in the form of LE and HE images. These LE and HE images are merged to create a high-contrast display image, as depicted in Figure 1 [30]. Our study aims to investigate the influence of three distinct types of high-contrast display images generated by DXA scans on the outcomes of DL. These high-contrast display images encompass images of bone density (*IBD*), an image of HE, LE log ratio (*ILR*), and a collage image (*CI*). The IBD image, which serves as the initial output, can be generated by following these steps:

$$IBD = \frac{R_{st} \times \frac{\log HE_i}{HE_0} - \frac{\log LE_i}{LE_0}}{u_l - u_h \times R_{st}}, \tag{1}$$

$$R_{st} = \frac{\log \frac{LE_0}{\frac{1}{n}\sum_{i=1}^{n} LE_i}}{\log \frac{HE_0}{\frac{1}{n}\sum_{i=1}^{n} HE_i}}, \tag{2}$$

where u_l, u_h, are the constant values of a specific *LE* X-ray and *HE* X-ray, respectively. HE_0 and LE_0 are incident energy at the source X-ray, and *HEi*, *LEi* are detector counts at a specific scan location (i.e., image pixel). The BMD value at the bone region is always higher than the soft tissue region; therefore, from Equation (1), we get a brighter bone and darker soft tissue region image.

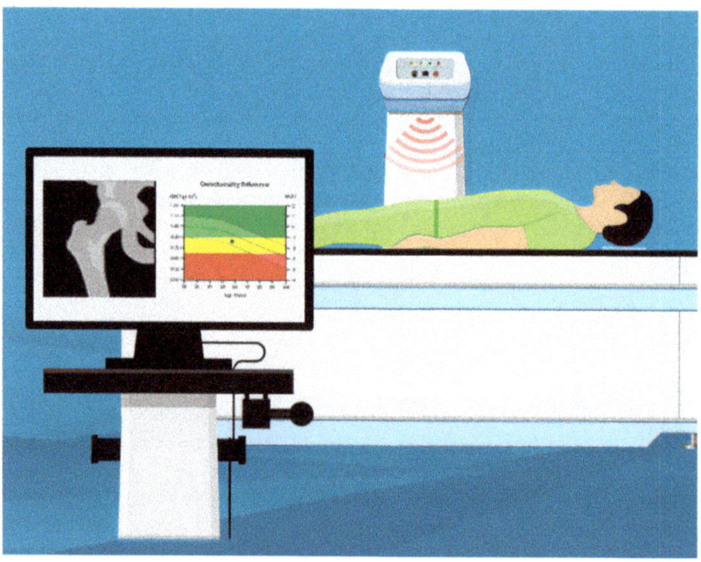

Figure 1. Acquire data with DXA machine scanning femur [30].

The second image we generate from raw DXA data is *ILR*, which can be generated as follows:

$$ILR = pow\left(T - \frac{\log(HE_i + LE_i)}{\log(u_l + u_h)}, B + C\right), \tag{3}$$

where *T* is a constant value and we used *T* = 2.0 as a constant value. Similarly, *B* and *C* represent image brightness and contrast enhancement. We used *B* and *C* as constant values to be determined from experiments. Using Equation (3), we get higher-contrast images with clear boundaries between bone, soft tissue, and air.

The third image we generate from DXA data is *CI*, which can be formed as follows:

$$CI = \left(\log\left(\frac{HE_i}{LE_i}\right) - \frac{\log\left(\frac{LE_i}{HE_i}\right)}{1 + e^{-\delta \times (\log(\frac{LE_i}{HE_i}) - \mu)}} \right) \times (B + C) - logLE_i, \quad (4)$$

where 'δ' is the standard deviation (STD) of the log (LE/HE) ratio and 'μ' is the mean of the log (*LE/HE*). The values of B and C represent image brightness and contrast values to be obtained from experiments. From Equation (4), we get a very high-contrast image with clear boundaries between different regions (i.e., bone, soft tissues, and air) compared to *IBD* and *ILR*. As we see from Figure 2, some information (most probably an artifact), indicated by red arrows in Figure 2(a2,a3), is hidden which appears in the *CI* and *IBD* images, as indicated by red arrows in Figure 2(c1,c3). So, *CI* gives a very high-contrast image with detailed DXA scan information. One drawback of the *ILR* image is that when an image contains an artifact, the image appears very black, as shown in Figure 2(c2). This problem does not affect *CI* images, as we see in Figure 2(c3). We normalize the intensities of the image from 0 to 255, and the final image is saved in '.png' format to be used in the *DL* model.

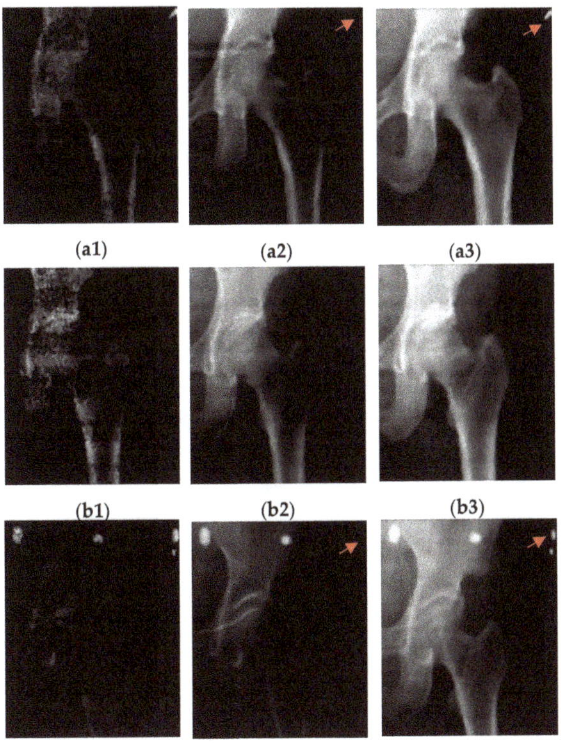

(a1) (a2) (a3)
(b1) (b2) (b3)
(c1) (c2) (c3)

Figure 2. Generated high-contrast images from DXA scan. In this figure (**a1**,**b1**,**c1**) are IBD images, (**a2**,**b2**,**c2**) represent ILR images, and (**a3**,**b3**,**c3**) represent high contrast *CI* images. As indicated by the arrows, some information is hidden at the arrow positions in the ILR images, while this information is revealed in the *CI* images. We primarily use *CI* for the final segmentation of DXA images. These collage images are created by combining high-energy (HE) and low-energy (LE) images, enhancing the contrast and improving segmentation performance. Bone mineral density is calculated from the HE and LE images using different algorithms as in Equations (1) and (2).

2.3. Noise in DXA Images

DXA images, while widely used for bone density assessment, are susceptible to various forms of noise. These noises can originate from factors such as equipment imperfections, patient movement during imaging, or interference from surrounding objects. Such noise can obscure important anatomical details and hinder accurate segmentation.

The presence of noise in DXA images can significantly affect the performance of DL models for femur segmentation. Noise introduces inconsistencies and irregularities in the image data, making it challenging for the model to distinguish between bone structures and background noise. As a result, DL algorithms may produce inaccurate segmentation, leading to erroneous clinical interpretations. Various preprocessing techniques can be employed to mitigate the adverse effects of noise on DL performance in femur segmentation. Additionally, data augmentation strategies such as random rotations, translations, and elastic deformations can help improve the robustness of DL models to noise.

Denoising images is a very important preprocessing step in image classification and segmentation [28,29]. S. Calderon et al. [28] studied the impact of denoising and contrast enhancement using a deceived non-local means (DNLM) filter in a CNN-based approach for age estimation using digital hand X-ray images. The results they obtained suggest that both image contrast enhancement and denoising can remarkably improve the results in a CNN-based model [28]. G. B. P. Costa et al. conducted a study about the effect of noise in image classification. Their study revealed that the image classification task was improved by the denoising process [29]. To improve the accuracy of image segmentation, noise reduction in both HE and LE images is an essential preprocessing procedure in DXA imaging. We employed and tested several denoising techniques, such as Non-local Mean Filter (NLMF), Gaussian filtering, and wavelet-based denoising to check the improvement in segmentation and BMD calculation using DL-based segmentation. The NLMF eliminates noise by comparing the similarity of patches through pixel neighborhoods. Gaussian filtering involves convolving the DXA image with a Gaussian kernel. Similarly, wavelet-based filtering in DXA imaging involves using wavelet transforms to analyze and process images. The denoising methods were configured with specific settings tailored to their respective algorithms. For instance, NLMF utilized optimal parameters such as a search window, a patch size, and a filtering strength (h), for effective noise reduction. Gaussian filtering involved setting parameters such as a standard deviation (sigma) and a kernel size. Wavelet-based denoising used a Daubechies (db2) wavelet, with a decomposition level, and applied thresholding parameter values.

2.3.1. Non-Local Mean Filter

The NLMF denoising technique is commonly used in medical imaging, including DXA imaging. In DXA imaging, where the images are often affected by noise due to factors such as low X-ray dose and scatter, denoising techniques like NLMF can be essential for improving image quality and enhancing the accuracy of subsequent analysis. Kim Kyuseok et al., (2020) [31] demonstrated through visual assessment and quantitative analysis that the NLM algorithm outperforms existing methods in processing casting images, offering an efficient solution to mitigating noise in high-energy X-ray imaging systems and potentially enhancing image restoration through accompanying software. Seungwan Lee et al., (2022) [32] introduce a newly improved non-local means (INLM) denoising algorithm tailored for X-ray images, accounting for the thickness of aluminum (Al) filters commonly used in X-ray systems, demonstrating its efficacy in reducing image noise. Results indicate that the proposed INLM algorithm, particularly when applied to X-ray images with a 5 mm Al filter thickness, exhibits superior performance in noise reduction and image evaluation compared to conventional methods, highlighting its importance for optimizing image processing applications in photon-counting X-ray systems.

The NLMF works by averaging the pixel values in an image, with the averaging process weighted based on the similarity between patches of pixels in the image. Unlike traditional filters that rely on local information, the NLM filter considers non-local similarities, meaning it compares patches of pixels from different image regions to determine the weighting for averaging. The approach of NLMF effectively removes noise while preserving image details, making it particularly suitable for medical imaging applications where preserving fine structures and details is crucial for accurate diagnosis and analysis. NLMF can help reduce the effects of noise, leading to clearer and more accurate DXA images, which in turn can improve the reliability of measurements such as BMD calculations and segmentation of bone structures.

The optimized NLMF was tested and verified with the DXA images of the uniform femur phantom and real human femur images. Preliminary results showed that the signal-to-noise ratio (SNR) for high- and low-energy femur DXA images improved by 15.26% and 13.55%, respectively. Key parameters used include a search window size of 21 × 21 pixels, a patch size of 7 × 7 pixels, and a filtering strength (h) set to 10. Figure 3 shows some of the denoised results of femur DXA images using NLMF. More details about our NLMF work for denoising DXA images are available in references [33,34].

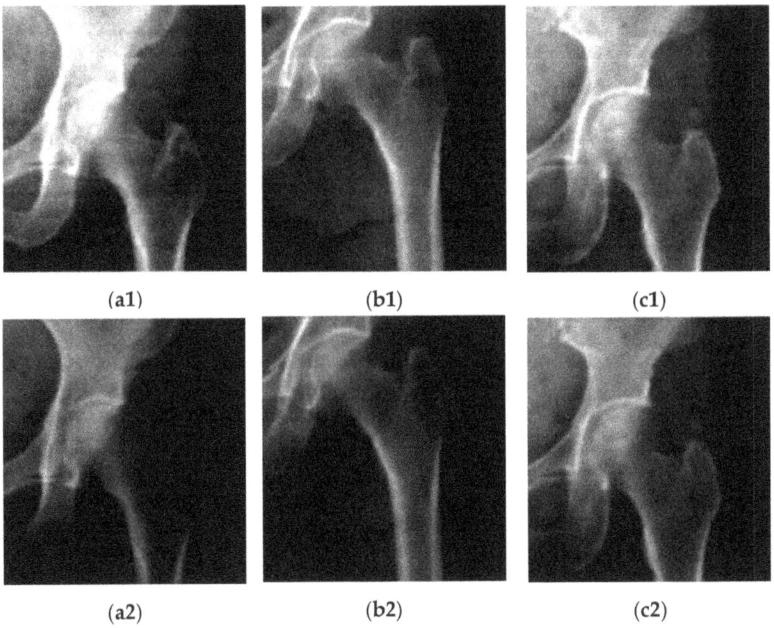

Figure 3. Denoising DXA images using NLMF. In this figure, (**a1**,**a2**,**a3**) represent three cases of high contrast *CI* images without using a denoising filter, while (**a2**,**b2**,**c2**) show the same images after applying NLMF denoising.

2.3.2. Gaussian Filtering

Gaussian filtering (GF) is a common technique used for image processing and noise reduction. Gaussian filtering involves convolving the DXA image with a Gaussian kernel. This kernel is essentially a Gaussian distribution centered at the pixel of interest, with the values of neighboring pixels weighted according to their distance from the center. The Gaussian kernel is defined by its standard deviation, which determines the amount of smoothing applied to the image [35–38].

However, it is important to note that GF may also blur fine details in the image, so the choice of the filter's parameters (such as the standard deviation of the Gaussian kernel) should be carefully considered based on the specific requirements of the application. The equation for GF in DXA image processing can be expressed as:

$$G(i,j) = \frac{1}{2\pi\sigma^2} e^{-\frac{i^2+j^2}{2\sigma^2} \times I(i,j)} \tag{5}$$

where *G(i,j)* is the filtered image, *I(i,j)* is the original image, and σ is the standard deviation of the Gaussian distribution. This equation represents the convolution of the original image *I(i,j)* with a 2D Gaussian kernel. The Gaussian kernel is centered at *(i,j)* with values determined by the Gaussian function, where σ controls the width of the kernel and hence the amount of smoothing applied to the image. In our study, the Gaussian filter was implemented with a standard deviation (σ) of 1.5 and a kernel size of 5 × 5 pixels.

Nazia Fathima et al. (2020) proposed a novel approach for accurately measuring BMD from X-ray images using a modified U-Net with an attention unit [39]. The proposed method includes preprocessing steps, including Gaussian filtering, to enhance image quality. Results demonstrate improved segmentation accuracy and high classification accuracy for osteoporosis detection, validating the effectiveness of the approach.

2.3.3. Wavelet-Based Methods

Wavelet-based denoising is often used in medical imaging. Wavelet-based denoising decomposes medical images into different frequency components using wavelet transforms. High-frequency noise in the decomposed image is then suppressed through thresholding methods, while preserving important diagnostic features. The denoised image is reconstructed, resulting in reduced noise and improved image quality suitable for clinical diagnosis and analysis [40,41]. G. Elaiyaraja et al., (2022) [42] proposed an optimal wavelet threshold-based denoising filter that effectively removes adaptive Gaussian noise from medical images, offering superior performance and efficiency. Farah Deeba et al., (2020) [43] proposed a wavelet-based Mini-grid Network Medical Image Super-Resolution (WMSR) method that enhances low-resolution medical images by leveraging the stationary wavelet transform (SWT). By combining wavelet sub-band images and utilizing sub-pixel layers, the model achieves superior performance in image reconstruction speed and quality.

Wavelet-based filtering in DXA imaging involves using wavelet transforms to analyze and process images [44,45]. Weiya Xie et al., (2021) [46] explored the application of photoacoustic time–frequency spectral analysis (PA-TFSA) for assessing bone mineral density (BMD) and structure. Utilizing wavelet transform-based PA-TFSA, simulations and experiments on bone samples revealed significant associations between frequency components and bone characteristics. Parameters derived from PA-TFSA, particularly mid band-fit and slope, demonstrated sensitivity in distinguishing between osteoporotic and normal bones.

Wavelet transforms are powerful tools for image processing because they can represent both frequency and spatial information simultaneously. The wavelet transform decomposes an image into different frequency components. It consists of two main components: the scaling function (approximation) and the wavelet function (detail). The wavelet transform can be applied in either one dimension (1D) or two dimensions (2D). The continuous wavelet transform (CWT) of a signal or image $f(x)$ is defined as

$$W(a,b) = \int_{-\infty}^{\infty} f(x)\varnothing_{a,b}(x)dx \tag{6}$$

where $\varnothing_{a,b}(x)dx$ is the wavelet function, scaled by a and translated by b.

In practice, continuous wavelet transformation is often computationally expensive. Therefore, the discrete wavelet transform (DWT) is commonly used. It decomposes the image into different scales and orientations, resulting in a multi-resolution representation of the image. DWT is applied iteratively to decompose the image into approximation (low-frequency) and detail (high-frequency) coefficients. This can be represented mathematically as

$$W(j,k) = \sum_{n} s_{j,k}(n).\varnothing(n-k) \quad (7)$$

where $W(j,k)$ are the wavelet coefficients at scale j and position k, $s_{j,k}(n)$ are the scalling cofficents, and $\varnothing(n-k)$ is the wavelet function.

Wavelet-based denoising involves thresholding the wavelet coefficients to remove noise while preserving image details. This is achieved by setting coefficients below a certain threshold to zero. Common thresholding methods include hard thresholding and soft thresholding.

Hard Thresholding (HT):

$$T_H(x) = \begin{cases} x, & if\ |x| > T \\ 0, & otherwise \end{cases} \quad (8)$$

Soft Thresholding (ST):

$$T_S(x) = sign(x).(|x| - T)_+ \quad (9)$$

where x represents the wavelet coefficient, T is the threshold value, and (.)+ denotes the positive part of the function. If the absolute value of x is less than T, the coefficient is set to zero. This approach is straightforward and leads to sparse representations by eliminating coefficients that are considered noise.

Hard thresholding is advantageous in scenarios where the noise level is well-defined or known, as it can effectively suppress noise without overly distorting the image features. However, it can sometimes introduce artifacts in regions where the signal is weak but not strictly zero, leading to a loss of subtle details in the reconstructed image. In contrast to soft thresholding, which applies a smoother shrinkage to coefficients, hard thresholding offers a more aggressive noise reduction by eliminating coefficients below the threshold. Its effectiveness depends on selecting an appropriate threshold T, which balances noise suppression with the preservation of significant signal components. Using the Daubechies (db2) wavelet, with a decomposition level of 3, the soft thresholding achieved optimal denoising results.

2.4. Data Augmentation

DL algorithms necessitate a substantial volume of data for effective training. However, the limited size of medical datasets presents a significant challenge in this regard. To meet the large dataset requirements of deep neural network training, the small data size is increased using data augmentation [47,48]. In this study, we addressed the data scarcity by randomly selecting 350 images and applying transformations such as image translations, and horizontal and vertical reflections, along with their subsequent scaling. These augmentations resulted in the expansion of our dataset to 1800 images, facilitating more robust training of deep neural networks.

2.5. Depp Learning Architecture

Figure 4 provides an overview of DXA image analysis employing DL methodologies. In this investigation, we introduce the U-Net, SegNet, and FCN approaches for femur segmentation from DXA images with preprocessing and postprocessing steps. The integration of the pre- and post-processing steps has been demonstrated to be crucial in significantly enhancing the efficacy of diverse tasks and applications. Through the utilization of pre-processing methodologies such as noise reduction, image standardization, and data enhancement, the original input data can be improved and optimized before its utilization in the model. These pre-processing approaches are employed to alleviate the impact of noise, variability, and discrepancies in the data to amplify the flexibility and precision of subsequent analyses or forecasts. Similarly, post-processing techniques like filtering, smoothing, and calibration can further perfect the outcomes generated by the model to ensure their alignment with specific quality or criteria. In essence, the incorporation of the pre- and post-processing stages plays a pivotal role in maximizing the efficiency and efficacy of the entire data processing pipeline to attain exceptional performance outcomes.

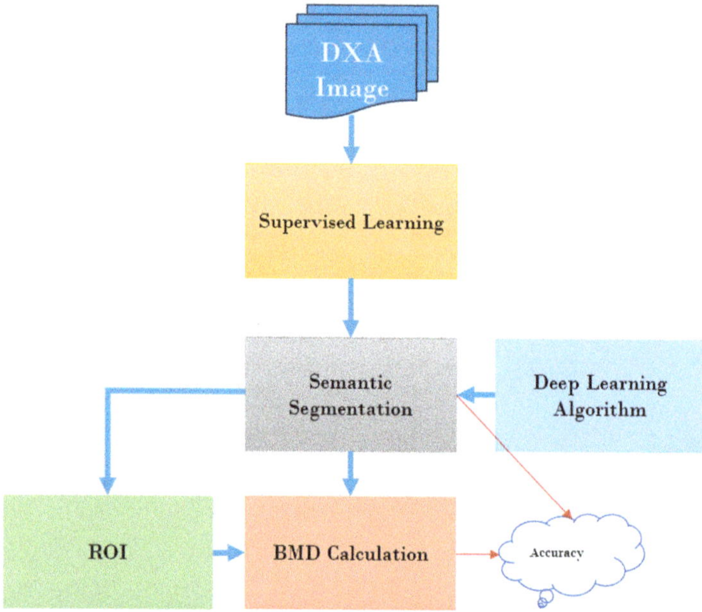

Figure 4. Overview of DXA image analysis using deep learning.

FCNs are adept at generating dense predictions from input data of variable dimensions. Notably, both training and inference processes operate on the entire image concurrently through dense feedforward computation and backpropagation [49]. Unlike conventional CNN models featuring fully connected layers (FCL), FCNs exclusively consist of convolutional layers without any FCL, as indicated by their names. Within the suggested DL segmentation networks, we employ the sigmoid activation function in the activation layer to classify each pixel within the femur image into bone, soft tissue, or air categories. Further elaboration on the U-Net, SegNet, and FCN methodologies can be found in references [49–51]. The Adadelta optimization algorithm was employed to train all segmentation models, utilizing a batch size of 25. Initially, the learning rate was set to 0.2, with dynamic adjustments made throughout 200 epochs to optimize training. Weighted cross-entropy was utilized to compute the loss, aiming to minimize the overall loss H during the training phase as follows:

$$H = -\sum_{c=1}^{M} y_c log(\hat{y}_c) \tag{10}$$

where y represents the ground truth labels and $\hat{y}$ represents the predicted map of segmentation. M represents the number of classes and 'c' represents classes (bone, tissue, air). The research was implemented in Python utilizing the Keras framework, with TensorFlow serving as the backend library.

2.6. Post-Processing

Unlike DL models, conventional semantic segmentation techniques typically necessitate the use of boundary smoothing filters to refine the femur bone boundaries delineated by the segmentation model and eliminate imperfections. In a prior investigation, we employed a binary smoothing filter to eradicate inaccuracies from segmented DXA images [23,24]. Addressing imperfections introduced during pixel labeling in image segmentation using machine learning is challenging. Morphological image processing (MIP) effectively handles these issues by considering the image's shape and structure. In instances where sometimes DL labeling of femur boundaries results in non-smooth contours, binary smoothing proves effective in refining femur object boundaries by eliminating small-scale noise while preserving large-scale features. For additional information regarding binary smoothing, please refer to our previous work cited in [23,24].

2.7. Evaluation and Performance Analysis

To evaluate the performance of DL-based predictions with preprocessing noise filters compared to ground truth using a range of evaluation metrics, these metrics—including True Positives (TP), True Negatives (TN), False Positives (FP), and False Negatives (FN)—are recorded for each segmented pixel in the DXA image to quantify the accuracy of the models over the total number of pixels (n = TP + TN + FP + FN) in an image. These metrics were calculated over the entire dataset of test images.

One of the key evaluation metrics used in the study was Intersection over Union (IOU), which is a measure commonly used to assess the accuracy of segmented objects compared to ground truth. IOU is calculated as the ratio of the area of intersection between the segmented object and ground truth to the area of their union. This metric indicates how well the segmented object aligns with the ground truth, with higher IOU values indicating better alignment.

Additionally, Sensitivity (True Positive Rate, TPR) and Specificity (True Negative Rate, TNR) were used to measure the proportion of positive and negative pixels accurately identified, respectively. Sensitivity measures the proportion of bone tissue pixels correctly identified in the femur DXA image, while Specificity measures the proportion of soft tissue pixels correctly identified. The False Positive Rate (FPR) and False Negative Rate (FNR) were also calculated to quantify the rate of misclassification of positive and negative pixels, respectively.

We employed a threshold-based approach to assess the overall accuracy of the segmentation methods. A segmentation method was considered successful if IOU was >0.92, Sensitivity was >95%, or Specificity was ≥93%. These thresholds were chosen to ensure that the segmentation methods accurately identified both bone and soft tissue regions in the DXA images.

We used a five-fold cross-validation approach, by dividing the dataset as 80% for training and the remaining 20% for independent testing. Test data were swapped with training data during cross-validation to ensure a robust evaluation of the segmentation methods. Finally, the segmentation methods' retrieved ground truths (RGT) and manual ground truths (GT) were applied to the test data in each cross-validation fold to compare their performance. For more details, visit our previous work referenced in [23,24,26], as the same evaluation and performance analysis strategy was adopted to evaluate the accuracy of this work.

$$Accuracy = 100 \times \frac{1}{N}\sum_{i=1}^{N} x_i \; where, \; f(x_i) = \begin{cases} 0, & if\ JI < 0.92 \,||\varepsilon < 95\,||\vartheta < 93 \\ 1, & else \end{cases} ,JI = \frac{|A_c \cap B_c|}{|A_c|+|B_c|}. \quad (11)$$

where JI is the Jaccard index or IOU, epsilon (ε) is the image segmentation sensitivity, and ϑ is the image segmentation specificity. $|A_c|$ is the set of pixels predicted as class c and $|B_c|$ is the set of pixels in the ground truth labeled as class c (i.e., bone, soft tissue, or air). $|A_c \cap B_c|$ is the number of pixels common to both sets (true positives).

Furthermore, we also calculated the average Dice score (*DISC*) for image segmentation involving three regions (bone, soft tissue, and air). We compute the Dice score separately for each class and then average them. The *DISC* was calculated using the following formula:

$$DISC = 2 \times \frac{|A_c \cap B_c|}{|A_c| + |B_c|}, \tag{12}$$

We calculated the mean Dice score by averaging the Dice scores for all classes:

$$DISC_{Mean_i} = \frac{1}{N}\sum_{c=1}^{N} D_c, \tag{13}$$

where N is the number of classes (in this case, $N = 3$). Finally, we calculated the average *DISC* for the overall test dataset.

3. Results

3.1. Data

We used the same dataset used in our previous studies [23,24], with some additional femur images acquired on a DXA scanner (OsteoPro MAX, Yozma BMTech Worldwide Co., Ltd., Seongnam-si, Gyeonggi-do, Republic of Korea). Radiology experts manually segmented femur images as the "ground truth". We extracted manually annotated images from the DXA system in the ".png" format, alongside high-contrast images, to train and test our DL models. Each and every pixel in the femur image was annotated and assigned a class label (i.e., either bone, soft tissue, or air).

3.2. Noise Reduction

The proposed denoising techniques via NLMF, Gaussian filtering, and wavelet-based methods for DXA images were evaluated experimentally with femur data. The performances of denoising filters was evaluated quantitatively in terms of mean-to-standard deviation ratio (MSR), signal-to-noise ratio (SNR), and contrast-to-noise ratio (CNR). The proposed noise reduction techniques significantly improve the quality of DXA images and image segmentation results, while preserving the fine details of anatomical structures. Compared to other denoising methods, wavelet-based filtering has shown higher performance in the case of the average improvement ratio of MSR, SNR, and CNR and improves the quality of DXA images. Table 1 presents the effects of NLMF, GF, and CWT filters. MSR, or mean-to-standard deviation ratio, quantifies the quality of DXA images by comparing the mean pixel value across the entire image to the standard deviation within a specified region of interest (ROI) of 10×10 pixels, as outlined in [34,38,52]. A higher MSR value indicates superior image quality, reflecting clearer and more defined anatomical details. SNR, measured in decibels, was calculated for both original and denoised DXA images. It evaluates the signal strength relative to the noise level, with μo representing the mean pixel value of the bone object and σROIbg denoting the standard deviation in a background ROI of equivalent size. Higher SNR values indicate effective noise reduction and improved image clarity. Similarly, CNR in decibels was estimated to assess the contrast between the bone and background regions, providing insights into the image's diagnostic utility and overall enhancement achieved through denoising techniques. A sample of original and denoised images with NLMF is shown in Figure 5.

Table 1 illustrates the effectiveness of various denoising techniques applied to femur DXA images. The metrics MSR, SNR, and CNR are critical for evaluating image quality and noise reduction. The results indicate that wavelet-based methods, particularly Continuous Wavelet Transform—Soft Thresholding (CWT-ST), provide the best performance in enhancing image quality and reducing noise, followed by Continuous Wavelet Transform—Hard Thresholding (CWT-HT), NLMF, and GF. These improvements are essential for accurate segmentation and reliable BMD calculations.

Table 1. Denoising technique performance analysis for femur DXA images. The results are presented in this table as mean values from all evaluated images.

Technique	Index	Image	
		Original	Denoised
NLMF	MSR	73.31	77.01
	SNR	5.26	5.74
	CNR	27.54	31.11
GF	MSR	73.31	74.15
	SNR	5.26	5.54
	CNR	27.54	29.26
CWT-ST	MSR	73.31	**78.21**
	SNR	5.26	**6.27**
	CNR	27.54	**32.63**
CWT-HT	MSR	73.31	77.81
	SNR	5.26	6.25
	CNR	27.54	31.41

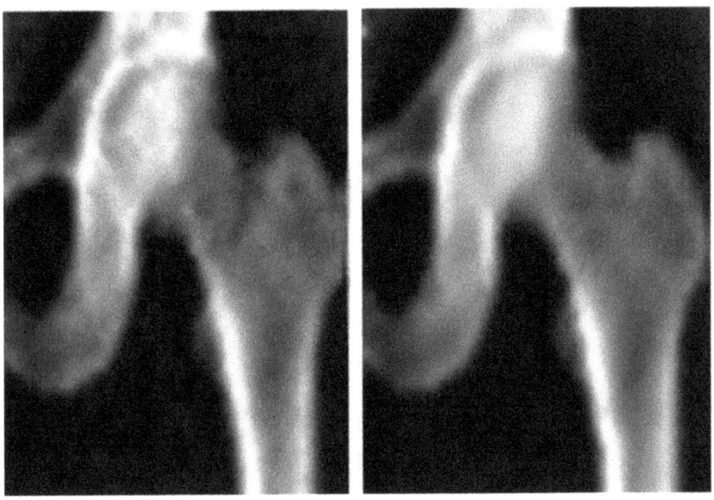

(**a**) (**b**)

Figure 5. Image denoised with NLMF: (**a**) original and (**b**) denoised image.

3.3. Segmentation

This section presents the performance of U-Net, SegNet, and FCN approaches using test data (i.e., 250 femur images). Some of the selected results from U-Net, SegNet, and FCN output with non-smooth contours, and binary smoothed contours are shown in Figures 6 and 7, respectively. Figure 6 presents the raw output of DL models without any additional post-processing applied. This means that the segmentation results shown in Figure 6 are directly generated by the DL algorithms (SegNet, U-Net, and FCN) without any further enhancement or adjustment.

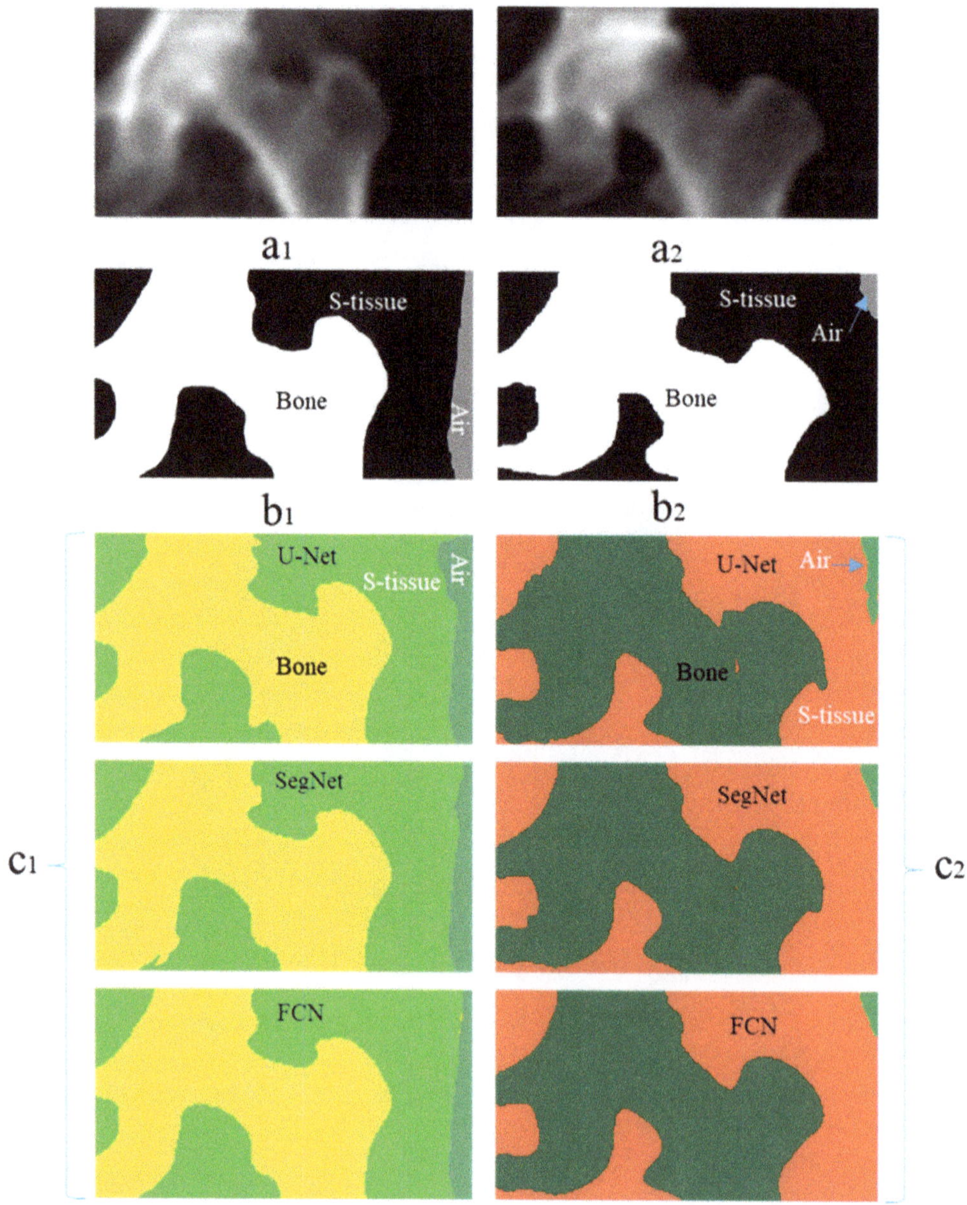

Figure 6. Selected results of two cases segmented with SegNet, U-Net, and FCN without smoothing filters. In this figure, (**a1**,**a2**) are the original images, while (**b1**,**b2**) represent the ground truths. (**c1**) on left side and (**c2**) on right side depict the segmentation masks of the original images (**a1**,**a2**) segmented using various DL algorithms without any post-processing applied. These two images were acquired using two distinct DXA imaging devices. The blue arrows indicate air labels in the images.

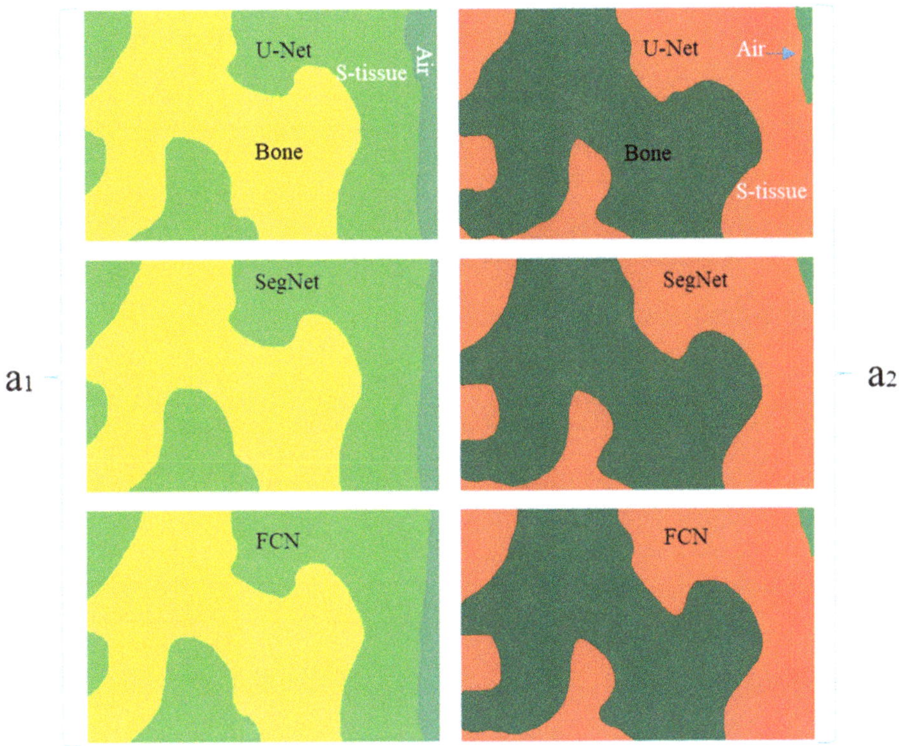

Figure 7. A binary smoothing filter was applied to the DL output in Figure 6. The results depict two cases segmented using SegNet, U-Net, and FCN with the addition of a binary smoothing filter as post-processing. In this figure, (**a1**) on the left side shows the segmentation mask corresponding to a1 in Figure 6, while (**a2**) on the right side displays the segmentation mask corresponding to (**a2**) in Figure 6. These masks were generated using various DL algorithms (U-Net, SegNet, and FCN, each labeled accordingly) and then smoothed with a binary smoothing filter. The blue arrows indicate air labels in the images.

In contrast, Figure 7 illustrates the results of the same DL models after applying a binary smoothing filter as post-processing. The binary smoothing filter is applied to the segmentation masks obtained from the DL models to refine and smooth out the boundaries and contours of the segmented regions. This post-processing step aims to improve the visual clarity and accuracy of the segmentation results by reducing pixel-level noise and inconsistencies in the DL outputs.

Thus, Figure 7 provides a comparison to Figure 6 by demonstrating how the application of a binary smoothing filter can potentially enhance the segmentation outcomes produced by the DL models, particularly in terms of achieving smoother and more coherent boundaries for the identified regions of interest. Table 2 shows the segmentation performance results in terms of average accuracy computed using the JI, sensitivity, and specificity of all test images. In addition, the average Dice score was calculated for bone, soft tissue, and air region in each image, and finally, a combined average Dice score for overall test data of each segmentation model was calculated. A couple of the predicted segmentation contours by different segmentation models using NLMF are shown in Figure 8.

Table 2. Segmentation performance of different methods on the test dataset.

DL Model	Trainable Parameters	Loss Function	Optimization Algorithm	Pre-Processing (Denoising Method)	Segmentation Accuracy (%)	DISC (%)
SegNet	13.3M	Cross-Entropy	Adam	No pre-processing	88.12	75.24
				NLMF	92.99	84.98
				GF	92.35	83.70
				CWT-ST	96.63	92.26
				CWT-HT	93.68	86.36
U-Net	12.7M	Dice Loss	Adam	No pre-processing	86.32	71.64
				NLMF	91.56	82.12
				GF	89.98	78.96
				CWT-ST	94.14	87.28
				CWT-HT	91.89	82.78
FCN	134.5M	Cross-Entropy	Adam	No pre-processing	89.36	77.72
				NLMF	94.76	88.52
				GF	93.89	86.78
				CWT-ST	**98.84**	**96.68**
				CWT-HT	94.97	88.94

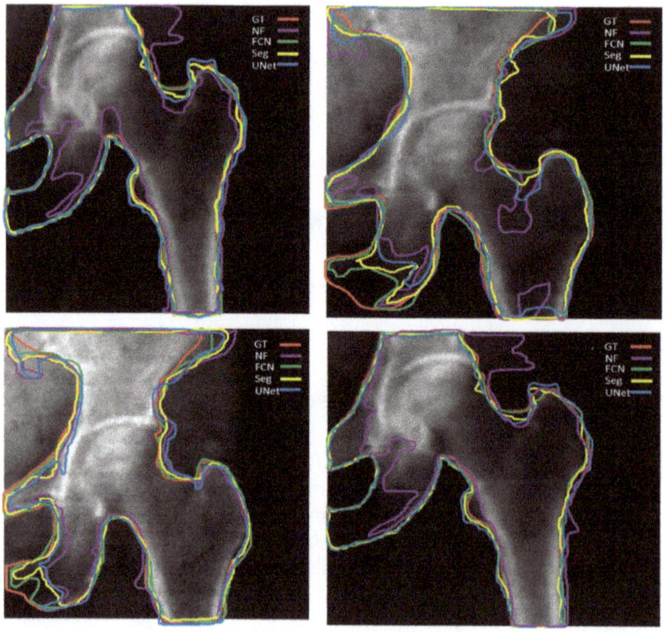

Figure 8. Predicted femur boundaries using SegNet, U-Net, and FCN segmentation models. Ground truth femur boundaries are outlined in red, while the predicted boundaries are represented by yellow (SegNet), blue (U-Net), green (FCN), and purple (FCN with no filter). The above segmentation results were obtained following a preprocessing with NLMF.

Table 2 highlights the segmentation performance of three deep learning models—SegNet, U-Net, and FCN—under various pre-processing conditions applied to femur DXA images. The table reveals that the FCN, particularly when combined with CWT-ST and a binary smoothing filter, achieves the highest segmentation accuracy (98.84%) and DISC (96.68%), significantly outperforming the other models. This suggests that the FCN's architecture, coupled with effective denoising through wavelet-based soft thresholding, is particularly well-suited for this task. Moreover, the table underscores the positive impact of pre-processing techniques, with all models showing improved performance when any form of denoising is applied. Among these techniques, CWT-ST and CWT-HT consistently deliver superior results compared to NLMF and Gaussian filtering. This performance boost can be attributed to the wavelet methods' ability to preserve fine details while effectively reducing noise.

Our evaluation of 250 test femur images using DL models with a noise reduction pre-processing step demonstrated superior performance compared to simple DL methods, particularly in high-contrast femur sections (femur head and shaft) and challenging areas (greater and smaller trochanteric and ischium). The data were collected on multiple devices, and models covered the diversity well and performed competently. These findings emphasize the importance of selecting both advanced pre-processing strategies and robust model architectures to achieve optimal segmentation accuracy in femur DXA imaging.

3.4. BMD Analysis

We conducted a comparison of BMD between manually segmented images and those segmented by our models. Initially, a set of 100 femur images was randomly chosen and given to three individuals to manually segment the femur, select regions of interest (ROIs), and calculate BMD at three distinct regions: the femur neck, ward, and greater trochanter (G.T.). Subsequently, the average BMD value was computed from the three expert readings for each region, and these estimations were then compared to those obtained through model-based segmentation to assess consistency. Finally, we conducted a statistical analysis, calculating the correlation (R^2) to evaluate the correlation of BMD measurements between different segmentation methods and manual segmentation. The FCN segmentation method with a CWT as preprocessing and binary smoothing of object boundaries as a post-processing step exhibited the highest correlation, as indicated in Table 3.

The findings underscore the superior "sensitivity, specificity, and accuracy" of the FCN method with preprocessing and postprocessing steps compared to other models, providing a robust solution for segmentation challenges in DXA imaging. While CNN is acknowledged as an innovative segmentation method, its reliance on large training datasets poses a limitation. To address this, we implemented transfer learning to enhance the training efficiency of DL models using small femur DXA images. Leveraging the weights of the pre-trained VGG-16 network from the extensive ImageNet dataset, our study on femur segmentation demonstrates that current DL models outperform previously utilized methods [23–26].

Training CNN-based networks with limited data poses a significant challenge in DL. Without optimized techniques and data augmentation, achieving satisfactory performance can be daunting. Therefore, employing appropriate optimization methods, data augmentation strategies, and transfer learning can greatly aid in training a reliable segmentation network. Transfer learning involves fine-tuning a deep network that has been pre-trained on either medical images or general datasets. When combined with data augmentation, transfer learning offers an additional distinct solution with numerous parameters. To tackle this challenge effectively, we successfully implemented transfer learning in our DXA image analysis, leveraging pre-trained models from ImageNet to enhance performance.

Table 3 presents the R^2 between BMD measurements obtained from manual segmentation and various DL models across different denoising methods. This correlation was assessed at three distinct regions of the femur: the neck, ward, and greater trochanter (G.T.). The results indicate that the FCN model, particularly when using Continuous CWT-ST as a preprocessing method and binary smoothing as a postprocessing step, achieved the highest overall mean correlation ($R^2 = 0.9928$). This high correlation suggests that the FCN model with CWT-ST preprocessing and binary smoothing most accurately replicates expert manual segmentation.

Table 3. Correlation record of BMD measurements between different segmentation methods to manual segmentation.

DL Model	Denoising Method	R^2			
		Neck	**Ward**	**G.T.**	**Mean**
SegNet	Without noise filter	0.9029	0.9108	0.9192	0.9109
	NLMF	0.9819	0.9798	0.9592	0.9736
	GF	0.9231	0.9307	0.9382	0.9306
	CWT-ST	0.9331	0.9467	0.9481	0.9426
	CWT-HT	0.9288	0.9348	0.9421	0.9352
U-Net	Without noise filter	0.8929	0.9018	0.8892	0.8946
	NLMF	0.9811	0.9896	0.9560	0.9755
	GF	0.8989	0.9102	0.9013	0.9034
	CWT-ST	0.9601	0.9634	0.9418	0.9551
	CWT-HT	0.9599	0.9595	0.9388	0.9527
FCN	Without noise filter	0.9413	0.9349	0.9324	0.9362
	NLMF	0.9698	0.9789	0.9814	0.9767
	GF	0.9479	0.9398	0.9404	0.9427
	CWT-ST	**0.9913**	**0.9949**	**0.9924**	**0.9928**
	CWT-HT	0.9734	0.9829	0.9816	0.9795

In comparison, other models and denoising techniques showed varied performance. For instance, SegNet and U-Net with NLMF also demonstrated strong correlations (mean R^2 of 0.9736 and 0.9755, respectively), but not as high as the FCN with CWT-ST. Models without any noise filtering consistently exhibited lower correlations, highlighting the importance of denoising in improving segmentation accuracy. Further investigations may discover the rational serviceability of the FCN and other DL models in the clinical diagnosis of osteoporosis and the prediction of fracture risk. All DL-based models have proven to have better performance than previously applied techniques. The study has shown that convolutional networks can effectively be utilized with high performance on a small clinical data set using transfer learning in semantic segmentation.

4. Discussion

The study investigated the influence of noise and image quality on the performance of DL models for femur segmentation from DXA images. A dataset comprising DXA images with varying levels of noise and image quality was utilized to assess the effectiveness of different denoising techniques in improving segmentation accuracy.

Firstly, the impact of noise on segmentation performance was evaluated. DXA images with simulated Gaussian noise at different levels were subjected to segmentation using a baseline CNN model. Results demonstrated a noticeable decrease in segmentation accuracy with increasing noise levels. However, when denoising techniques such as Gaussian filtering, CWT filtering, and Non-local Mean Filter were applied as preprocessing steps, the segmentation accuracy improved significantly across all noise levels. Particularly, the Wavelet-based Mean Filter exhibited the highest efficacy in reducing noise and enhancing segmentation accuracy.

Secondly, the influence of image quality on segmentation performance was analyzed. DXA images with varying levels of blur and contrast were utilized to assess the robustness of the CNN model to image quality variations. It was observed that images with low contrast and blur presented challenges for the CNN model, leading to decreased segmentation accuracy. Nevertheless, by applying preprocessing techniques such as contrast enhancement and wavelet-based filtering, the model demonstrated improved performance in segmenting femur structures from low-quality images.

Furthermore, the combined effect of noise and image quality on segmentation performance was investigated. DXA images with simultaneous variations in noise and image quality were used to simulate real-world scenarios. The results indicated that denoising techniques played a crucial role in mitigating the adverse effects of both noise and low image quality on segmentation accuracy. Wavelet-based methods, in particular, exhibited robust performance in preserving image details while reducing noise, resulting in improved segmentation accuracy even in challenging imaging conditions.

Image denoising plays a crucial role in enhancing the performance of CNN-based DL models, particularly in medical image analysis tasks such as femur segmentation from DXA images. Various denoising techniques, including Non-local Mean Filter, Gaussian filtering, median filtering, and wavelet-based methods, contribute to improving the robustness and accuracy of CNN models by reducing the impact of noise on image data.

Incorporating FCN-based segmentation with a wavelet-based preprocessing filter, and a binary smoothing filter as a postprocessing step, we observed a significant enhancement in femur segmentation accuracy and BMD calculation in DXA images. Our findings demonstrated that the FCN, when coupled with a wavelet-based filter, outperformed alternative segmentation techniques, achieving a notable accuracy rate. The application of wavelet-based filtering as a preprocessing step effectively reduced noise and enhanced image quality, thereby facilitating more precise femur segmentation in DXA imaging. Furthermore, the combination of FCN and awavelet-based filter exhibited robust performance across a range of imaging conditions, including challenging regions like the greater and smaller trochanteric areas, as well as high-contrast regions such as the femur head and shaft. These results underscore the potential of FCN-based segmentation with wavelet-based filtering as a reliable approach for accurate femur segmentation in DXA imaging, offering superior sensitivity, specificity, and overall accuracy compared to traditional methods.

The results highlight the significance of incorporating denoising techniques into CNN-based DL models for femur segmentation from DXA images. By effectively mitigating the influence of noise and image quality variations, these techniques enhance the accuracy and reliability of segmentation, facilitating more precise diagnosis and monitoring of osteoporosis.

5. Conclusions

Segmentation of the femur in DXA images poses challenges due to factors like reduced contrast, noise, and variations in bone shape. This study investigated the impact of noise on DL techniques for femur segmentation, incorporating noise reduction techniques to enhance DL-based models' accuracy. By applying CNN to DXA images with and without noise reduction filters, we observed that the FCNN outperformed the use of noise reduction algorithms before model training, resulting in precise bone density calculation and improved osteoporosis diagnosis. The study demonstrated a higher accuracy of 98.84% for different segmentation methods and a significantly higher correlation (R2 = 0.9928) for BMD measurement compared to manual segmentation. This research contributes to advancing DXA imaging segmentation, enhancing diagnostic accuracy, and stimulating further inquiry into medical imaging and DL applications. The incorporation of Non-local Mean Filter, Gaussian filtering, and wavelet-based methods for image denoising in CNN-based DL models significantly contributes to improving the performance and robustness of femur segmentation from DXA images, ultimately enhancing the quality of osteoporosis diagnosis and patient care.

6. Future Directions

In the realm of medical imaging, future directions for femur segmentation from DXA images encompass several promising avenues. One avenue involves delving into advanced denoising techniques, such as deep learning-based models or adaptive algorithms, to further refine noise reduction strategies and enhance the accuracy of segmentation. Additionally, exploring the fusion of DXA images with other modalities like MRI or CT scans could unlock synergistic benefits, improving segmentation accuracy by leveraging complementary information from diverse imaging sources. Transfer learning and domain adaptation techniques offer another promising pathway, enabling model training on diverse datasets and enhancing generalization to different imaging conditions and patient populations. Moreover, developing methods for uncertainty quantification could provide valuable insights into the reliability of segmentation results, facilitating more informed decision-making in clinical settings. Clinical validation studies are essential for assessing the real-world performance and clinical utility of DL-based segmentation models, necessitating collaboration with medical professionals and institutions for rigorous validation against ground truth annotations and clinical outcomes. User-friendly software tools and platforms for DXA image segmentation should be developed to streamline integration into existing medical workflows and promote widespread adoption. Longitudinal studies exploring the prognostic value of segmentation-based biomarkers for predicting clinical outcomes, such as fracture risk or treatment response, offer valuable insights into personalized risk assessment and treatment planning in osteoporosis management. Lastly, addressing ethical considerations and potential biases in DXA image segmentation is paramount for ensuring equitable and unbiased healthcare delivery, requiring robust strategies for bias mitigation, fairness assessment, and transparency in model development and deployment.

Author Contributions: Methodology, D.H.; Validation, D.H.; Investigation, Dildar Hussain; Writing—original draft, D.H.; Project administration, Y.H.G.; Funding acquisition, Y.H.G. All authors have read and agreed to the published version of the manuscript.

Funding: This work was supported by the Institute of Information & Communications Technology Planning & Evaluation (IITP) grant funded by the Korean government (MSIT) (No. 1711160571, MLOps Platform for Machine learning pipeline automation).

Institutional Review Board Statement: Not applicable.

Informed Consent Statement: Informed consent was obtained from all subjects involved in the study.

Data Availability Statement: No new data were created or analyzed in this study.

Acknowledgments: We extend our appreciation to ChatGPT for its assistance and insights during the preparation of this article. Its contributions have helped enhance the clarity and coherence of our work [53]. Additionally, we acknowledge Sejong University for the support and resources that made this research possible, and we thank the participants involved in the study for their cooperation and contribution to advancing scientific knowledge in this field.

Conflicts of Interest: The authors declare no conflict of interest.

References

1. Karlsson, K.M.; Sernbo, I.; Obrant, K.; Redlund-Johnell, I.; Johnell, O. Femoral neck geometry and radiographic signs of osteoporosis as predictors of hip fracture. *Bone* **1996**, *18*, 327–330. [CrossRef] [PubMed]
2. Holvik, K.; Ellingsen, C.; Solbakken, S.; Finnes, T.; Talsnes, O.; Grimnes, G.; Tell, G.; Søgaard, A.; Meyer, H. Cause-specific excess mortality after hip fracture: The Norwegian Epidemiologic Osteoporosis Studies (NOREPOS). *BMC Geriatr.* **2023**, *23*, 201.
3. Ghalenavi, E.; Mirfeizi, Z.; Hashemzadeh, K.; Sahebari, M.; Joker, M. Diagnostic Value of Radiographic Singh Index Compared to Dual-Energy X-ray Absorptiometry Scan in Diagnosing Osteoporosis: A Systematic Review. *Arch. Bone Jt. Surg.* **2024**, *12*, 1–11. [PubMed]
4. Dendere, R.; Potgieter, J.H.; Steiner, S.; Whiley, S.P.; Douglas, T.S. Dual-Energy X-ray Absorptiometry for Measurement of Phalangeal Bone Mineral Density on a Slot-Scanning Digital Radiography System. *IEEE Trans. Biomed. Eng.* **2015**, *62*, 2850–2859. [CrossRef] [PubMed]

5. Peel, N.; Johnson, A.; Barrington, N.; Smith, T.; Eastell, D.R. Impact of anomalous vertebral segmentation on measurements of bone mineral density. *J. Bone Miner. Res.* **1993**, *8*, 719–723. [CrossRef] [PubMed]
6. Stolojescu-Crisan, C.; Holban, S. A comparison of X-ray image segmentation techniques. *Adv. Electr. Comput. Eng.* **2013**, *13*, 85–92. [CrossRef]
7. Fathima, S.; Tamilselvi, R.; Beham, M.; Nagaraj, A. A deep learning approach on segmentation of bone for bmd measurement from dexa scan images. In Proceedings of the 2020 Sixth International Conference on Bio Signals, Images, and Instrumentation (ICBSII), Chennai, India, 27–28 February 2020.
8. Sanchez, M.; Sánchez, M.; Vidal, V.; Verdu, G.; Verdú, G.; Mayo, P.; Rodenas, F. Medical image restoration with different types of noise. In Proceedings of the 2012 Annual International Conference of the IEEE Engineering in Medicine and Biology Society, San Diego, CA, USA, 28 August–1 September 2012.
9. Goyal, B.; Dogra, A.; Agrawal, S.; Sohi, B.S. Noise issues prevailing in various types of medical images. *Biomed. Pharmacol. J.* **2018**, *11*, 1227–1237. [CrossRef]
10. Vijaya, K.V.; Kalpana, V. Effect of noise on segmentation evaluation parameters. In *Soft Computing: Theories and Applications: Proceedings of SoCTA 2019*; Springer: Singapore, 2020.
11. Tenbrinck, D.; Jiang, X. Image segmentation with arbitrary noise models by solving minimal surface problems. *Pattern Recognit.* **2015**, *48*, 3293–3309. [CrossRef]
12. Chen, T.; Wang, C.; Shan, H. Berdiff: Conditional bernoulli diffusion model for medical image segmentation. In *International Conference on Medical Image Computing and Computer-Assisted Intervention*; Springer Nature: Cham, Switzerland, 2023.
13. Khalid, F.; Hanif, M.; Rehman, S.; Qadir, J.; Shafique, M. Fademl: Understanding the impact of pre-processing noise filtering on adversarial machine learning. In Proceedings of the 2019 Design, Automation & Test in Europe Conference & Exhibition (DATE), Florence, Italy, 25–29 March 2019; IEEE: Piscataway, NJ, USA, 2019.
14. Maharana, K.; Mondal, S.; Nemade, B. A review: Data pre-processing and data augmentation techniques. *Glob. Transit. Proc.* **2022**, *3*, 91–99. [CrossRef]
15. Dhar, T.; Dey, N.; Borra, S.; Sherratt, R. Challenges of deep learning in medical image analysis—Improving explainability and trust. *IEEE Trans. Technol. Soc.* **2023**, *4*, 68–75. [CrossRef]
16. Naylor, K.E.; McCloskey, E.V.; Eastell, R.; Yang, L. Use of DXA-based finite element analysis of the proximal femur in a longitudinal study of hip fracture. *J. Bone Miner. Res.* **2012**, *28*, 1014–1021. [CrossRef] [PubMed]
17. Burkhart, T.A.; Arthurs, K.L.; Andrews, D.M. Manual segmentation of DXA scan images results in reliable upper and lower extremity soft and rigid tissue mass estimates. *J. Biomech.* **2009**, *42*, 1138–1142. [CrossRef] [PubMed]
18. Yasufumi, H.; Kichizo, Y.; Toshinobu, F.M.I.; Kichiya, T.; Yasuho, N. Assessment of bone mass by image analysis of metacarpal bone roentgenograms: A quantitative digital image processing (DIP) method. *Radiat. Med.* **1990**, *8*, 173–178.
19. Matsumoto, C.; Kushida, K.; Yamazaki, K.; Imose, K.; Inoue, T. Metacarpal bone mass in normal and osteoporotic Japanese women using computed X-ray densitometry. *Calcif. Tissue Int.* **1994**, *55*, 324–329. [CrossRef] [PubMed]
20. Wilson, J.P.; Mulligan, K.; Fan, B.; Sherman, J.L.; Murphy, E.J.; Tai, V.W.; Powers, C.L. Shepherd, Dual-energy X-ray absorptiometry–based body volume measurement for 4-compartment body composition. *Am. J. Clin. Nutr.* **2012**, *95*, 25–31. [CrossRef] [PubMed]
21. Roberts, M.; Cootes, T.; Pacheco, E.; Adams, J. Quantitative vertebral fracture detection on DXA images using shape and appearance models. *Acad. Radiol.* **2007**, *14*, 1166–1178. [CrossRef] [PubMed]
22. Sarkalkan, N.; Weinans, H.; Zadpoor, A.A. Statistical shape and appearance models of bones. *Bone* **2014**, *60*, 129–140. [CrossRef] [PubMed]
23. Hussain, D.; Han, S.-M. Computer-aided osteoporosis detection from DXA imaging. *Comput. Methods Programs Biomed.* **2019**, *173*, 87–107. [CrossRef] [PubMed]
24. Hussain, D.; Al-Antari, M.A.; Al-Masni, M.A.; Han, S.-M.; Kim, T.-S. Femur segmentation in DXA imaging using a machine learning decision tree. *J. X-ray Sci. Technol.* **2018**, *26*, 727–746. [CrossRef]
25. Hussain, D.; Han, S.-M.; Kim, T.-S. Automatic hip geometric feature extraction in DXA imaging using regional random forest. *J. X-ray Sci. Technol.* **2019**, *27*, 207–236. [CrossRef]
26. Hussain, D.; Naqvi, R.A.; Loh, W.-K.; Lee, J. Deep learning in DXA image segmentation. *Comput. Mater. Contin.* **2021**, *66*, 2587–2598. [CrossRef]
27. Samuel, D.; Karam, L. Understanding how image quality affects deep neural networks. In Proceedings of the 2016 Eighth International Conference on Quality of Multimedia Experience (QoMEX), Lisbon, Portugal, 6–8 June 2016.
28. Calderon, S.; Fallas, F.; Zumbado, M.; Tyrrell, P.N.; Stark, H.; Emersic, Z.; Meden, B.; Solis, M. Assessing the Impact of the Deceived Non Local Means Filter as a Preprocessing Stage in a Convolutional Neural Network Based Approach for Age Estimation Using Digital Hand X-ray Images. In Proceedings of the 2018 25th IEEE International Conference on Image Processing (ICIP), Athens, Greece, 7–10 October 2018.
29. Costa, G.B.P.; Contato, W.A.; Nazare, T.S.; Neto, J.E.; Ponti, M. An empirical study on the effects of different types of noise in image classification tasks. *arXiv* **2016**, arXiv:1609.02781.
30. Chehab, M.E. Illinois Bone & Joint Institute. (n.d.). Osteoporosis Podcasts. Available online: https://www.ibji.com/podcasts/osteoporosis/ (accessed on 6 March 2024).
31. Kim, K.; Choi, J.; Lee, Y. Effectiveness of non-local means algorithm with an industrial 3 mev linac high-energy X-ray system for non-destructive testing. *Sensors* **2020**, *20*, 2634. [CrossRef] [PubMed]

32. Lee, S.; Lee, Y. The impact of improved non-local means denoising algorithm on photon-counting X-ray images using various Al additive filtrations. *Nucl. Instrum. Methods Phys. Res. Sect. A Accel. Spectrometers Detect. Assoc. Equip.* **2021**, *1027*, 166244. [CrossRef]
33. Al-antari, M.A.; Al-masni, M.A.; Metwallyb, M.; Hussain, D.; Parkb, S.J.; Shinb, J.S.; Hana, S.M.; Kima, T.S. Denoising images of dual energy X-ray absorptiometry using non-local means filters. *J. X-ray Sci. Technol.* **2018**, *26*, 395–412. [CrossRef] [PubMed]
34. Al-Antari, M.A.; Al-Masni, M.A.; Metwally, M.; Hussain, D.; Valarezo, E.; Rivera, P.; Gi, G.; Park, J.M.; Kim, T.Y.; Park, S.J.; et al. Non-local means filter denoising for DXA images. In Proceedings of the 2017 39th Annual International Conference of the IEEE Engineering in Medicine and Biology Society (EMBC), Jeju Island, Republic of Korea, 11–15 July 2017.
35. Ilango, G.; Marudhachalam, R. New hybrid filtering techniques for removal of Gaussian noise from medical images. *ARPN J. Eng. Appl. Sci.* **2011**, *6*, 8–12.
36. Mayasari, R.; Heryana, N. Reduce noise in computed tomography image using adaptive Gaussian filter. *arXiv* **2019**, arXiv:1902.05985.
37. Sagheer, S.V.M.; George, S.N. A review on medical image denoising algorithms. *Biomed. Signal Process. Control.* **2020**, *61*, 102036. [CrossRef]
38. Kwon, J.W.; Cho, S.I.; Ahn, Y.B.; Ro, Y.M. Noise reduction in DEXA image based on system noise modeling. In Proceedings of the 2009 International Conference on Biomedical and Pharmaceutical Engineering, Singapore, 2–4 December 2009.
39. Fathima, S.N.; Tamilselvi, R.; Beham, M.P.; Sabarinathan, D. Diagnosis of Osteoporosis using modified U-net architecture with attention unit in DEXA and X-ray images. *J. X-ray Sci. Technol.* **2020**, *28*, 953–973. [CrossRef]
40. Ziyad, S.R.; Radha, V.; Vaiyapuri, T. Noise removal in lung LDCT images by novel discrete wavelet-based denoising with adaptive thresholding technique. *Int. J. E-Health Med. Commun.* **2021**, *12*, 1–15. [CrossRef]
41. Zavala-Mondragón, L.A.; de With, P.H.; van der Sommen, F. Image noise reduction based on a fixed wavelet frame and CNNs applied to CT. *IEEE Trans. Image Process.* **2021**, *30*, 9386–9401. [CrossRef] [PubMed]
42. Elaiyaraja, G.; Kumaratharan, N.; Rao, T.C.S. Fast and efficient filter using wavelet threshold for removal of Gaussian noise from MRI/CT scanned medical images/color video sequence. *IETE J. Res.* **2019**, *68*, 10–22. [CrossRef]
43. Deeba, F.; Kun, S.; Dharejo, F.A.; Zhou, Y. Wavelet-based enhanced medical image super resolution. *IEEE Access* **2020**, *8*, 37035–37044. [CrossRef]
44. Khan, S.U.; Khan, I.U.; Ullah, I.; Saif, N.; Ullah, I. A review of airport dual energy X-ray baggage inspection techniques: Image enhancement and noise reduction. *J. X-ray Sci. Technol.* **2020**, *28*, 481–505. [CrossRef] [PubMed]
45. Fathima, S.N.; Tamilselvi, R.; Beham, M.P. A Survey on Osteoporosis Detection Methods with a Focus on X-ray and DEXA Images. *IETE J. Res.* **2020**, *68*, 4640–4664. [CrossRef]
46. Xie, W.; Feng, T.; Zhang, M.; Li, J.; Ta, D.; Cheng, L.; Cheng, Q. Wavelet transform-based photoacoustic time-frequency spectral analysis for bone assessment. *Photoacoustics* **2021**, *22*, 100259. [CrossRef] [PubMed]
47. Kaur, S.; Hooda, R.; Mittal, A.; Akashdeep; Sofat, S. Deep CNN-based method for segmenting lung fields in digital chest radiographs. In Proceedings of the Advanced Informatics for Computing Research: First International Conference, ICAICR 201, Jalandhar, India, 17–18 March 2017; Springer: Singapore, 2017.
48. Krizhevsky, A.; Sutskever, I.; Hinton, G.E. Imagenet classification with deep convolutional neural networks. *Commun. ACM* **2017**, *60*, 84–90. [CrossRef]
49. Shelhamer, E.; Long, J.; Darrell, T. Fully convolutional networks for semantic segmentation. *IEEE Trans. Pattern Anal. Mach. Intell.* **2017**, *39*, 640–651. [CrossRef] [PubMed]
50. Ronneberger, O.; Fischer, P.; Brox, T. U-net: Convolutional networks for biomedical image segmentation. In *International Conference on Medical Image Computing and Computer-Assisted Intervention*; Springer International Publishing: Munich, Germany, 2015.
51. Badrinarayanan, V.; Kendall, A.; Cipolla, R. Segnet: A deep convolutional encoder-decoder architecture for image segmentation. *arXiv* **2015**, arXiv:1505.07293. [CrossRef]
52. Wang, L.; Lu, J.; Li, Y.; Yahagi, T.; Okamoto, T. Noise removal for medical X-ray images in wavelet domain. *Electr. Eng. Jpn.* **2008**, *163*, 37–46. [CrossRef]
53. OpenAI. ChatGPT-3.5. Microsoft Corporation. Available online: https://chat.openai.com/ (accessed on 30 March 2024).

diagnostics

MDPI

Article

Performance Comparison of Convolutional Neural Network-Based Hearing Loss Classification Model Using Auditory Brainstem Response Data

Jun Ma [1], Seong Jun Choi [2], Sungyeup Kim [3] and Min Hong [4,*]

1 Department of Software Convergence, Soonchunhyang University, Asan 31538, Republic of Korea; ringring369@gmail.com
2 Department of Otorhinolaryngology—Head and Neck Surgery, College of Medicine, Soonchunhyang University Cheonan Hospital, Cheonan 31151, Republic of Korea; akas9238@hanmail.net
3 Insitute for Artificial Intelligence and Software, Soonchunhyang University, Asan 31538, Republic of Korea; sungyeup.kim@gmail.com
4 Department of Computer Software Engineering, Soonchunhyang University, Asan 31538, Republic of Korea
* Correspondence: mhong@sch.ac.kr

Abstract: This study evaluates the efficacy of several Convolutional Neural Network (CNN) models for the classification of hearing loss in patients using preprocessed auditory brainstem response (ABR) image data. Specifically, we employed six CNN architectures—VGG16, VGG19, DenseNet121, DenseNet-201, AlexNet, and InceptionV3—to differentiate between patients with hearing loss and those with normal hearing. A dataset comprising 7990 preprocessed ABR images was utilized to assess the performance and accuracy of these models. Each model was systematically tested to determine its capability to accurately classify hearing loss. A comparative analysis of the models focused on metrics of accuracy and computational efficiency. The results indicated that the AlexNet model exhibited superior performance, achieving an accuracy of 95.93%. The findings from this research suggest that deep learning models, particularly AlexNet in this instance, hold significant potential for automating the diagnosis of hearing loss using ABR graph data. Future work will aim to refine these models to enhance their diagnostic accuracy and efficiency, fostering their practical application in clinical settings.

Keywords: auditory brainstem response; ABR; deep learning; VGG16; VGG19; DenseNet121; Densenet201; Alexnet; image processing; hearing loss

1. Introduction

Convolutional Neural Networks (CNNs) are a specialized category of deep learning algorithms predominantly utilized in the fields of image and video recognition. Characteristically, CNNs automate the process of learning and classifying image features through a structured network comprising convolutional layers, pooling layers, and fully connected layers. The convolutional layer primarily serves to extract pertinent features from images, while the pooling layer reduces computational load by diminishing the spatial dimensions of the data. The fully connected layer then performs the final task of classification. These networks are extensively applied across various tasks in computer vision, including image classification, object detection, and face recognition, due to their robustness in handling complex visual inputs [1,2]. In the realm of medical imaging, CNNs assume a critical role given the intricate nature of most medical datasets. They provide effective mechanisms for processing and interpreting such data swiftly, which is indispensable in clinical settings. Consequently, CNNs are employed in diverse medical imaging applications encompassing disease classification, tissue categorization, and image segmentation, among others [3]. This versatility underlines the significance of CNNs in advancing medical image analysis and improving diagnostic methodologies.

The Auditory Brainstem Response (ABR) is an electrophysiological measurement reflecting the brainstem's activity in response to auditory stimuli. This response involves the transmission of a neuroelectric signal from the cochlea through the auditory pathways to the auditory cortex of the brain. The ABR test, a diagnostic procedure used to assess hearing functionality, measures the waveform of this electrical response. This test is particularly valuable in clinical settings for evaluating hearing impairment. Its non-invasive nature and independence from patient consciousness—being unaffected by sleep or anesthesia—make it particularly suitable for use in populations unable to provide reliable auditory feedback. These include newborns, infants, young children, the elderly, and individuals with congenital disabilities. Consequently, the ABR test provides a robust and objective method for assessing auditory function across a diverse patient demographic [4].

ABR measurement is a neurophysiological method used to record changes in brain waves triggered by auditory stimulation. This technique involves the application of click sounds at intervals of approximately 0.8 ms combined with energy modulation during auditory transmission to stimulate brain wave activity. When auditory stimuli ranging from 10 dB to 100 dB are administered, typically, five to seven waves are detectable. In normal adults, a waveform responsive to the stimulus typically emerges within approximately 10 milliseconds following the onset of the click sound [5]. Among the identifiable waves, wave number 5 (V wave) is particularly significant for clinical assessments. The threshold of hearing is determined by analyzing the latency periods of the V wave across various dB levels. Hearing loss is subsequently diagnosed based on these latency values within the specified dB range [6].

During the ABR testing procedure, as illustrated in Figure 1, small electrodes marked by red circles are affixed to the subject's forehead and behind the ears. These electrodes detect electrical activity within the auditory nerve and brainstem in response to auditory stimuli. The test involves the administration of a series of click sounds delivered through an eartip inserted into the ear, with the brain's responses—essentially, brain waves—being detected and automatically recorded by a computer system. This method offers an objective assessment of hearing functionality, contrasting with other methods that rely on subjective patient responses. In the process of measurement, the audiologist identifies and records the V wave, which is critical for hearing and occurs between 6 ms and 8 ms, from among wave information labeled from 1 to 5, corresponding to each decibel (dB) stimulus level. The measurement concludes once the waveforms for all dB stimulus levels have been successfully recorded [7].

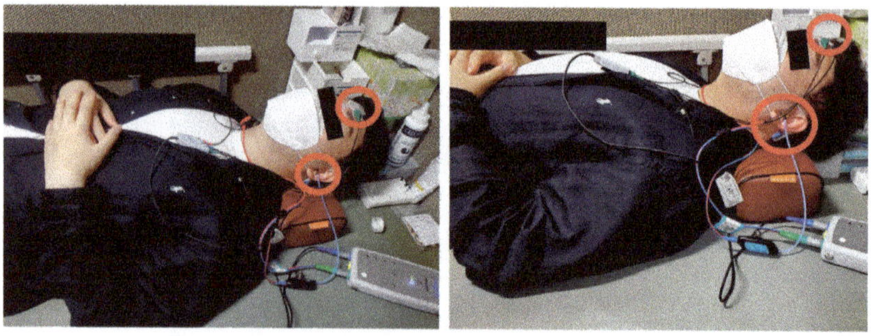

Figure 1. Auditory brainstem response test scene.

In our previous study [8], we conducted preprocessing to standardize the auditory brainstem response (ABR) graph outputs across various manufacturers. ABR graph data from five different manufacturers—Audera, Navigator, Eclipse, Viking Select, and Interacoustics—were collected. Each ABR graph was normalized as depicted in Figure 2, resulting in a dataset comprising 10,000 data entries. Furthermore, during the preprocessing phase, a total of 2010 images were filtered out due to significant reductions in graph resolution or improper graph outputs. Consequently, the analysis was conducted on 7990 images using normalized ABR data from the remaining valid datasets.

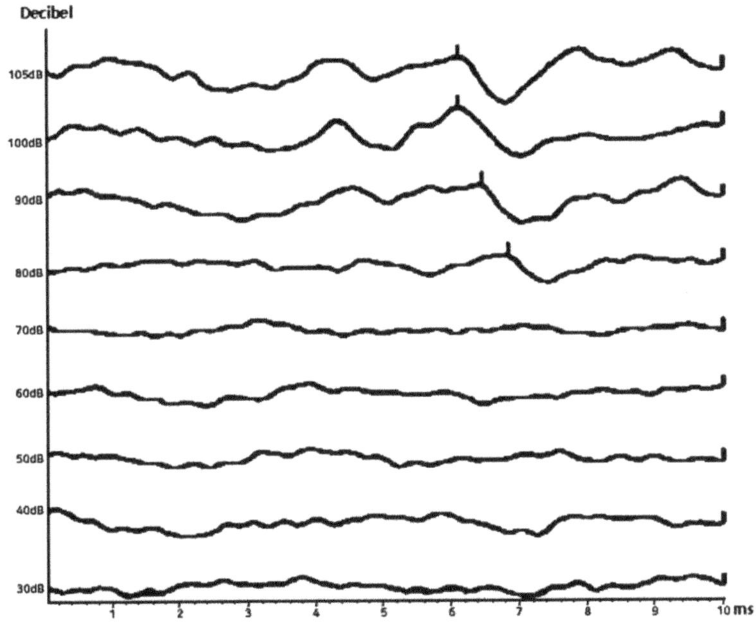

Figure 2. Auditory brainstem response data after pre-processing.

2. Materials and Methods

2.1. Auditory Brainstem Response Data

The ABR graph serves as a crucial diagnostic tool for evaluating the auditory system by visualizing changes in electrical activity over time. An analysis of the ABR graph can reveal significant insights into the auditory system's condition through various characteristic features described below.

Multi-wave form: the ABR graph typically displays a series of waves, each sequentially representing electrical activity in different parts of the brain at specific times.

The size and spacing of waves: the initial wave usually appears as the largest and most distinct wave, with subsequent waves diminishing progressively in size and becoming closer together.

Latency and amplitude: these parameters are critical; latency refers to the timing of the wave's occurrence post stimulus, and amplitude denotes the wave's magnitude.

Baseline: the baseline of the graph indicates normal brain activity levels; any deviation from this baseline may suggest abnormalities in the auditory pathway.

Axes: the ABR graph is typically oriented with time (ms) on the horizontal axis and amplitude (μV) on the vertical axis, where sound pressure levels (decibels, dB) may also be considered.

By systematically assessing these features—particularly changes in latency and amplitude—anomalies such as hearing impairment or disorders within the central auditory pathway can be detected. Thus, the ABR graph not only aids in assessing the patient's hearing status but also contributes to the formulation of an appropriate treatment plan [9–12]. The methodology for analyzing an ABR graph, as depicted in Figure 3, is essential for comprehensively evaluating both hearing function and the broader state of the auditory system.

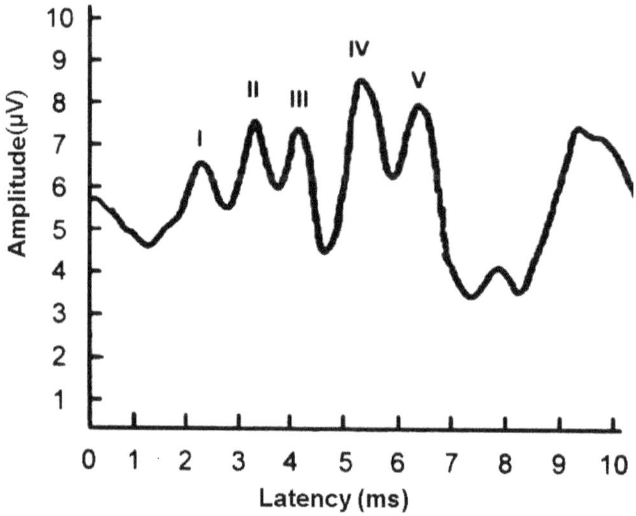

Figure 3. Auditory brainstem response example graph.

Wave I: This initial wave is indicative of the acoustic signal's arrival at the auditory nerve. The latency for Wave I, representing the time taken for the stimulus to reach the auditory nerve, typically ranges from 1 to 4 ms.

Wave II: occurring in the auditory brainstem, specifically in the region associated with the auditory pathway's initial processing stages, the latency from Wave I to Wave II measures the transmission time to the auditory brainstem and is generally observed between 2 and 4 ms.

Wave III: this wave is generated in the cochlear nuclei—the Dorsal Cochlear Nucleus (DCN) and Ventral Cochlear Nucleus (VCN)—located in the lower auditory brainstem. The latency from Wave II to Wave III, which measures the passage of stimulus through the auditory brainstem, typically ranges between 3 and 5 ms.

Wave IV: Representing signals generated en route to the Medial Superior Olive (MSO) at the upper part of the auditory brainstem; the latency between Wave III and Wave IV usually spans 4 to 5.5 ms.

Wave V: The largest of the acoustic signals, Wave V emanates from the output region of the auditory brainstem, reaching the Inferior Colliculus (IC). The latency from Wave IV to Wave V is noted between 5.5 and 7 ms.

Wave latency: the latency of each wave quantifies the time required for its generation. Within a normal auditory system, these latencies fall within specific ranges; however, abnormalities may manifest as delayed latencies.

Wave amplitude: The amplitude of each wave reflects its magnitude. Typically, a healthy auditory system produces waves of a large and consistent amplitude. Reduced amplitude may indicate auditory abnormalities.

Interpeak latency: This metric illustrates the latency differences between consecutive waves, reflecting the conduction time along the central auditory pathway. Normal auditory systems exhibit consistent interpeak latencies, whereas increased values may suggest central auditory pathway dysfunction.

These metrics provide a comprehensive framework for assessing the integrity and functionality of the auditory system through ABR testing, facilitating the identification and characterization of potential auditory impairments.

Hearing loss: Hearing loss is characterized as a reduction in the ability to perceive or interpret sounds; the loss is attributable to anomalies within the auditory system, which may involve the external or internal ear structures or the auditory nerve. This condition can be either temporary or permanent and may affect one or both ears. Within the context of this study, the severity of hearing loss is assessed based on the detection of the V wave in the ABR graph data, as illustrated in Figure 4. Typically, V waves are elicited by sound stimuli ranging from 10 to 100 dB, presented in 10 dB increments. An absence of V waves in waveforms at or below 40 dB typically leads otolaryngologists to diagnose hearing loss.

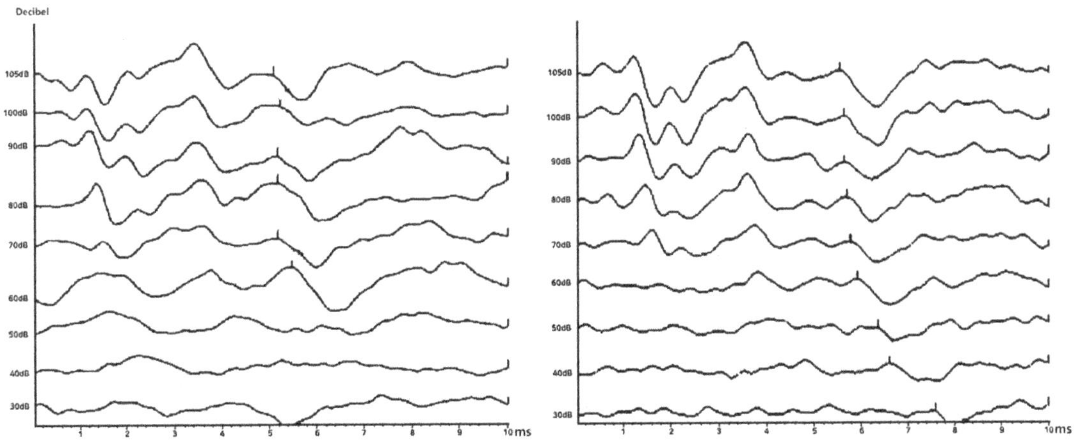

Figure 4. An ABR graph of a patient with hearing loss (**left**) and an ABR graph of a normal person (**right**).

The analysis of an ABR graph involves a detailed evaluation of various parameters to ascertain the auditory system's status. Key indicators include the latency and amplitude of waves; a delay in wave latency or a reduction in amplitude may signal auditory impairments or issues within the central auditory pathway. By examining these attributes, clinicians can effectively gauge a patient's hearing condition and devise appropriate treatment strategies. Such diagnostic practices are crucial for the early detection and management of hearing loss, thereby enhancing the quality of life and communication abilities of affected individuals [13–16].

2.2. CNN Classification Model

In our previous study, we evaluated the classification of hearing loss using solely the VGG16 model [8]. Expanding upon this initial approach, the current study incorporates a broader array of convolutional neural network models, specifically VGG16, VGG19, DenseNet121, DenseNet201, AlexNet, and InceptionV3, to perform more comprehensive learning and classification tests. Each model, recognized for its unique strengths and limitations, has previously been utilized across a variety of medical image classification tasks. For the purposes of this study, we tailored the hyperparameters, specifically the batch size and layer configurations, to optimize the learning process for ABR image classification. This part details the architectural nuances and characteristics of each model and describes the specific modifications made to the hyperparameters to enhance model performance for this application. These adjustments are pivotal in refining our approach to accurately classifying hearing loss through deep learning techniques.

2.2.1. VGG16 and VGG19

The VGG model, developed by the Visual Geometry Group at Oxford, includes two primary configurations: VGG16 and VGG19. VGG19 extends the architecture of VGG16 by adding three additional convolutional layers positioned before the 3rd, 4th, and 5th max pooling layers, enhancing its depth and complexity. In our experiments with the VGG16 model, the original images, sized 573 × 505 pixels, were initially resized to 224 × 224 pixels. Subsequently, the images were further scaled down to dimensions of 286 × 252, 143 × 126, 71 × 63, 35 × 31, and 17 × 15 to facilitate object recognition. The learning process involved adjustments in the dense layer configurations, with neuron counts set to 1024, 512, and 2, optimizing the network's ability to discern features at various scales. The VGG19 model, with its additional convolutional layers, retains the number of layers for sizes 286 × 252 and 143 × 126 but adds a layer each at smaller scales (71 × 63 and smaller), comprising a total of 19 layers to enhance detail recognition [17–21]. In a related study conducted by Dey et al., a pneumonia detection model utilizing VGG19 applied to chest X-ray images demonstrated a high classification accuracy of up to 97.94% [22]. Similarly, Mateen et al. reported that the VGG19 model was effectively utilized in medical image analysis, achieving an impressive classification accuracy of 98.13% in a retinopathy classification system using fundus images [23]. For the purpose of this study, which was tailored to ABR image classification, the VGG19 model and the VGG16 model were enhanced by incorporating two additional dense layers and a dropout layer in each configuration to prevent overfitting. The learning framework was structured with dense layers of 1024, 512, 256, 128, and 2 neurons, with a batch size of 8, serving as a robust hyperparameter setup. This architecture was designed to maximize the model's ability to accurately classify ABR images, leveraging deeper layers for more nuanced feature extraction.

2.2.2. DenseNet121 and DenseNet201

DenseNet is a CNN model engineered to enhance training efficiency by integrating the concept of shorter connections. This design enables direct links between the input and output layers, fostering a deeper and structurally more efficient network capable of delivering precise performance outcomes. Unlike traditional CNNs, which feature connections primarily to the immediately subsequent layer, DenseNet boasts a comprehensive connection structure with L(L + 1)/2 direct connections, greatly enriching the flow of information across the network. To efficiently manage down-sampling, the architecture is segmented into three distinct dense blocks. Each block is separated by a transition layer which performs both convolution and pooling operations, thus maintaining the network's depth while progressively reducing its dimensionality [24–28]. In a related study by Chauhan et al., a DenseNet model was employed to differentiate COVID-19 patients from healthy individuals using chest X-ray images, achieving an impressive accuracy rate of 98.45% [29]. Within the context of this paper, the DenseNet121 and DenseNet201 models, comprising 121 and 201 layers, respectively, were utilized. Tailored specifically for ABR image classification, the learning process was conducted using dense layers configured with 256 and 2 neurons and a hyperparameter setting of a batch size of 8. This configuration was designed to optimize the network's capability for high-accuracy classification in ABR imaging.

2.2.3. AlexNet

AlexNet, a CNN model, significantly impacted the field of deep learning after securing victory in the 2012 ImageNet Large Scale Visual Recognition Challenge (ILSVRC). Named after Alex Krizhevsky, the lead author of the seminal paper "ImageNet Classification with Deep Convolutional Neural Networks", AlexNet's architecture has been instrumental in advancing CNN development. Its structure features a sequential layout comprising an input layer and five convolutional layers—each accompanied by a max pooling layer and a normalization layer—culminating in a dense layer dedicated to classification tasks [30–33]. In research conducted by Chen et al., the efficacy of various models including 3DAlexNet,

ResNet50, and InceptionV4 was evaluated for the classification of magnetic resonance images to diagnose prostate cancer, yielding classification accuracies of 92.1%, 87.6%, and 85.7%, respectively [34]. Another study by Titoriya and Sachdeva utilized the AlexNet model to classify breast cancer tissue images, demonstrating a high classification accuracy of 95.7%, thereby underscoring its potential for medical imaging applications [35]. In this study, modifications were made to the original AlexNet architecture to enhance its suitability for Auditory Brainstem Response (ABR) image classification. Adjustments included the integration of max pooling layers at the first and fifth convolutional layers and the insertion of dropout layers within each dense layer to mitigate overfitting. The learning process was optimized by setting the hyperparameters of the dense layers to 4096, 4096, and 2, with a batch size of 8, facilitating an improved learning rates and robust classification performance in ABR image analysis.

2.2.4. InceptionV3

Generally, there is a correlation between increased model size and both accuracy and computational effort. For instance, the DenseNet architecture enhances performance by deepening the model with skip connections, yet this also escalates computational demands, resulting in longer training durations due to the increased depth. Similarly, enlarging model size augments computational requirements, which presents a limitation when operating within memory constraints. The Inception model, devised by Google, addresses this challenge by employing convolutional decomposition to expand the model size while minimizing computational costs. The InceptionV3 model, which was utilized in this research, stands out among the Inception series with its 42-layer deep network, which is optimized to maintain a balance between a low parameter count and computational efficiency, akin to that of the VGG models [36,37]. In research conducted by Wang, Cheng, et al., the InceptionV3 model was applied to develop a classification system for lung nodules using chest X-ray images, achieving a classification accuracy of up to 86.4% [38]. In the context of this study, the InceptionV3 model was adapted for ABR image classification. Modifications were made to the model's configuration, setting the hyperparameters of the dense layers to 256 and 2, and the batch size to 8, to tailor the learning process specifically for ABR image analysis. This strategic adjustment aims to leverage the model's efficiency and deep learning capabilities for precise ABR image classification.

3. Results

Model Training and ABR Data Classification Results

Using 7990 ABR data excluding impure data, learning and classification tests were conducted with 4794, 1598, and 1598 train, validation, and test data at a ratio of 6:2:2, respectively. The accuracy, loss results, and test classification confusion matrix results of each model's learning are as follows.

The results in Figure 5 highlight the performance metrics for the VGG16 model during training, showing an accuracy of 91.58% and a loss of 6.52%. The confusion matrix for the test dataset for this model indicated 769 true negatives (tn), 52 false negatives (fn), 707 true positives (tp), and 70 false positives (fp). The VGG19 model demonstrated improved training performance, achieving an accuracy of 94.84% and a loss of 4.64%. The test data for this model produced a confusion matrix with 770 tn, 12 fn, 735 tp, and 81 fp. Additionally, the DenseNet121 model recorded a training accuracy of 92.52% and a loss of 5.77%, with its confusion matrix displaying 727 tn, 49 fn, 753 tp, and 69 fp.

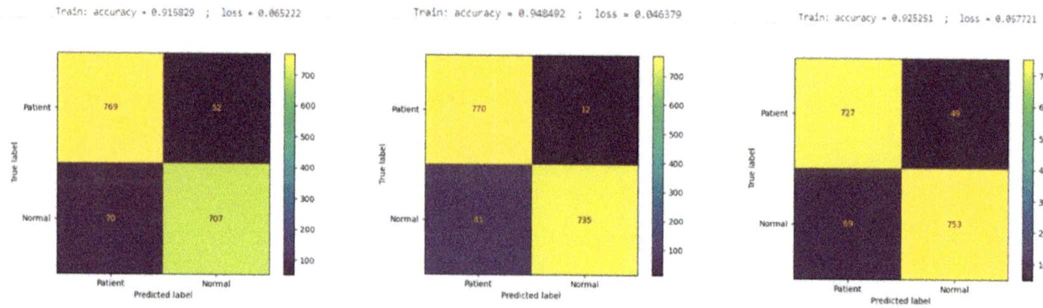

Figure 5. Learning results of VGG16 model (**left**), VGG19 model (**middle**), and DenseNet121 model (**right**).

And the results in Figure 6, the DenseNet201 model's training performance exhibited an accuracy of 93.09% and a loss of 5.19%, with the confusion matrix for the test dataset indicating 752 tn, 34 fn, 739 tp, and 73 fp. The AlexNet model, on the other hand, achieved a training accuracy of 96.54% and a remarkably lower loss of 2.99%. The corresponding confusion matrix demonstrated its high precision with 748 tn, 51 fn, 785 tp, and only 14 fp, underscoring its efficacy in accurately classifying the conditions with minimal misclassifications. Lastly, the InceptionV3 model registered a training accuracy of 91.64% and a loss of 6.59%, with its test data confusion matrix revealing 760 tn, 56 fn, 685 tp, and 97 fp.

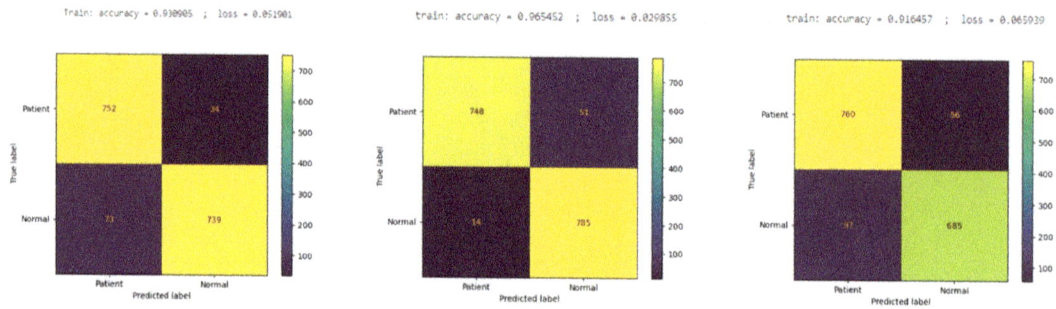

Figure 6. Learning results of DenseNet201 model (**left**), AlexNet model (**middle**), and InceptionV3 model (**right**).

Based on the confusion matrix results from the test data for each model, various performance metrics were calculated and systematically tabulated. These metrics include accuracy, the true negative rate (TNR), the true positive rate (TPR), the false positive rate (FPR), the false negative rate (FNR), precision, and the F1 score. The formulas for each of these metrics are outlined below, with their respective results being derived from Equations (1)–(7):

$$\text{Accuracy} = \frac{\text{tp} + \text{tn}}{\text{tp} + \text{tn} + \text{fp} + \text{fn}} \tag{1}$$

Accuracy: this is calculated as the ratio of correctly predicted observations (both true positives and true negatives) to the total observations in the dataset.

$$\text{TNR} = \frac{\text{tn}}{\text{tn} + \text{fp}} \tag{2}$$

True negative rate (TNR), also known as specificity: this measures the proportion of actual negatives that are correctly identified.

$$\text{TPR} = \frac{\text{tp}}{\text{tp} + \text{fn}} \tag{3}$$

True positive rate (TPR), also known as sensitivity or recall: this metric indicates the proportion of actual positives that are correctly identified.

$$\text{FPR} = \frac{\text{fp}}{\text{fp} + \text{tn}} \tag{4}$$

False positive rate (FPR): this is calculated as the ratio of the number of false positives to the sum of the false positives and true negatives.

$$\text{FNR} = \frac{\text{fn}}{\text{fn} + \text{tp}} \tag{5}$$

False negative rate (FNR): this measures the proportion of positives which yield negative test outcomes with the model.

$$\text{Precision} = \frac{\text{tp}}{\text{tp} + \text{fp}} \tag{6}$$

Precision is also known as the positive predictive value: this is the ratio of true positives to the combined total of true positives and false positives.

$$\text{F1 score} = 2 \times \frac{\text{precision} \times \text{TPR}}{\text{precision} + \text{TPR}} \tag{7}$$

F1 score: this is the harmonic mean of Precision and Recall, providing a balance between the two when their rates may vary.

The calculated values for these metrics are compiled in a Table 1 within this paper, providing a comprehensive assessment of each model's performance on the test dataset. This structured approach allows for a detailed comparison and evaluation of the effectiveness of each classification model in the context of hearing loss detection.

Table 1. Total data validity calculation results.

	Accuracy	TNR	TPR	FPR	FNR	Precision	F1 Score
VGG16	92.37%	93.67%	90.99%	6.33%	9.01%	93.15%	0.9206
VGG19	94.18%	98.47%	90.07%	1.53%	9.93%	98.39%	0.9405
DenseNet121	92.62%	93.69%	91.61%	6.31%	8.39%	93.89%	0.9273
DenseNet201	93.30%	95.67%	91.01%	4.33%	8.99%	95.60%	0.9325
AlexNet	95.93%	93.62%	98.25%	6.38%	1.75%	93.90%	0.9602
InceptionV3	90.43%	93.14%	87.60%	6.86%	12.40%	92.44%	0.8995

The result of the classification models utilized in this research were evaluated based on outputs of classification scores. Figure 7 presents the classification score results for the AlexNet model, which demonstrated the highest accuracy among the models tested. This figure displays image data that was randomly selected from the test dataset. It includes the filename of the data—where "napa" denotes an image associated with hearing loss and "tupa" indicates a normal hearing image. Additionally, the outcomes of the classification process are indicated, with the number 0 representing hearing loss and the number 1 representing normal hearing. The scores leading up to these classifications are also documented to provide a comprehensive view of the model's performance in distinguishing between the two categories. This detailed display of results facilitates an understanding of the model's efficacy in accurately classifying auditory conditions based on ABR image data.

These findings illustrate the varying levels of performance and accuracy across the models tested, offering insights into their respective strengths and areas for improvement in the classification of conditions based on the training and validation datasets.

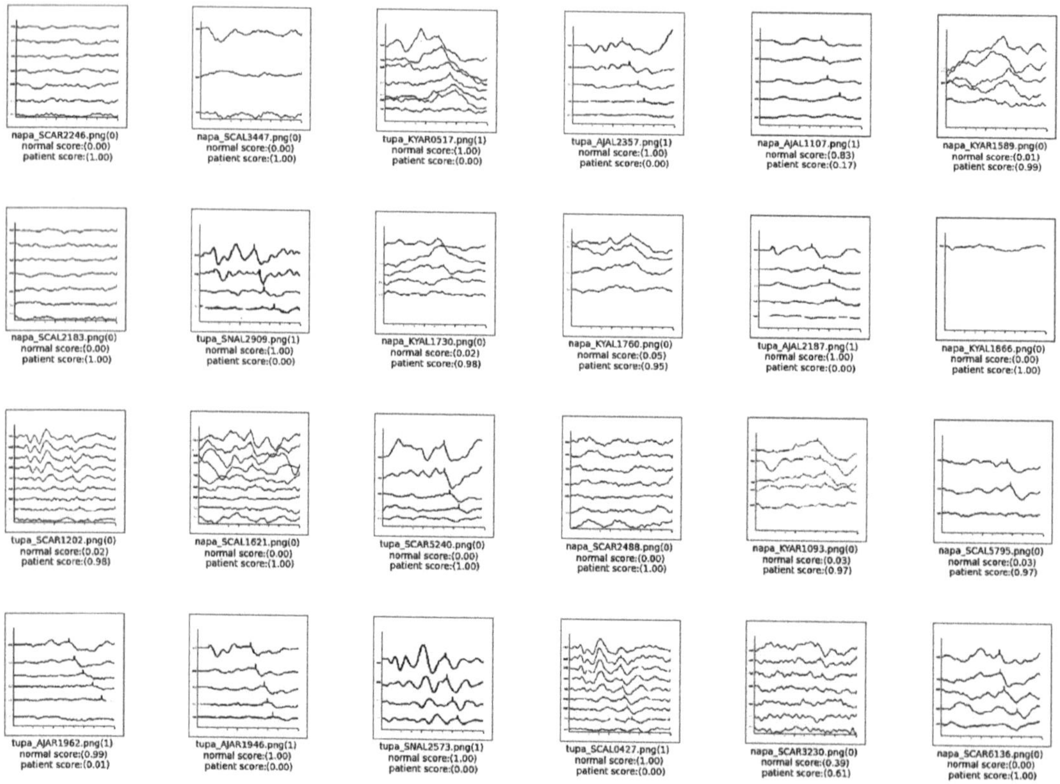

Figure 7. Classification result scores of AlexNet model.

4. Discussion

4.1. Classification Data Analysis

In the current study, the classification learning revealed that AlexNet achieved the highest overall accuracy, recording 95.93%. However, when focusing specifically on the accuracy of hearing loss classification, as measured by the TNR, VGG19 excelled with a TNR of 98.47%, making it the most effective model for this particular objective. Given that the primary aim of this research is to accurately classify hearing loss, the VGG19 model emerges as the superior performer in this context. Nevertheless, it is important to consider the structural differences between the models. AlexNet, with its comparatively shallower layer depth, consumes fewer temporal resources during the learning classification process. Thus, for applications involving the classification of ABR data on a scale larger than the current dataset of 7990 cases, AlexNet presents a viable option due to its efficiency in handling larger datasets without a significant increase in computational demand. This balance between accuracy and efficiency is crucial for scaling the application of these models to larger datasets in future studies.

4.2. Analysis of ABR Data That Are Not Classified Correctly

In this part, we will discuss the results that were not classified correctly among the classification results from various models.

4.2.1. False Negative: In Case the Data Are actually Normal but Are Classified as a Patient with Hearing Loss

Figure 8 presents a selection of misclassified cases identified through the application of the AlexNet, VGG19, and VGG16 models within the classification analyses of this study. These instances involve subjects who, despite having normal hearing, were erroneously classified as suffering from hearing loss. The graph on the left represents ABR data from a 1-year-old infant, with V waves detected at 60 dB, 40 dB, and 30 dB. The analysis suggests that the model's misclassification may stem from the limited number of data points in this sample, contrasting with the more comprehensive ABR data typically gathered from the general population, which is measured in 10 dB increments from 30 dB to 90 dB. The middle graph displays ABR results for a 52-year-old individual; no V wave was detected at 30 dB. This subject, diagnosed with normal hearing by a medical professional, was analyzed in comparison to typical public ABR data, which usually exhibits V waves across all tested decibels. The absence of a V wave at 30 dB in this case led to a model misrecognition, highlighting a deviation from expected patterns observed in broader datasets. The graph on the right documents ABR measurements for an 18-year-old individual. The analysis determined that the model misclassification occurred because the final 30 dB graph plotted very close to the x-axis. This proximity likely influenced the model's perception, causing it to misidentify the presence of a V wave, which deviates from the normative data where V waves are consistently present across all measurements. These illustrations underscore challenges with model accuracy when faced with atypical data representations, emphasizing the need for the continuous refinement of classification algorithms to enhance diagnostic precision in clinical settings.

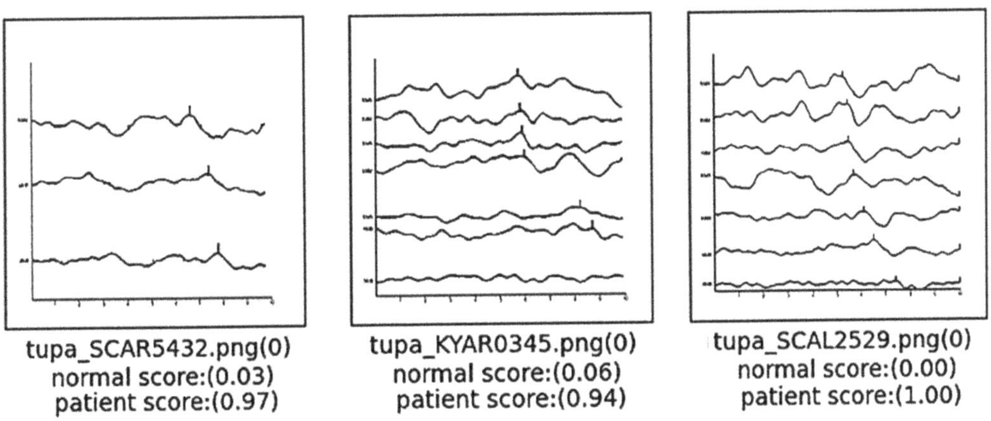

Figure 8. Cases of false negatives.

4.2.2. False Positive: In Case the Data Represent Patients with Actual Hearing Loss but Classified as Normal

Figure 9 presents images indicative of hearing loss that were inaccurately classified as normal by the model. For the images on the left and right, prior to the existing preprocessing steps, the resolution was significantly compromised, necessitating further preprocessing to enhance resolution quality. Despite these efforts, the resolution remained comparatively lower than that of typical ABR graphs, which likely impeded the model's ability to accurately classify these cases. Concerning the image in the middle, the analysis suggests that the misclassification occurred due to the proximity of the wave in the bottom graph to the x-axis and its closeness to the graph directly above it. This spatial arrangement may have confused the model, leading to an incorrect interpretation of the data. This observation underscores the sensitivity of classification models to variations in graphical representation and highlights the need for robust preprocessing techniques to ensure consistent image quality across all data inputs. Such improvements are critical for enhancing the accuracy of diagnostic models in clinical applications.

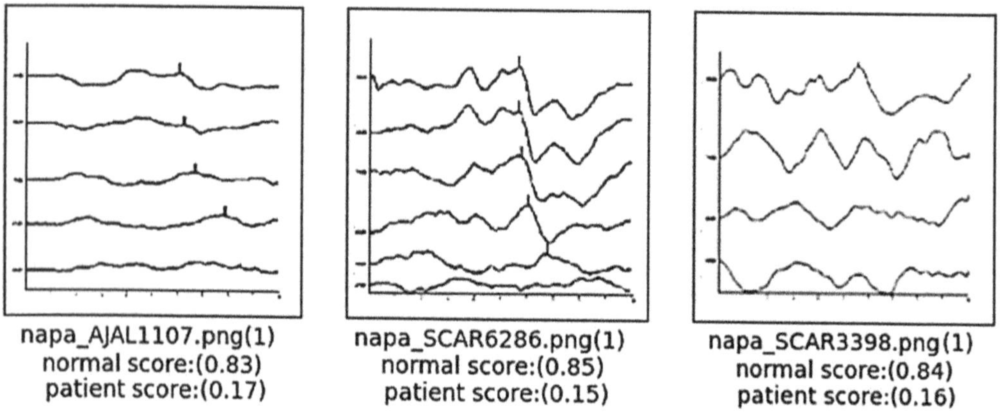

Figure 9. Cases of false positives.

5. Conclusions

In this study, we developed multiple models to classify hearing loss using preprocessed ABR graph data and evaluated their comparative performances. The AlexNet model exhibited the highest accuracy with a value of 95.93%, while the VGG19 model demonstrated the best TNR at 98.47%. Among the six evaluated models, AlexNet showed the quickest learning speed, followed by InceptionV3, VGG16, VGG19, DenseNet121, and DenseNet201 in terms of processing time. For instances in which images are incorrectly classified, future work will involve exploring further supplementation and preprocessing strategies. These will aim to enhance image quality without compromising the integrity of the original data including measures such as increasing resolution, augmenting the X and Y axes, and adjusting the wave positioning for each decibel level.

Previously, the process of diagnosing hearing loss using ABR involved audiologists and otolaryngologists manually reviewing each ABR graph, which was time-consuming. This study was conducted to address this issue and improve the efficiency of the diagnostic process. The findings of this research pave the way for the development of a robust model that can support the preliminary automatic classification of ABR data, assisting in the prediagnostic stages before clinical evaluation by a physician. Additionally, the study plans to extend into the creation of an automatic V-latency detection algorithm which will be designed for universal application across various devices rather than being confined to specific equipment. This advancement is anticipated to simplify the diagnosis of hearing loss and related auditory conditions, thereby enhancing patient care and diagnostic efficiency.

Author Contributions: Conceptualization, M.H. and S.J.C.; methodology, J.M. and M.H.; software, J.M.; validation, J.M., S.K., and S.J.C.; formal analysis, J.M. and S.K.; investigation, J.M.; resources J.M. and S.J.C.; data curation, J.M., S.K., and S.J.C.; writing original draft preparation, J.M.; writing review and editing, M.H., S.K., and S.J.C.; visualization, J.M.; supervision, M.H.; project administration, S.J.C.; funding acquisition, M.H. and S.J.C. All authors have read and agreed to the published version of the manuscript.

Funding: This research was funded by the BK21 FOUR (Fostering Outstanding Universities for Research) Grant, No. 5199990914048, and was supported by the Soonchunhyang University Research Fund.

Institutional Review Board Statement: This study was conducted according to the guidelines of the Declaration of Helsinki and approved by the Institutional Review Board of Soonchunhyang University Cheonan Hospital (Cheonan, Korea) (IRB number: 2021-06-040; approval date: 4 June 2021).

Informed Consent Statement: Because of the retrospective design of the study, patient consent was waived.

Data Availability Statement: Data sharing is not applicable.

Conflicts of Interest: The authors declare no conflicts of interest.

References

1. Li, Q.; Cai, W.; Wang, X.; Zhou, Y.; Feng, D.D.; Chen, M. Medical image classification with convolutional neural network. In Proceedings of the 2014 13th International Conference on Control Automation Robotics & Vision (ICARCV), Singapore, 10–12 December 2014; pp. 844–848.
2. Mu, R.; Zeng, X. A review of deep learning research. *KSII Trans. Internet Inf. Syst. (TIIS)* **2019**, *13*, 1738–1764.
3. Yadav, S.S.; Jadhav, S.M. Deep convolutional neural network based medical image classification for disease diagnosis. *J. Big Data* **2019**, *6*, 1–18. [CrossRef]
4. Eggermont, J.J. Auditory Brainstem Response. In *Handbook of Clinical Neurology*, 3rd ed.; Elsevier: Amsterdam, The Netherlands, 2019; Volume 160, pp. 451–464.
5. Sun, J.; Liu, H.; Wang, X.; Li, Y.; Ni, X. Application of auditory brainstem response to different types of hearing loss in infants. *J. Clin. Otorhinolaryngol. Head Neck Surg.* **2022**, *36*, 120–125.
6. Aldè, M.; Binda, S.; Primache, V.; Pellegrinelli, L.; Pariani, E.; Pregliasco, F.; Berardino, F.D.; Cantarella, G.; Ambrosetti, U. Congenital cytomegalovirus and hearing loss: The state of the art. *J. Clin. Med.* **2023**, *12*, 4465. [CrossRef]
7. Elberling, C.; Parbo, J. Reference data for ABRs in retrocochlear diagnosis. *Scand. Audiol.* **1987**, *16*, 49–55. [CrossRef]
8. Ma, J.; Seo, J.H.; Moon, I.J.; Park, M.K.; Lee, J.B.; Kim, H.; Ahn, J.H.; Jang, J.H.; Lee, J.D.; Choi, S.J.; et al. Auditory Brainstem Response Data Preprocessing Method for the Automatic Classification of Hearing Loss Patients. *Diagnostics* **2023**, *13*, 3538. [CrossRef] [PubMed]
9. Hood, L.J. Principles and applications in auditory evoked potentials. *Ear Hear.* **1996**, *17*, 178. [CrossRef]
10. Sininger, Y.S. Auditory brain stem response for objective measures of hearing. *Ear Hear.* **1993**, *14*, 23–30. [CrossRef]
11. Sininger, Y.S.; Abdala, C.; Cone-Wesson, B. Auditory threshold sensitivity of the human neonate as measured by the auditory brainstem response. *Hear. Res.* **1997**, *104*, 27–38. [CrossRef] [PubMed]
12. Aiyer, R.G.; Parikh, B. Evaluation of auditory brainstem responses for hearing screening of high-risk infants. *Indian J. Otolaryngol. Head Neck Surg.* **2009**, *61*, 47–53. [CrossRef]
13. Verhulst, S.; Jagadeesh, A.; Mauermann, M.; Ernst, F. Individual differences in auditory brainstem response wave characteristics: Relations to different aspects of peripheral hearing loss. *Trends Hear.* **2016**, *20*, 2331216516672186. [CrossRef] [PubMed]
14. Galambos, R.; Despland, P.A. The auditory brainstem response (ABR) evaluates risk factors for hearing loss in the newborn. *Pediatr. Res.* **1980**, *14*, 159–163. [CrossRef] [PubMed]
15. McCreery, R.W.; Kaminski, J.; Beauchaine, K.; Lenzen, N.; Simms, K.; Gorga, M.P. The impact of degree of hearing loss on auditory brainstem response predictions of behavioral thresholds. *Ear Hear.* **2015**, *36*, 309. [CrossRef] [PubMed]
16. Stapells, D.R.; Gravel, J.S.; Martin, B.A. Thresholds for auditory brain stem responses to tones in notched noise from infants and young children with normal hearing or sensorineural hearing loss. *Ear Hear.* **1995**, *16*, 361–371. [CrossRef] [PubMed]
17. Simonyan, K.; Zisserman, A. Very deep convolutional networks for large-scale image recognition. *arXiv* **2014**. [CrossRef]
18. Qassim, H.; Verma, A.; Feinzimer, D. Compressed residual-VGG16 CNN model for big data places image recognition. In Proceedings of the 2018 IEEE 8th Annual Computing and Communication Workshop and Conference (CCWC), Las Vegas, NV, USA, 8–10 January 2018; pp. 169–175.
19. Mascarenhas, S.; Agarwal, M. A comparison between VGG16, VGG19 and ResNet50 architecture frameworks for Image Classification. In Proceedings of the 2021 International Conference on Disruptive Technologies for Multi-Disciplinary Research and Applications (CENTCON), Bengaluru, India, 19–21 November 2021; Volume 1, pp. 96–99.
20. Carvalho, T.; De Rezende, E.R.; Alves, M.T.; Balieiro, F.K.; Sovat, R.B. Exposing computer generated images by eye's region classification via transfer learning of VGG19 CNN. In Proceedings of the 2017 16th IEEE International Conference on Machine Learning and Applications (ICMLA), Cancun, Mexico, 18–21 December 2017; pp. 866–870.
21. Yin, J.; Qu, J.; Huang, W.; Chen, Q. Road Damage Detection and Classification based on Multi-level Feature Pyramids. *KSII Trans. Internet Inf. Syst.* **2021**, *15*, 786.
22. Dey, N.; Zhang, Y.D.; Rajinikanth, V.; Pugalenthi, R.; Raja, N.S.M. Customized VGG19 architecture for pneumonia detection in chest X-rays. *Pattern Recognit. Lett.* **2021**, *143*, 67–74. [CrossRef]
23. Mateen, M.; Wen, J.; Nasrullah; Song, S.; Huang, Z. Fundus image classification using VGG-19 architecture with PCA and SVD. *Symmetry* **2018**, *11*, 1. [CrossRef]
24. Huang, G.; Liu, Z.; Van Der Maaten, L.; Weinberger, K.Q. Densely connected convolutional networks. In Proceedings of the IEEE Conference on Computer Vision and Pattern Recognition, Honolulu, HI, USA, 21–26 July 2017; pp. 4700–4708.

25. Singh, D.; Kumar, V.; Kaur, M. Densely connected convolutional networks-based COVID-19 screening model. *Appl. Intell.* **2021**, *51*, 3044–3051. [CrossRef]
26. Jaiswal, A.; Gianchandani, N.; Singh, D.; Kumar, V.; Kaur, M. Classification of the COVID-19 infected patients using DenseNet201 based deep transfer learning. *J. Biomol. Struct. Dyn.* **2021**, *39*, 5682–5689. [CrossRef]
27. Chhabra, M.; Kumar, R. A Smart Healthcare System Based on Classifier DenseNet 121 Model to Detect Multiple Diseases. In *Mobile Radio Communications and 5G Networks, Proceedings of the Second MRCN 2021*; Springer Nature: Singapore, 2022; pp. 297–312.
28. Frimpong, E.A.; Qin, Z.; Turkson, R.E.; Cobbinah, B.M.; Baagyere, E.Y.; Tenagyei, E.K. Enhancing Alzheimer's Disease Classification using 3D Convolutional Neural Network and Multilayer Perceptron Model with Attention Network. *KSII Trans. Internet Inf. Syst.* **2023**, *17*, 2924–2944.
29. Chauhan, T.; Palivela, H.; Tiwari, S. Optimization and fine-tuning of DenseNet model for classification of COVID-19 cases in medical imaging. *Int. J. Inf. Manag. Data Insights* **2021**, *1*, 100020. [CrossRef]
30. Krizhevsky, A.; Sutskever, I.; Hinton, G.E. Imagenet classification with deep convolutional neural networks. *Commun. ACM* **2017**, *60*, 84–90. [CrossRef]
31. Yuan, Z.W.; Zhang, J. Feature extraction and image retrieval based on AlexNet. In Proceedings of the 8th International Conference on Digital Image Processing (ICDIP 2016), Chengdu, China, 20–22 May 2016; SPIE: Bellingham, WA, USA, 2016; Volume 10033, pp. 65–69.
32. Alippi, C.; Disabato, S.; Roveri, M. Moving convolutional neural networks to embedded systems: The alexnet and VGG-16 case. In Proceedings of the 2018 17th ACM/IEEE International Conference on Information Processing in Sensor Networks (IPSN), Porto, Portugal, 11–13 April 2018; pp. 212–223.
33. Abd Almisreb, A.; Jamil, N.; Din, N.M. Utilizing AlexNet deep transfer learning for ear recognition. In Proceedings of the 2018 Fourth International Conference on Information Retrieval and Knowledge Management (CAMP), Kota Kinabalu, Malaysia, 26–28 March 2018; pp. 1–5.
34. Chen, J.; Wan, Z.; Zhang, J.; Li, W.; Chen, Y.; Li, Y.; Duan, Y. Medical image segmentation and reconstruction of prostate tumor based on 3D AlexNet. *Comput. Methods Programs Biomed.* **2021**, *200*, 105878. [CrossRef]
35. Titoriya, A.; Sachdeva, S. Breast cancer histopathology image classification using AlexNet. In Proceedings of the 2019 4th International Conference on Information Systems and Computer Networks (ISCON), Mathura, India, 21–22 November 2019; pp. 708–712.
36. Szegedy, C.; Liu, W.; Jia, Y.; Sermanet, P.; Reed, S.; Anguelov, D.; Erhan, D.; Vanhoucke, V.; Rabinovich, A. Going deeper with convolutions. In Proceedings of the IEEE Conference on Computer Vision and Pattern Recognition, Boston, MA, USA, 7–12 June 2015; pp. 1–9.
37. Szegedy, C.; Vanhoucke, V.; Ioffe, S.; Shlens, J.; Wojna, Z. Rethinking the inception architecture for computer vision. In Proceedings of the IEEE Conference on Computer Vision and Pattern Recognition, Las Vegas, NV, USA, 26 June–1 July 2016; pp. 2818–2826.
38. Wang, C.; Chen, D.; Hao, L.; Liu, X.; Zeng, Y.; Chen, J.; Zhang, G. Pulmonary image classification based on inception-v3 transfer learning model. *IEEE Access* **2019**, *7*, 146533–146541. [CrossRef]

diagnostics

MDPI

Article

Deep Learning Detection and Segmentation of Facet Joints in Ultrasound Images Based on Convolutional Neural Networks and Enhanced Data Annotation

Lingeer Wu, Di Xia, Jin Wang, Si Chen, Xulei Cui *, Le Shen and Yuguang Huang

Department of Anesthesiology, Peking Union Medical College Hospital, Chinese Academy of Medical Sciences and Peking Union Medical College, Beijing 100006, China; 13426365656@126.com (L.W.); xiad08@163.com (D.X.); wangjin05@163.com (J.W.); chensi93@pumch.cn (S.C.); pumchshenle@aliyun.com (L.S.); garybeijing@163.com (Y.H.)

* Correspondence: cui.xulei@aliyun.com

Abstract: The facet joint injection is the most common procedure used to release lower back pain. In this paper, we proposed a deep learning method for detecting and segmenting facet joints in ultrasound images based on convolutional neural networks (CNNs) and enhanced data annotation. In the enhanced data annotation, a facet joint was considered as the first target and the ventral complex as the second target to improve the capability of CNNs in recognizing the facet joint. A total of 300 cases of patients undergoing pain treatment were included. The ultrasound images were captured and labeled by two professional anesthesiologists, and then augmented to train a deep learning model based on the Mask Region-based CNN (Mask R-CNN). The performance of the deep learning model was evaluated using the average precision (AP) on the testing sets. The data augmentation and data annotation methods were found to improve the AP. The AP50 for facet joint detection and segmentation was 90.4% and 85.0%, respectively, demonstrating the satisfying performance of the deep learning model. We presented a deep learning method for facet joint detection and segmentation in ultrasound images based on enhanced data annotation and the Mask R-CNN. The feasibility and potential of deep learning techniques in facet joint ultrasound image analysis have been demonstrated.

Keywords: ultrasound image; deep learning; facet joint; convolutional neural network; enhanced data annotation; ventral complex

1. Introduction

Pain is a common medical condition. Pain diseases ranked first among all diseases according to the Global Burden of Disease study published by the Lancet in 2018, with lower back pain being the primary cause of motion limitation, causing loss of labor ability, in most countries [1]. Up to 80% of people experience chronic neck pain and lower back pain during their lifetime [2]. Facet joint conditions are the most common causes of chronic spinal-derived lower back pain [3,4].

Each facet joint is located at the junction of the pedicle and the laminae, which consist of the upper facet joint and lower facet joint of the adjacent vertebrae. Facet joints are the only synovial joints in the spine, and their surfaces are covered by hyaline cartilage. The capsule contains synovial fluid. The articular surface consists of the outer fibrous capsule and the inner synovial capsule. There are non-myelinated notional receptors and myelinated mechanoreceptors distributed in each joint capsule, which means that the facet joints play an essential role in maintaining spinal stability and regular physiological activity.

In clinical practice, 15–40% of lower back pain is caused by degeneration of the lumbar facet joints. Conservative treatment for articular pain of the lumbar facet joints is mainly used, such as hot compresses, short-wave ultrasounds, and oral nonsteroidal anti-inflammatory drugs. Although lower back pain can be temporarily alleviated, the long-term effect is not so satisfactory. With the development of minimally invasive techniques, interventional therapy has become a safer and more effective method for the treatment of lower back pain caused by facet joint conditions. Facet joint injections are one of the most common procedures performed by pain management anesthesiologists [5]. Injections can be directed by fluoroscopy, computed tomography (CT), and palpation or loss-of-resistance techniques [6]. Techniques such as fluoroscopy and CT have significant disadvantages. For example, the patients will be exposed to ionizing radiation, which means that these techniques may be not appropriate for patients who are pregnant. In addition, it has been reported that the palpation and loss-of-resistance techniques for epidural injections have a failure rate of 6% to 20% [7]. Alternatively, ultrasound, a real-time nonionizing imaging technique, has the capacity to visualize soft tissues and bony surfaces, and has been increasingly utilized for facet joint injections in recent years [8,9]. For instance, Overnauer et al. [10] compared ultrasound-guided and CT-guided facet joint injections. Their results revealed that injections guided by ultrasound were faster than and presented the same therapeutic effects as CT-guided injections [10]. Additionally, Wang et al. [11] compared image guidance technologies for interventional pain procedures. Their study revealed that although ultrasound guidance is beneficial in spinal injections, the success rate of the procedure still depends greatly on the experience of the anesthesiologists [11]. It requires a long learning curve for most pain specialists to be familiar with ultrasound guidance techniques [6]. That means the visualization and identification of ultrasound imaging, especially for the facet joints, remains a problem for anesthesiologists. Artificial intelligence (AI) is an emerging technology for addressing the above issues.

AI has become one of the most popular tools for medical image analysis. The convolutional neural networks (CNNs) are among the AI techniques [12,13] which have been applied to ultrasound images to identify and recognize different target objects, such as neural vascular structures [14], left ventricles [15], breast tumors [16], and the spine [17]. CNNs may gradually become able to interpret low-level features as if they were high-level features, which are mainly applied to object detection and recognition in image and video analysis. Deep learning network models based on object detection are increasingly used in medical image processing, especially in spinal ultrasounds, including spinal image recognition, disease detection, and disease prediction. However, deep learning network models based on object detection need to be based on a large amount of data. In fact, the current medical image training data scale is significantly smaller than the public data sets used in other fields such as natural image understanding, which results in lower prediction performance. The current CNN model is mainly limited to the single region features of fixed morphology. However, the facet joints are often shown in ultrasonic images with different scales, unclear edges, and irregular shapes, etc., which cannot be thoroughly characterized by single region features.

The ventral complex, located in the ventral dural space, is a complex composed of tissues such as the ventral dural membrane and the anterior longitudinal ligament, showing a high echo zone on ultrasound. The ventral complex plays an essential role in the indication of ultrasound-guided intraspinal puncture. Its presence usually indicates that an ultrasound beam can pass through the tissue, thus indicating that the plane is suitable for an ultrasound-guided puncture approach. Therefore, the ventral complex plays an important localization role in lumbar ultrasound. We hypothesized that CNNs might assist in the detection and segmentation of facet joints in ultrasound images. In this study, we proposed a deep learning method for recognizing ultrasound images of facet joints based on enhanced data annotation and CNNs. In the enhanced data annotation, the facet joint was considered as the first target and the posterior vertebral body as the second or auxiliary target.

2. Materials and Methods

Figure 1 shows the proposed deep learning-based ultrasonic image detection and segmentation method for facet joints. The facet joints were labeled as the primary target and the ventral complex was used as the auxiliary target. Preprocessing and data augmentation were conducted for each input ultrasound image. The deep learning network, the Mask Region-based CNN (Mask R-CNN), refs. [18,19] was used, in which ResNet101 and the feature pyramid network (FPN) were used as the backbone. The FPN was used to extract image features. The region proposal network (RPN) was used to find the region of interest (ROI). The ROI Align was used to transform all the proposed ROIs generated in the RPN process into a feature map of the same size, which was then reshaped into a one-dimensional vector, so as to facilitate the subsequent generation of masks, coordinates, and classifications for facet joint detection and segmentation [20].

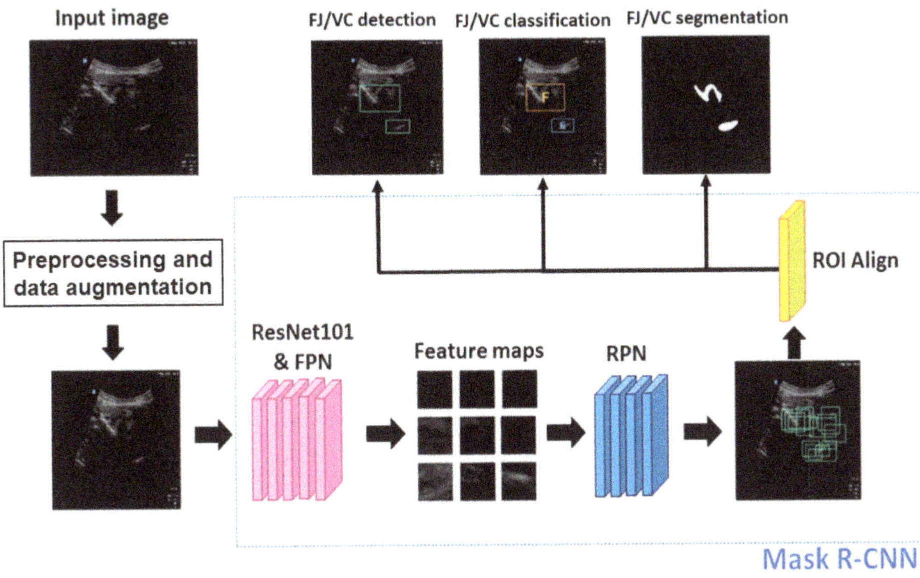

Figure 1. Flow chart of facet joint (FJ) and ventral complex (VC) detection and segmentation using the proposed method. In FJ/VC classification, the orange box represents the detected FJ (denoted "F") and the blue box indicates the detected VC (denoted "E"). FPN = feature pyramid network; RPN = region proposal network; ROI = region of interest.

2.1. Ultrasound Image Data

2.1.1. Data Collection

In this study, the clinical data were collected from 300 patients. Most of these patients suffered from lower back pain and were undergoing pain treatment at the Department of Anesthesiology of Peking Union Medical College Hospital (PUMCH), Chinese Academy of Medical Sciences. The study protocol was numbered K22C2241 and approved by the Ethics Review Committee of PUMCH. All the patients received ultrasound scanning before and after pain treatment. The ultrasound images were captured by one or two expert sonographers using a SonoSite X-Porte scanner (Fujifilm, Tokyo, Japan), with a 2–5 MHz curved-array transducer (C60xp/5-2), a scanning depth of 7 cm, and a gain of 50%. The pixels of the original images in this study were 960 $\times$ 720. The collected data had the same scanning angle and a similar scanning field (Figure 2).

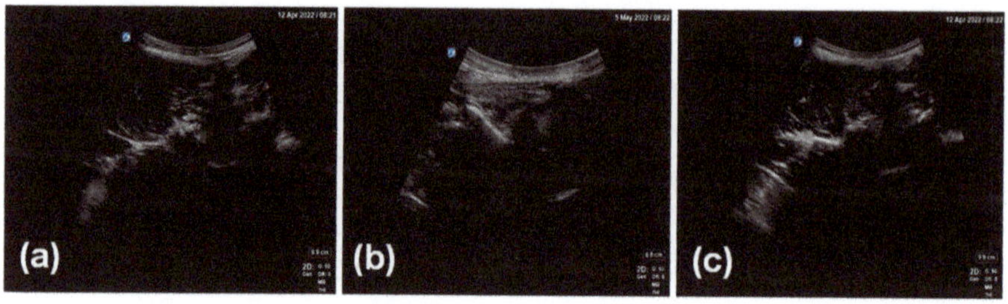

Figure 2. Examples of collected ultrasound images (**a**–**c**).

2.1.2. Enhanced Data Annotation

Two professional anesthesiologists confirmed the inclusion of a facet joint and ventral complex in all the image data and processed the ultrasound image data to remove the patient's name, physician's unit, and physician's name. Under the review and proofreading of the senior anesthesiologist, the annotator marked the outline of the facet joint and the ventral complex in each image. The marking software used in this study was LabelMe (version 4.5.13), an open annotation tool. In this study, we proposed an enhanced data annotation method for the facet joint using the facet joint as the first target and the ventral complex as the second target (Figure 3). In order to evaluate the influence of data annotation on deep learning, we considered two labeling manners: (i) a local labeling method (Figure 3b), which only involved facet joints and ventral association; (ii) a full labeling method (Figure 3a), which involved transverse processes, facet joints and even bone structures in the median line and ventral association.

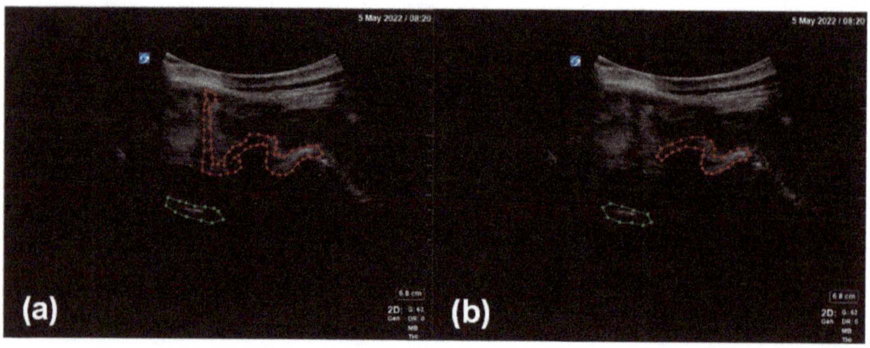

Figure 3. Two methods of enhanced data annotation: (**a**) full labeling method; (**b**) local labeling method. The red annotation indicates the facet joint, and the green annotation indicates the ventral complex.

2.1.3. Data Cleaning

The process of data cleaning was to screen out abnormal data, including (1) images and marked data that were damaged; (2) data naming formats that did not meet the requirements; (3) incorrect marks; (4) incorrect association of the original image with the marked image; (5) images that did not contain a target object. In this study, a total of 391 original ultrasonic images remained after the data cleaning.

2.1.4. Data Classification

The purpose of data classification was to divide the data into training sets and testing sets for the deep learning model. The 391 original ultrasonic images were divided into a training set composed of 356 images and a testing set composed of 35 images. Note that all images from the same patient were assigned to the same subset (training or testing).

2.2. Image Preprocessing and Data Augmentation

Image preprocessing was conducted to the input image, including image scaling and data augmentation. Since the input of the CNN must be images with the same width and height, one of the preprocessing tasks was to scale the original image to 256 × 256 pixels. In addition, the data augmentation method of horizontal flipping was used in this study according to medical knowledge. Generally, data augmentation should conform to certain physical meaning. Otherwise, inappropriate data augmentation may reduce the recognition accuracy. Specifically, each image in the training set was flipped left to right, so the number of images in the training set were doubled.

2.3. The Mask R-CNN

In this study, the ventral complex (denoted "E") and the facet joint (denoted "F") needed to be segmented in ultrasonic images. In theory, semantic segmentation and instance segmentation are both acceptable. Considering that, in some cases, E and F could be multiple and overlap, instance segmentation was adopted in this work. Mask R-CNN [18] is the most typical instance segmentation algorithm, in which target segmentation is achieved on the basis of target detection. Compared with U-Net [21] and its improved algorithms that can only achieve semantic segmentation, Mask R-CNN is more suitable for instance segmentation where the number of targets is small and the target occupies a low proportion of image pixels. This is because when the number of targets is small and the proportion of pixels is low, U-Net takes the background as an independent segmentation target and the number of network layers is low. In the training process, U-Net often quickly converged on a local optimality, ignoring all the targets and identifying the whole image as the background. In addition, the objective of this study was to detect and segment facet joints at the same time, for which Mask R-CNN was well-suited, while U-Net was only suitable for image segmentation. For these reasons, we considered that Mask R-CNN would be more suitable for the facet joint detection and segmentation task in this work.

2.3.1. The FPN

The FPN is the backbone network of the Mask R-CNN [18], which is mainly used to extract image features (Figure 4). The FPN was divided into five layers in order to extract image features. Low-level features usually contain more details, such as textures and edges, but they may also contain a lot of noise. High-level features generally contain more semantic information (shape, position, etc.), but the spatial resolution is small, and the information loss is severe. Suppose the size of the original image is $H \times W$ pixels. The FPN extracted 5 features of different levels from low to high, whose feature sizes were $H/2 \times W/2, H/4 \times W/4, H/8 \times W/8, H/16 \times W/16$, and $H/32 \times W/32$, respectively. In this study, H = 256 and W = 256. Standard ResNet [22] networks are usually used along with FPNs, such as the ResNet50 and ResNet101 networks. In this paper, the ResNet101 network was used.

2.3.2. The RPN

The RPN found ROIs based on image features extracted from the FPN. The RPN can be understood as fulfilling two tasks. One is a classification task, and the other is a regression task. The regression task is to obtain the coordinates of the candidate box (the coordinates of the upper left and lower right of the candidate box), and the classification task is to determine whether there is a target in the candidate box (the probability of having a target). After the two tasks were completed, each candidate box with a probability score greater than 0.7 for an object was retained as a proposal.

2.3.3. The ROI Align

The ROI Align was mainly used to transform all the proposed ROIs generated in the RPN process into feature maps of the same size. The feature maps are then reshaped into a one-dimensional vector. The size of the vector was 49 (i.e., 7 × 7) for facet joint detection and 196 (i.e., 14 × 14) for facet joint segmentation. Finally, the Mask R-CNN [15] produced segmentation results (mask), recognition results (coordinates), and categories of images through three independent fully connected networks (Figure 1).

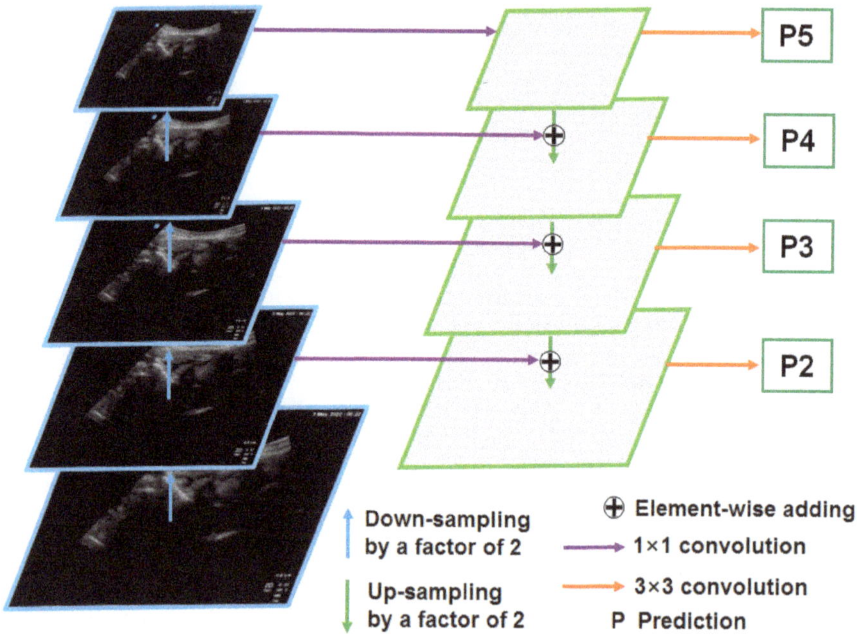

Figure 4. Structure of the feature pyramid network.

2.4. Evaluation Metrics

During the testing phase, the results of the deep learning model were compared with the labeled data. The performance of the model was evaluated using the following metrics.

(1) True Positive (*TP*) represents the correct prediction of positive data.
(2) True Negative (*TN*) represents the correct prediction of negative data.
(3) False Positive (*FP*) represents the incorrect prediction of positive data.
(4) False Negative (*FN*) represents the incorrect estimate of negative data.

The Intersection over Union (IoU) score is a standard performance measure for object segmentation tasks. Given a set of images, the IoU metric gives the similarity between the predicted region and the ground truth region of the objects presented in the set of images, and is defined by

$$\text{IoU} = \frac{TP}{FP + TP + FN} \text{ or } \text{IoU} = \frac{A \cap B}{A \cup B} \quad (1)$$

where A is the area of model-based segmentation and B is the area of the ground truth.

With the deep learning algorithms, the success of the model depends on the result of the confusion matrix. The success rate of each algorithm for detecting facet joints was determined. The average precision (AP) was used to evaluate the facet joint detection and segmentation performance of the proposed method. The AP values of all categories were averaged to produce mAP as well. Specifically, AP50 and AP@50:5:95 were used. AP50 is defined as the AP when the IoU equals 50%. AP@50:5:95 is defined as the mean value of those APs corresponding to IoUs from 50% to 95% with a step of 5%.

2.5. Experimental Setup

The training platform for the experiments was a self-built server, based on a single 32G V100 PCIe GPU. The number of training iterations was set at 25,000, and completing all mini-batch trainings took approximately 8 min 50 s multiplied by 25,000 and divided by 20 (batch size = 16). That was approximately 219,166 s or about 61 h. The initial learning rate was set at 0.0001. The loss function of the Mask R-CNN was a multi-task loss L, $L = L_C + L_D + L_S$, where the subscripts 'C', 'D', and 'S' denote classification, detection, and segmentation, respectively [18]. L_S is the average binary cross-entropy loss [18]. The testing sets were input to the trained deep learning model. A total of 421 ultrasound images were included in the dataset, and 420 valid data were obtained after data cleaning. Among them, 29 images did not contain any objects, so were deemed to constitute a negative set, and the remaining 391 image data were divided into a training set composed of 356 images and a validation set composed of 35 images. Validation sets and training sets were annotated in coco format.

2.6. Statistical Analysis

Categorical variables were expressed as frequencies and percentages, which were compared using *t*-tests. Statistical analyses were conducted using SPSS 20.0 (SPSS Inc., Chicago, IL, USA)

3. Results

In Figure 5, the green area is the inspected and segmented facet joint structure, and the orange area is the inspected and segmented ventral complex. The figure shows that the results for the facet joint structures inspected and segmented using the proposed method are close to the results of manual annotations by human experts, indicating the good performance of the enhanced data annotation method and the deep learning model adopted in this work.

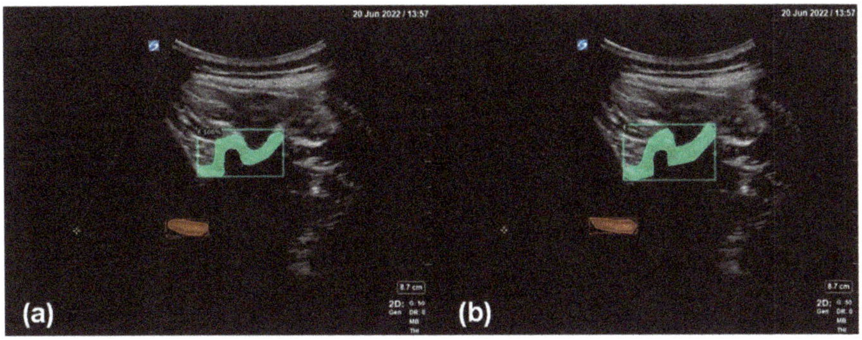

Figure 5. (**a**) Detection and segmentation of the facet joint (green) and the ventral complex (orange) using the proposed method. (**b**) The facet joint (green) and the ventral complex (orange) as manually annotated by human experts.

Table 1 shows the effects of data augmentation on the AP of facet joint detection and segmentation using the proposed method. Compared to using no data augmentation, using horizontal flipping for data augmentation improves the AP50 and AP@50:5:95 values in both detection and segmentation. Based on the symmetry of the facet joints, we considered that horizontal flipping is an effective method for data augmentation.

Table 1. Effects of data augmentation on the average precision (AP) of facet joint detection and segmentation using the proposed method.

Data Augmentation	Detection		Segmentation	
	AP50	AP@50:5:95	AP50	AP@50:5:95
None	92.17%	54.73%	82.86%	36.71%
Horizontal flip	92.88%	54.95%	90.01%	39.86%
p	<0.001	<0.001	<0.001	<0.001

Table 2 shows the effects of data annotation methods on the AP of facet joint detection and segmentation using the proposed method. Although it is generally believed that using the full labeling method reduces the false detection rate and improves the recognition accuracy of the target, our results show that, for the task of facet joint detection and segmentation in ultrasound images, using the local labeling method produces higher AP50 and AP@50:5:95 values in facet joint segmentation using the proposed method.

Table 2. Effects of data annotation methods on the average precision (AP) of facet joint detection and segmentation using the proposed method.

Data Annotation Area	Detection		Segmentation	
	AP50	AP@50:5:95	AP50	AP@50:5:95
Full labeling method	92.17%	54.73%	82.86%	36.71%
Local labeling method	98.57%	63.86%	90.75%	44.01%
p	<0.001	<0.001	<0.001	<0.001

4. Discussion

The application of AI in ultrasound imaging is currently a hot topic, especially in the fields of liver, cardiovascular, thyroid, and musculoskeletal systems [23–26]. AI techniques include conventional machine learning methods and deep learning methods. CNNs are types of deep learning techniques, biologically inspired neural networks that mimic the physiology of the visual cortex by responding differently to specific features [27]. CNNs are composed of a series of convolutional layers, followed by a pooling layer, and finally a fully connected layer. CNNs have been applied to ultrasound images to identify and recognize different target objects, such as neural vascular structures [28] (12), left ventricles [29,30], breast tumors, and the spine [31]. However, CNNs and other deep learning techniques have not been applied to ultrasound image analysis for facet joint detection and segmentation.

In recent years, musculoskeletal ultrasound has been widely used in the field of rehabilitation, anesthesiology, orthopedics, and other fields for puncture positioning and real-time guidance. It is well known that the bony structures of the lumbar spine, including spinous processes, vertebral arch plates, facet joints and transverse processes, have typical ultrasound characteristics. Ultrasound scans of the lumbar spine structure can be applied to minimally invasive treatments such as spinal or epidural anesthesia, lumbar nerve blocks, quadratus lumborum plane blocks, and endoscopic foraminal surgery. Among them, the recognition of the bone structure of the facet joint is particularly of interest. On one hand, the posterior medial branch of the lumbar nerve is close to the rear of the facet joint. On the other hand, the blocking point of the posterior medial branch is the recess between the outer side of the upper facet and the proximal edge of the transverse process. The structure of the facet joint can be accurately identified using the characteristic ultrasound images of the upper vertebral arch plate, the lower facet, the facet joint, and the transverse process. Real-time ultrasound guidance can accurately locate the position of the puncture needle and avoid intravascular injection.

In the early stage of this study, we supposed that the bone structure of the facet joint would be connected to the transverse process and spinous process, so we labeled all the relevant bone structures in the ultrasound images, i.e., using the full labeling method. But soon we found that this full labeling approach resulted in poor model accuracy. The main reason may be that the range of bone structures is large in full labeling, and the boundary is not so clear, resulting in a large boundary error in deep learning. After recognizing this, we chose a local labeling method, that is, only the bony structure of the facet joint area was labeled. With this method, the boundary can be relatively clear, and the area that needs to be labeled is relatively small. It turns out that the local labeling method has better accuracy. Although we generally believed that labeling more features would reduce the false detection rate of the target and improve the recognition accuracy of the target, our experimental results showed that a larger scope of labeling is not better. We should meet the requirements of medical scenarios and computer vision algorithms.

In this study, we investigated an approach using CNNs and enhanced data annotation methods for facet joint detection and segmentation in ultrasound images. Specifically, the Mask R-CNN and enhanced annotation of the facet joint and ventral complex yields satisfying detection and segmentation performance. To the best of our knowledge, this work is the first to demonstrate the feasibility of deep learning models in detecting and segmenting facet joints in ultrasound images.

The considerations of the enhanced data annotation method proposed in this work are discussed. The ventral complex can assist in the detection and recognition of the facet joint. For instance, if there are facet joint–transverse process objects appearing in ultrasound images, with no dura mater appearing, then the facet joint–transverse process object detection may be true, or false. However, if there are facet joint–transverse process objects appearing in ultrasound images, with dura mater appearing, then facet joint–transverse process object detection should be true. The dura mater is the structure inside the spinal canal. If the dura mater appears in the ultrasound image, it means that the ultrasonic scanning plane must be at the level of the intervertebral space, and the facet joint–transverse process is also at the level of the intervertebral space. Therefore, the facet joint–transverse process must be true at the level where the dura mater can appear. However, if the facet joint–transverse process appears in the absence of dura mater, it may be that the ossification and calcification of intervertebral tissue has caused a failure to recognize the dura mater in ultrasound images, or it may be due to other tissues whose structures are similar to the facet joint–transverse process. With the enhanced data annotation method and combined identification of the facet joint and ventral complex, the negative sets can be correctly recognized (Figure 6).

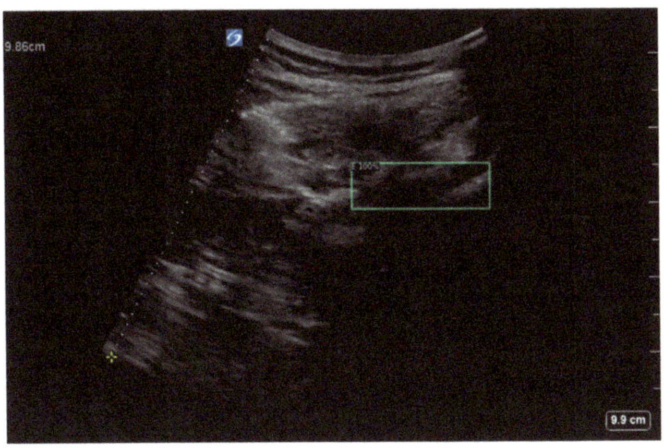

Figure 6. Detection of a negative set using the proposed method.

It should be noted that facet joint detection and segmentation in ultrasonic images is not the final goal. Clinicians are concerned about the accuracy of the puncture of each target.

Based on the segmentation results from our deep learning model, the search inflection point of the segmentation (binary images) can be computed. As the final detection result of the target, the positioning error of the facet joint–transverse process can be controlled within an available range (typically 5 mm). This can effectively improve the accuracy of each puncture. Specifically, after the target detection and segmentation results are obtained, the search inflection points of the recognition coordinates can be obtained according to the recognition coordinates of the lumbar facet joint–transverse process and the dura mater, and the search inflection point can be added to the recognition coordinates to obtain the target recognition coordinates.

Finally, this study has some limitations. First, the size of the ultrasound image samples was relatively limited. It is worth noting that after horizontal flipping was performed on the training set, the detection AP50 of the test set increased from 93.8% to 94.2%, and the segmentation AP50 of the test set increased from 66.2% to 69.9%. Our experiments showed that the accuracy of the model can be increased by enlarging the amount of training sets in different ways. That is to say, the size of training sets should also be increased to get better experimental results. In addition, the full labeling method performs better for detection, and the local labeling method for segmentation (Table 2), so combining the respective neural network layers/branches for each model may be considered in future work. Last but not least, the images were collected at a single center with a single scanner. In future work, more images collected at different centers with different scanners may be used to further validate and improve the performance of deep learning models in facet joint detection and segmentation in ultrasound images.

5. Conclusions

In conclusion, this study is the first to present a deep learning method for facet joint detection and segmentation in ultrasound images based on enhanced data annotation and the Mask R-CNN. The feasibility and potential of deep learning techniques in facet joint ultrasound image analysis have been demonstrated. In the future, the proposed method may be used in the field of pain management and medical education.

Author Contributions: Conceptualization, L.W.; Data curation, D.X.; Investigation, J.W., X.C. and Y.H.; Methodology, J.W., S.C. and X.C.; Project administration, L.S.; Resources, L.W., D.X., S.C. and X.C.; Software, X.C.; Visualization, L.S. and Y.H.; Writing—original draft, L.W.; Writing—review & editing, X.C. All authors have read and agreed to the published version of the manuscript.

Funding: This work was supported in part by the Chinese Academy of Medical Sciences Innovation Fund for Medical Sciences and Health (Grant No. 2021-I2M-C&T-B-015) and National High Level Hospital Clinical Research Funding (No. 2022-PUMCH-A-149).

Institutional Review Board Statement: The study was conducted in accordance with the Declaration of Helsinki, and approved by the Peking Union Medical College (PUMC) Hospital Institutional Review Board (protocol code K22C2241, protocol time: 2 December 2022).

Informed Consent Statement: Not applicable.

Data Availability Statement: The datasets generated and analyzed and the code during the current study are available from the corresponding author on reasonable request.

Acknowledgments: The authors thank the anonymous reviewers for their valuable comments and suggestions.

Conflicts of Interest: The authors declared they have no competing interests. The authors alone are responsible for the content and writing of the paper.

References

1. GBD 2017 Disease and Injury Incidence and Prevalence Collaborators. Global, regional, and national incidence, prevalence, and years lived with disability for 354 diseases and injuries for 195 countries and territories, 1990–2017: A systematic analysis for the Global Burden of Disease Study 2017. *Lancet* **2018**, *392*, 1789–1858. [CrossRef]
2. Rubin, D.I. Epidemiology and risk factors for spine pain. *Neurol. Clin.* **2007**, *25*, 353–371. [CrossRef]

3. Bodor, M.; Murthy, N.; Uribe, Y. Ultrasound-guided cervical facet joint injections. *Spine J.* **2022**, *22*, 983–992. [CrossRef]
4. Chan, J.J.I.; Gan, Y.Y.; Dabas, R.; Han, N.R.; Sultana, R.; Sia, A.T.H.; Sng, B.L. Evaluation of association factors for labor episodic pain during epidural analgesia. *J. Pain Res.* **2019**, *12*, 679–687. [CrossRef]
5. Manchikanti, L.; Boswell, M.V.; Singh, V.; Pampati, V.; Damron, K.S.; Beyer, C.D. Prevalence of facet joint pain in chronic spinal pain of cervical, thoracic, and lumbar regions. *BMC Musculoskelet. Disord.* **2004**, *5*, 15. [CrossRef]
6. Cui, X.; Zhang, D.; Zhao, Y.; Song, Y.; He, L.; Zhang, J. An open-label non-inferiority randomized trail comparing the effectiveness and safety of ultrasound-guided selective cervical nerve root block and fluoroscopy-guided cervical transforaminal epidural block for cervical radiculopathy. *Ann. Med.* **2022**, *54*, 2681–2691. [CrossRef]
7. Engel, A.; King, W.; Schneider, B.J.; Duszynski, B.; Bogduk, N. The Effectiveness of Cervical Medial Branch Thermal Radiofrequency Neurotomy Stratified by Selection Criteria: A Systematic Review of the Literature. *Pain Med.* **2020**, *21*, 2726–2737. [CrossRef]
8. Lin, T.L.; Chung, C.T.; Lan, H.H.; Sheen, H.M. Ultrasound-guided facet joint injection to treat a spinal cyst. *J. Chin. Med. Assoc.* **2014**, *77*, 213–216. [CrossRef] [PubMed]
9. Narouze, S.N. Ultrasound-guided cervical spine injections: Ultrasound "prevents" whereas contrast fluoroscopy "detects" intravascular injections. *Reg. Anesth Pain Med.* **2012**, *37*, 127–130. [CrossRef] [PubMed]
10. Obernauer, J.; Galiano, K.; Gruber, H.; Bale, R.; Obwegeser, A.A.; Schatzer, R.; Loizides, A. Ultrasound-guided versus Computed Tomography-controlled facet joint injections in the middle and lower cervical spine: A prospective randomized clinical trial. *Med. Ultrason.* **2013**, *15*, 10–15. [CrossRef] [PubMed]
11. Wang, D. Image Guidance Technologies for Interventional Pain Procedures: Ultrasound, Fluoroscopy, and CT. *Curr. Pain Headache Rep.* **2018**, *22*, 6. [CrossRef] [PubMed]
12. Maimon, A.; Netzer, O.; Heimler, B.; Amedi, A. Testing geometry and 3D perception in children following vision restoring cataract-removal surgery. *Front. Neurosci.* **2022**, *16*, 962817. [CrossRef] [PubMed]
13. Shin, Y.; Yang, J.; Lee, Y.H.; Kim, S. Artificial intelligence in musculoskeletal ultrasound imaging. *Ultrasonography* **2021**, *40*, 30–44. [CrossRef] [PubMed]
14. Huang, C.; Zhou, Y.; Tan, W.; Qiu, Z.; Zhou, H.; Song, Y.; Zhao, Y.; Gao, S. Applying deep learning in recognizing the femoral nerve block region on ultrasound images. *Ann. Transl. Med.* **2019**, *7*, 453. [CrossRef]
15. He, B.; Kwan, A.C.; Cho, J.H.; Yuan, N.; Pollick, C.; Shiota, T.; Ebinger, J.; Bello, N.A.; Wei, J.; Josan, K.; et al. Blinded, randomized trial of sonographer versus AI cardiac function assessment. *Nature* **2023**, *616*, 520–524. [CrossRef]
16. Lei, Y.; He, X.; Yao, J.; Wang, T.; Wang, L.; Li, W.; Curran, W.J.; Liu, T.; Xu, D.; Yang, X. Breast tumor segmentation in 3D automatic breast ultrasound using Mask scoring R-CNN. *Med. Phys.* **2021**, *48*, 204–214. [CrossRef]
17. Tang, S.; Yang, X.; Shajudeen, P.; Sears, C.; Taraballi, F.; Weiner, B.; Tasciotti, E.; Dollahon, D.; Park, H.; Righetti, R. A CNN-based method to reconstruct 3-D spine surfaces from US images in vivo. *Med. Image Anal* **2021**, *74*, 102221. [CrossRef]
18. He, K.; Gkioxari, G.; Dollar, P.; Girshick, R. Mask R-CNN. *IEEE Trans. Pattern. Anal Mach. Intell.* **2020**, *42*, 386–397. [CrossRef]
19. Kong, H.; Chen, P. Mask R-CNN-based feature extraction and three-dimensional recognition of rice panicle CT images. *Plant Direct* **2021**, *5*, e00323. [CrossRef]
20. He, F.; Liu, T.; Tao, D. Why ResNet Works? Residuals Generalize. *IEEE Trans. Neural Netw. Learn Syst.* **2020**, *31*, 5349–5362. [CrossRef]
21. He, Q.; Yang, Q.; Xie, M. HCTNet: A hybrid CNN-transformer network for breast ultrasound image segmentation. *Comput. Biol. Med.* **2023**, *155*, 106629. [CrossRef] [PubMed]
22. Ma, Z.; Qi, Y.; Xu, C.; Zhao, W.; Lou, M.; Wang, Y.; Ma, Y. ATFE-Net: Axial Transformer and Feature Enhancement-based CNN for ultrasound breast mass segmentation. *Comput. Biol. Med.* **2023**, *153*, 106533. [CrossRef] [PubMed]
23. Pesteie, M.; Abolmaesumi, P.; Ashab, H.A.; Lessoway, V.A.; Massey, S.; Gunka, V.; Rohling, R.N. Real-time ultrasound image classification for spine anesthesia using local directional Hadamard features. *Int. J. Comput. Assist. Radiol Surg.* **2015**, *10*, 901–912. [CrossRef] [PubMed]
24. Pesteie, M.; Lessoway, V.; Abolmaesumi, P.; Rohling, R. Automatic Midline Identification in Transverse 2-D Ultrasound Images of the Spine. *Ultrasound Med. Biol.* **2020**, *46*, 2846–2854. [CrossRef] [PubMed]
25. Yue, N.; Zhang, J.; Zhao, J.; Zhang, Q.; Lin, X.; Yang, J. Detection and Classification of Bronchiectasis Based on Improved Mask-RCNN. *Bioengineering* **2022**, *9*, 359. [CrossRef] [PubMed]
26. Wang, P.; Vives, M.; Patel, V.M.; Hacihaliloglu, I. Robust real-time bone surfaces segmentation from ultrasound using a local phase tensor-guided CNN. *Int. J. Comput. Assist. Radiol. Surg.* **2020**, *15*, 1127–1135. [CrossRef]
27. Wong, J.; Reformat, M.; Parent, E.; Lou, E. Convolutional Neural Network to Segment Laminae on 3D Ultrasound Spinal Images to Assist Cobb Angle Measurement. *Ann. Biomed. Eng.* **2022**, *50*, 401–412. [CrossRef] [PubMed]
28. Wang, J.C.; Shu, Y.C.; Lin, C.Y.; Wu, W.T.; Chen, L.R.; Lo, Y.C.; Chiu, H.C.; Özçakar, L.; Chang, K.V. Application of deep learning algorithms in automatic sonographic localization and segmentation of the median nerve: A systematic review and meta-analysis. *Artif. Intell. Med.* **2023**, *137*, 102496. [CrossRef] [PubMed]
29. Zeng, Y.; Tsui, P.H.; Pang, K.; Bin, G.; Li, J.; Lv, K.; Wu, X.; Wu, S.; Zhou, Z. MAEF-Net: Multi-attention efficient feature fusion network for left ventricular segmentation and quantitative analysis in two-dimensional echocardiography. *Ultrasonics* **2023**, *127*, 106855. [CrossRef]

30. Zhu, X.; Wei, Y.; Lu, Y.; Zhao, M.; Yang, K.; Wu, S.; Zhang, H.; Wong, K.K.L. Comparative analysis of active contour and convolutional neural network in rapid left-ventricle volume quantification using echocardiographic imaging. *Comput. Methods Programs Biomed.* **2021**, *199*, 105914. [CrossRef]
31. Hetherington, J.; Lessoway, V.; Gunka, V.; Abolmaesumi, P.; Rohling, R. SLIDE: Automatic spine level identification system using a deep convolutional neural network. *Int. J. Comput. Assist. Radiol. Surg.* **2017**, *12*, 1189–1198. [CrossRef] [PubMed]

MDPI AG
Grosspeteranlage 5
4052 Basel
Switzerland
Tel.: +41 61 683 77 34

Diagnostics Editorial Office
E-mail: diagnostics@mdpi.com
www.mdpi.com/journal/diagnostics

www.ingramcontent.com/pod-product-compliance
Lightning Source LLC
Chambersburg PA
CBHW040757250926
51850CB00007BA/1018
* 9 7 8 3 7 2 5 8 7 7 6 3 8 *